HOMEBUILDING DEBT-FREE

A GUIDE FOR THE OWNER-BUILDER

BY LYNN UNDERWOOD, C.B.O.

Homebuilding Debt-Free: A Guide for the Owner-Builder

Publication Date: January 2002
First Printing

ISBN 1-58001-078-4

Acquisitions Editor:	Mark Johnson
Cover Design:	Mary Bridges
Design and Production:	Lone Wolf Enterprises, Ltd.
Illustrator:	Anthony Guasp
Photography:	Robert Raynor
Production Coordinator:	Cindy Rodriguez
Printed and Bound by:	Copy-Rite Press

5360 Workman Mill Rd. Whittier, CA 90601-2298
The World's Leading Source in Code Publications
www.icbo.org

Comments on this publication are welcome and will be considered for future revisions.
Send to underwood737@cs.com

PRINTED IN THE U.S.A.

DEDICATION

Since October 23, 1974, I have been married to a beautiful, smart woman who tended to be nervous about virtually everything. When she was in her mid-thirties, she survived a life threatening surgery that forever changed her attitude. Every day in her life became precious. Problems became challenges and setbacks were less important. Her attitude was and continues to be inspiring. She is my wife, Sweet Cyndi. Be like her! Be grateful to be alive and to have the good fortune to solve problems. I would never have had the opportunity for this experience if it had not been for my supportive and understanding wife.

I could not have built our home without her and our children, Darius Alexander Orion and Cassandra Anastasia Andromeda, who helped me with everything from holding an elevation rod, to installing trusses, to gathering nails. They also endured abject poverty for three years during construction. They went without new clothes, toys, movies, and dining out. They helped inspire me with their attitude. All of my family were extremely supportive of my efforts during my construction projects, as well as my work in the development of this book.

TABLE OF CONTENTS

FOREWORD

By Dean Johnson

You are about to take the first steps on a path that can lead to financial freedom: building a house while remaining debt-free. And in the process you will achieve something tangible for yourself and your family—a home of your own. I understand your desire to increase the value of that investment by doing as much of the work yourself as possible. After all, on *Hometime* we've been showing people like you how to do that since 1986.

Chances are, you've been thinking about building your home for some time. You already have some construction skills, plus the confidence that you can pick up other skills as needed. That said, you probably also have the smarts to know that there are people out there who have a lot more building experience than you do—and that you could use their help.

That's why I can recommend Lynn Underwood's book to you. He's been there. He's been a certified building inspector, a plan reviewer, and a home builder. There is no better guide you could have to navigate through the construction and inspection process than someone who understands and appreciates both sides of the process: building and inspecting. In *Homebuilding Debt-Free* Lynn shares the knowledge he's earned through his years and years of experience. His book provides both insight and inspiration to those who wish to embark on the rewarding and challenging experience of building their own home.

From planning, to codes, framing, roofing, mechanicals, and trim-outs, Lynn provides a wealth of detailed information to help you build your own house—and build it well. You will find the checklists, tables, formulas, and the detailed text in this book invaluable as you work through the process of designing and then building your own home.

However, as Lynn would be the first to point out, this will not be an easy project. It will consume all of your energy and resources for some time to come. Remember to take breaks every now and again. Sometimes you need to step back and reflect to realize just how much you've accomplished and how good it makes you feel. Good luck with the project, and be sure and let Lynn and me know how it turns out!

Hometime

ACKNOWLEDGEMENTS

Although this inspirational guide may appear to have been produced by one person, many others shared in its development and overall appearance. Whereas my family was an inspiration, my personal and professional friends and coworkers have all had a hand in this work. I owe a debt of gratitude to the owner-builders whose work I inspected, or whose plans I reviewed, for their own dream homes. Several architects, designers, engineers, superintendents, contractors, subcontractors, financial advisors, lawyers, insurance, and safety experts have added to the thoroughness of this book. In addition, I must express appreciation and offer a special thanks to the following people for their unique contributions and help:

Terry Vosler
William Jones
David Mann, P.E.
Randy Schuler, P.E.
Sharon Bonesteel, A.I.A., C.B.O.
Jack Holden, C.B.O.
Leroy Sayre, C.B.O.
John C. Wiles
Lincoln Thomas
David H. Levy and Wendee Levy
Greg and Melody Simmons
Linus Kafka, P.A
Herrold Skidmore, P.E.
Dale Booth
Rod Cramer
Sebastian Kunnappilly
Barbara Piller
Larry Underwood, P.E.
David Eisenberg
Elizabeth Lassuy
Anthony Novelli

Alvaro Crespo
Sharon Ludwig
Dale Booth
Gary Bonesteel
Art Boehm and Jeff Ramos
Barbara Johnson, P.E.
John and Pat Cooper
Nancy and Dale Fish
Jim Herbig and Debbie Haas
Lynn McEvers
Keri and Rich Skinner
C.G. Gaither
Suzane Nunes
Carmel Gieson
Roland Morales, R.A.
Tom O'Brien
Jennifer Cover, P.E.
Roy Fewell
Kevin Ireton
Mary Bridges

An extra special thanks is extended to a few others. Anthony Guasp is a consummate professional in the building code enforcement community and is also a very accomplished artist. Tony has transformed this book with his unique skill. Robert Raynor is an attorney, a professional builder, as well as a very accomplished photographer. Robert has made a significant contribution toward the appearance and functionality of this book with his photography. Suzane Nunes worked tirelessly to build quality into this book. Barb Karg and Rick Sutherland are a genuine team of veteran experts who provided outstanding copyediting, typesetting, design, and layout. They also were very patient with a first-time author.

More than anyone else, Mark Johnson made this book what it is. Without the support, enthusiasm, and trust imparted by him, you would not be enjoying this book. Mark is the Director of Publications and Product Development with ICBO. He is also the most inspirational person I have ever met. His enthusiasm and zeal for this project was infectious—especially to me! I owe a huge debt of gratitude to Mark for his tireless help, hard work, late hours, and passion for excellence. Thanks Mark!

INTRODUCTION

Picture this: You're sipping lemonade, relaxing in the living room of a home you have built yourself. It's taken the better part of five years, but it's finished and you own it outright. That's right. No mortgage company has their hands on your domicile. Sound impossible? Are you convinced you haven't got the talent to build your own home? I'm here to remove those negative perceptions and introduce the concepts and practices of building your own home. You can achieve your goal of home ownership by building it yourself. I did, and hundreds of other people I know have done the same thing. They directed their ambition and took charge of their lives and their destiny.

It's frightening to consider that a residential mortgage will amount to as much as three times the purchase price of a house over the life of the payment schedule. A $100,000 house would cost more than $314,000 at 10 percent interest amortized over 30 years. Additionally, the monthly mortgage payment can amount to as much as half of a regular income. What would you do with an extra $873 every month?

Many families cannot afford to buy a new home, because the debt would overextend their economic stability. For those who need more space for a growing family, remodeling might be the answer. But, when the estimate comes back from the contractor, the price is usually too high. Because of this, many individuals feel locked into a mortgage payment, and as a result, more and more people are considering doing the work themselves. Becoming the owner-builder eliminates most labor and overhead expenses.

I can hear your reaction now: *"I can't possibly build a home! I may know some carpentry, but nothing about electrical, plumbing or roofing."* But read on, and you might find that it's a lot less daunting than it may appear.

In a larger sense, an achievement such as building your own home gives you a measure of control over your destiny. Consider the self-fulfillment aspect of building and owning the largest single investment you're likely to have in your lifetime. Consider how you'll feel knowing that you'll have something meaningful to pass on to your family. No insurance policy can compare to the indisputable value of a solid real estate investment such as a family home. The peace of mind you have knowing that the biggest expense in your life is paid for is well worth the effort.

There are many other reasons to build your own home. The idea is not new, nor does it belong exclusively to the "back to nature" culture. Building your own home is an expression of yourself and your loved ones. Most tract housing projects, for example, have three or four basic plans, with minor variations in floor layout or elevation views. Very little expression of individuality emanates from such a cookie cutter design.

Having the type of home you want will not only add comfort, but a sense of fulfillment to your life. The home you design and build will be customized for you and your family. Whether your particular interests include gardening, cooking, artistic painting, special collections, or movies, your house will be unique. A traditional design can never incorporate all of those personal aspects within a standard plan. Nor will a builder ever organize all of your inner thoughts into just the house you want. Your personality and life's interests, and those of your family, should unfold into the house of your dreams.

Your home should also match your lifestyle. Some individuals prefer large bedrooms, or a country kitchen, while others dream of having a full-length porch, or three-car garage. When you have a house built, a contractor charges for even small changes to the original plan. By doing it yourself, you'll have the flexibility to make changes along the way for next to nothing.

Another very important reason for building your own home is your family. I cannot tell you how much building a home will do to bring a family closer together. The common goal of constructing a living space for each other is very unifying, especially for a family dedicated to the goal of attaining the independence and self-reliance home ownership provides. The greatest reward is that each family member will recognize that their contribution made a difference. The value to your children's self-esteem will last a lifetime.

Building your own home will also teach you about yourself. You'll discover that you can perform new and challenging tasks and that every experience improves what you do on a daily basis. After a short time, it will become evident that one or more of the construction trades unfolds naturally for you. In the long run, you'll develop a respect for tradesmen you never appreciated before, and in doing so, will build lifelong friendships. You'll also become accomplished in the art of conservation and recycling, and adept at using and adapting myriad building materials. You'll find yourself working on your home with enthusiasm, be it day or night. You're going to have a great time!

The purpose of this book is that of an inspirational guide. It represents an effort to inspire and advise a prospective owner-builder. There is, however, no substitute for experience. I strongly recommend that you gain as much practical experience as possible before you begin work on your dream home. Building a home is a huge undertaking, especially if you've never built anything before. Enlist all the help you can and practice on something small, like a shed or playhouse. Learn all you can from experienced tradesmen. This book will not be enough to teach a novice all of the skills required to completely build a home without making mistakes along the way. Indeed, builders with vast experience still make mistakes on every project. Likewise, this book makes no representation that everyone can accomplish these goals. However, individuals with a good head on their shoulders, some experience in manual labor, and a willingness to learn, can achieve success.

As with any undertaking, building a home is marked by overcoming numerous setbacks. These are learning tools. General contractors, subcontractors, tradesmen, and laborers all experience setbacks, even after 30 years of experience. However, they don't look at these setbacks as failures, but as learning experiences. You need to take the same approach.

Familiarize yourself with building laws within your community. For starters, you must get a building permit, so consult with your local building department for any legal issues that affect your plans to build the home of your dreams. When looking for a site, be sure to review restrictive covenants that place conditions on the use of land. Ask pointed questions, learn your rights and obligations, and make informed decisions. Before purchasing land, consult a realtor or attorney experienced in real estate and zoning.

By definition, the term construction pertains to the assembly of manufactured products and natural materials to create a building or structure. As such, these materials are subject to the vagaries of nature and human error. Since it's not always possible to avoid problems in the construction field, I've created a section termed *"Uh, Oh"* that deals with corrections and fixes for common construction errors. You'll find these sections at the end of strategic chapters. In my years as a contractor and building inspector, I have encountered thousands of problems corrected with a safe alternative that will pass inspection. These troubleshooting sections contain some of the most commonly made mistakes. Depending upon local jurisdiction rules, your project must pass several building code inspections. There may be some aspects of your design that trouble an inspector, and it will be up to you to offer an alternative that the inspector can accept. These sections will help you begin the dialog with your inspector.

The scope of this book is limited to the first-time home builder. You can use this book as a guide for specific portions of a project, or as a cookbook for the entire project. I will provide a pattern for designing and building a small (less than 2000 square feet) conventionally shaped, light weight, wood frame building. This is the easiest house for most people to build, and the most forgiving of mistakes. It's also the easiest for a single person or small group to assemble and erect.

Read the entire book to get the overall picture before you begin your project. Part One includes a discussion of design principles, plan drawing, and permit acquisition. Part Two applies those designs to construction techniques including foundation, framing, utility installation, and finish trim work. Where possible, pictures of construction will help you understand construction concepts.

Chapter 3 contains examples of building plans, including site, foundation, floor, framing, electrical, plumbing, and mechanical plans, as well as elevation views. Use these as a guide to developing your own design. The foundation design can be either a monolithic (single pour) slab on grade, concrete masonry block (CMU), or concrete foundation wall. You may use an interior bearing foundation wall to allow for a reasonable span for wood floor joists, instead of a monolithic concrete slab. This would allow for the option of building a basement.

The wall frame I recommend is a wood stud type with double top and single bottom plates. This is the most popular, conventional, and simple wall framing system. In this case, the price is as much a factor as is ease of assembly. The framing I recommend is a 2 x 6 wood stud wall frame which must be sheathed with exterior grade plywood or hardboard siding. This will serve as the finish exterior wall surface, or as a substrate for a stucco or masonry veneer finish at a later date.

The floor framing system I recommend is solid wood joists or composite wood joists. Either is satisfactory, easy to install, and forgiving of mistakes. The solid wood joists are limited in their span by strength and size, while composite wood joists are a manufactured product with longer design spans.

The roof framing system I recommend is premanufactured wood trusses. Your first house is a tough place to learn to build a cut roof. Open-web, wood trusses are the mainstay of modern production framing because of the price and speed of installation. For the most part, they are easy to erect, quick to install, and you'll have fewer complications.

The plumbing system I recommend is plastic drainage and water supply pipe. Only specific kinds of plastic water pipe are approved for interior installations, so review the manufacturer's conditions carefully. A beginner can install each with a little training and observational experience. Pressure testing of the plumbing system will verify that the connections are water-tight. You'll know right away if there's a leak that will require repair.

I also recommend an electric heat pump for heating and cooling. If you live in a cold climate, for example, you may need to add a *heat strip* to ensure adequate heating. You may need help from a qualified mechanical equipment installer, but you may be able to do much of the work yourself.

Another option for heating your home is electric baseboard heaters. Electrical service equipment rated at or less than 200 amps will be sufficient for a small home such as this, even for an all electric house. An *all-in-one* electric service panel will provide a means of service disconnect along with a load center and circuit breakers in one box. Wiring with copper type NM cable is most commonly used in residences. An all electric house is easiest and cheapest for the first time home builder.

That's it for starters. There may be a lot to learn, but the end result is well-worth the effort. Keep in mind that a home is more than just a place to keep your stuff. It's a place that provides a safe haven from both natural and social danger. It's a legacy that can be handed down from generation to generation. As mobile as modern society has become, a home provides a sense of stability. Simply put, it's the heart of the family.

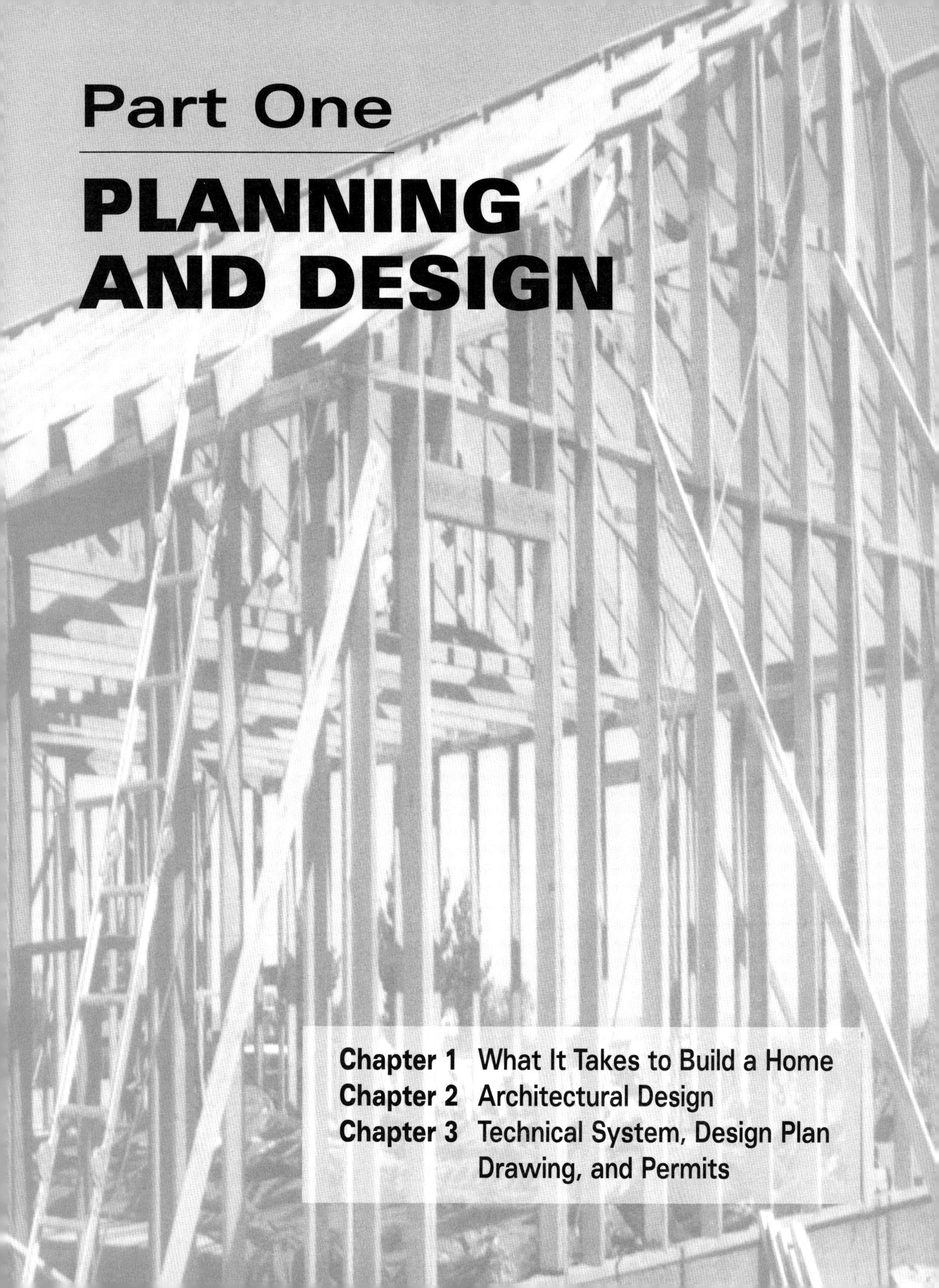

Part One

PLANNING AND DESIGN

Chapter 1

WHAT IT TAKES TO BUILD A HOME

Visualizing Your Work

I could start off by discussing tools, materials, labor, site selection, the importance of friends or your particular skills, but I consider something else more important. My high school shop teacher taught me that everything that is built has its origin in the human mind. If you intend to build a house, put it together mentally first. You must imagine it and construct it in your mind, one stud at a time. Mentally reduce each separate day's task into small increments. Try to work out all the possible problems before you begin each portion of a particular project. The times you select to build your project in your mind is up to you, but I found it desirable to begin envisioning a strategy near bedtime while drifting off to sleep.

Human resource directors know that some people do this naturally and some don't. Everyone is capable of developing this skill. My suggestion is that you begin assimilating this technique before you start your project. You will probably become a master of this before the project is finished. When you were young, you learned to imagine your assigned chores. You learned to mow the lawn, and planned your methods. Practice simple tasks like this to develop the skill of imagining.

Begin by trying to imagine all the problems you would encounter if you planned to drive to the store to get a newspaper. You might have a flat tire, run out of gas, have a wreck with another car or a tree, get lost, get a ticket, find out the store has sold out of papers by noon, forget your wallet, or forget your pants. How do you avoid these problems? You anticipate problems by thinking ahead. You always travel with a spare tire and enough gas. You practice defensive driving and remember to bring your driver's license with your wallet. You know the store usually sells out of papers by noon, and you choose to go earlier. You get cold without your pants, so you put them on. You see, you pretty much plan for these eventualities almost without thinking. That's what you need to do when you build a house. Learn to anticipate a problem and be prepared with a solution. Think the work through in your mind, one step at a time. If you are lacking some experience about what to plan for, this book will help you.

Available Time

You must be able to block out significant time in which to work on your project. If you are living on site, you will undoubtedly have time after work. However, you must plan on working during a significant number of weekends in order to bring the project to a relatively speedy fruition. You might also devote your vacations to the project for the first couple of years. You might try to stage certain aspects of the construction to give yourself a rest. First, pour foundation and floor slab. Wait a few weeks while saving enough money to purchase the lumber package for the wall framing. Then work to erect the wall frame and wait for a few weeks while you save money to purchase trusses, and so on.

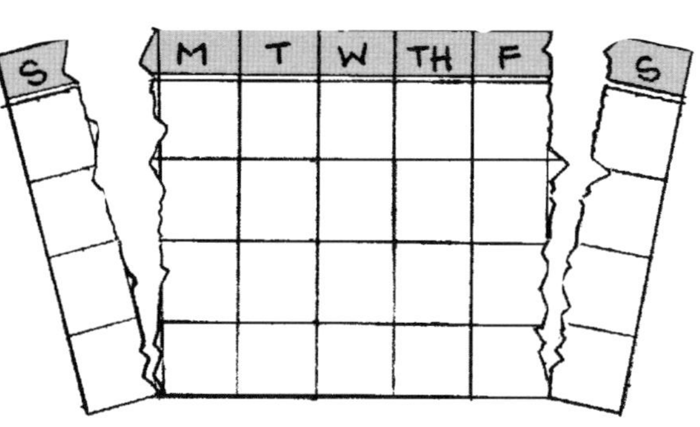

Patience and Commitment

The first trait that you must acquire is patience. Don't expect to achieve your goal of home building within a year, or even two. Remember where you were financially five years ago? Are you substantially different now than you were then? Even with a concerted effort on your part and help from this and other inspirational books, your path to home-ownership and financial independence will develop into a life-long journey. Don't expect immediate fulfillment. Remember that this may be a lifetime project; you really never completely finish your venture into home building. Even if you choose not to build another home, you will find ways to improve your newly-built home, and you'll have the disposable income to make those things happen. Your project might be a porch, a wine cellar, a work shop, a tennis court, a barbecue, or even a guest house for extended family. Once the building bug bites, there's really no cure!

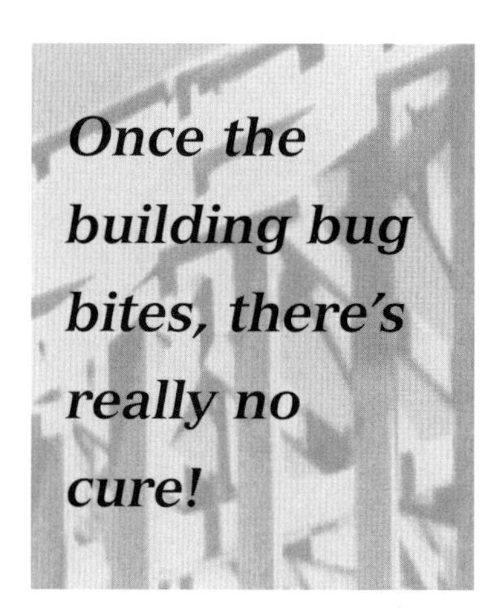

A second very important trait you must have is commitment. Commitment and patience should be your partners. They must also be your closest allies for the next few years. If you're accustomed to the 30 minute problem resolution established by modern sitcoms or even the 90 or 120 minute variety of motion pictures, you must change your paradigm. Think in terms of months or maybe even years for problem resolution. Sometimes you may not be able to do everything you want in a weekend. A rain storm may delay your well-laid plans for concrete work. A friend may have to postpone his offer to assist in a particularly complicated plumbing installation. Your school age kids may have their only chance to attend a sporting event during the one day when you can rent a grout mixer. Compromises and trade-offs will emerge, and priorities will be established by you as well as your family. Don't forget for whom you are doing this. My experience is that if the whole family develops a consensus, life is much less complicated. Think of this as a lesson in patience and learn to accept setbacks in stride. Remember, if it were easy, everyone would be doing it.

Are You Ready for This?

- ❑ Are you prepared for the upheaval that will inevitably result?
- ❑ Is your entire family as motivated as you are to do this?
- ❑ Do you enjoy watching and participating in construction work?
- ❑ Do you stop on the way to the store to watch a house being built?
- ❑ Do you do enjoy working by yourself? Can you physically lift heavy materials and tools?
- ❑ Do you think that you would enjoy building your own home?
- ❑ If you are not involved in construction, have you participated in a construction project?
- ❑ Do you know how to use some construction tools? Do you like to learn about others?
- ❑ Have you ever started projects, then lost interest? Will you be motivated to finish this project?
- ❑ Do you listen to advice of others? Do you act on that advice? Who gives you advice?
- ❑ Will this project bring joy into your life? Will you share your enthusiasm with others in your family?
- ❑ Why do you want to build your own home? Money? Joy? Family? Achievement?
- ❑ Are you expecting quick results? Does slow progress frustrate you?
- ❑ Can you see a project completed in your mind before you start?
- ❑ Can you predict at least some problems you will experience in a project before you begin?
- ❑ Are you highly organized and efficient in your daily life?
- ❑ Do you have a good memory? Do you remember what you ate for breakfast last Monday?
- ❑ How do you reach a decision? Do you discuss it with others? Do you brainstorm with others?
- ❑ What kind of people do you call your friends? Do you help each other out a lot?
- ❑ Do you like working crossword puzzles? How do you solve problems?
- ❑ Are you prone to emotional instability? Can you take highs and lows in stride?
- ❑ How do you deal with mistakes you make? Others make? Do you fix the problem—or fix the blame?
- ❑ How do you react to stressful situations? To confrontations?
- ❑ Are you confident about success in this endeavor?
- ❑ Do you know how to keep records? How much did the electric bill cost last May?
- ❑ Will you visit your neighbors often during your project?
- ❑ Are you safety conscious?
- ❑ Do you know how big a commitment you are making?

Is the Family Ready for This?

This is a project that extends to the entire family. Every member will alternately suffer some burden and experience some joy as a result. A project such as this will create joy as well as friction and stress among mutual participants. I will never forget one reaction I received when I was preparing to embark on this project. I decided to tell everyone whom I encountered about my plans in order to seek advice. One crusty old carpenter listened carefully as I explained my intent to put my wife and two adolescent children in a small RV for two years and build a 3000 square foot dream home with no mortgage, paycheck by paycheck. He carefully put his saw down on the work bench in the house he was framing, took his straw hat off, looked me directly in the eye, and delivered an admonition to me, which was his attempt at rescuing me from years of strife. He said simply, "Tell you what ...give your wife everything you own—car, land, money—and get a divorce now and save yourself the pain and suffering you're going to have from trying to do this."

His advice not only helped me to realize how a project like this impacts a marriage and a family, but how to avoid emotional upheaval by setting aside some time toward the end of each week to discuss feelings and problems caused by the turmoil of living within a construction project for an interminable period. Try to do something extra-special for the family at least once every month. Get away. Travel to a nearby city, stay in a motel room and visit a museum or carnival. Visit a farmer's market. Reward yourselves. You will have earned it.

Try to prepare your family emotionally for occasional setbacks. The cabinets won't fit. The truss manufacturer can't make the trusses for another three months. The hardware store must special order a faucet, which takes weeks. The paint is the wrong color. The delivery truck ran over your child's bicycle. Things will happen. How you react to those difficulties will define your character and your attitude toward your home.

Resources

In order to build a home with no mortgage, a regular apportionment of money is essential. Small periods of inactivity won't hurt some phases of construction, but will devastate others. Concrete or block work can be left temporarily without severe consequences, but wood framing left to the elements, even for a short time, can induce rot and invite vermin, including termites. You must develop discipline in funding your project. For the first several years just look at it as your own personal mortgage payment you pay to yourself. Emergencies will come up, and you will appreciate the flexibility your budget will allow, but do not let it become habitual.

Remember your goal. Structure your life in a manner that will encourage savings and frugal behavior. Every stretched meal is another box of nails. Every movie passed up is two sheets of drywall. Search for ways to reward yourself for your efforts. Take plenty of pictures and keep a diary. Plan for greater rewards after you have finished. As a family, collectively look forward to being able to make more opulent expenditures. Plan for your first real vacation.

Talk about remembering how hard it was digging footing trenches when, in the future, you're vacationing on the beach. In short, give your family and yourself something to look forward to. You must become adept at thrifty living. Buy a hair clipper and be your own barber for three or four years. Live on hamburger. Visit yard sales. Ask relatives and neighbors for their hand-me-down clothes for a few years. Plant a garden. In earlier times, families would partake in activities like hunting and fishing, which in addition to providing entertainment, would help improve their economic circumstances. Learn the joy and virtue of frugality. Your children will never forget the valuable lessons they will learn from this experience.

Taking on a project such as building your own home will require periodic need for moderate sums of money. While you could borrow money from a bank or mortgage company, it will add to the expense and you will remain in debt probably for the rest of your life. I want to encourage you to be debt-free.

Living on the Building Site

To build a debt-free house you must be able to buy the materials and install them yourself. In order to have enough money to buy the materials, you will have to make some lifestyle changes. If you are earning an income and living on a budget, your rent or mortgage payment represents the largest expense you have. Use that to your advantage. Most owner-builders start by selling their home, using the equity to purchase land and move onto their property. This eliminates their mortgage payment, freeing up at least $800-$1,000 per month. You must have a place to live on-site during the construction stage. Some owner-builders live in RVs (recreational vehicles), and some with larger families live in a small, older mobile home. Ask your local zoning department about the possibility of temporarily living on-site during construction. Although they may be prohibited from permanent use, some zoning departments will typically allow the temporary use of a mobile home or RV to serve as living quarters until the primary dwelling is complete, at which time the RV or mobile home would be removed. Some owner-builders build a garage or other outbuilding first and make it habitable, then live in it while completing the main house. Again, ask your local zoning department about this concept. They may assign the primary land use to the temporary structure despite its intended future use. Some owner-builders have used phased construction and built different portions of their overall house across several years. Their ultimate plan was too large for them all at one time. I have even seen some owner-builders live in larger tents for brief periods, though this frugal lifestyle may not meet with everyone's concept of personal sacrifice.

We did some of each. We first moved a small travel trailer onto our property and connected to permanent utilities while we started our project. Nine months later, the travel trailer we were living in was too small for the family, and the main house project wasn't as far along as we had expected. We decided to build a small, ancillary guest house, which we could live in comfortably until the main house was habitable. We lived there for a year and a half quite comfortably. After nine months in a travel trailer, even a small guest house seemed like a mansion.

However you plan your temporary settlement, ensure basic habitation necessities such as water, sewage, electric power, shelter, and possibly such amenities as a washing machine drain, an outdoor porch (for tools and expanded living), an outdoor sink, and outdoor electrical outlets for construction. In any case, you need to extricate yourself from a portion of your monthly income to have sufficient operating capital to fund this construction project.

Where Will the Money Come From?

Compose a monthly budget analysis. Look at what you spend now. Be honest. While every household has their share of extravagances that can be pared, don't overdo it. Try to avoid a totally Spartan lifestyle.

This will cause friction and may cause the family to lose interest. Allow for some luxuries. If you plan to live on-site, consider the fixed expenses you will incur setting up the temporary shelter and monthly expenses. While your savings or proceeds from the sale of your current home may pay for the land and initial shelter, you must plan for regular funds to apply to the building project. Use the following worksheet to estimate what you can hope to spend on your project on a monthly basis.

Monthly Budget Analysis

Monthly Income

1. Monthly family income $________
2. Bonuses, commissions, dividends $________
3. Retirement income/social security $________
4. Other income $________

Total income $________

Monthly Expenses

1. Electricity $________
2. Water $________
3. Sewer fees (if any) $________
4. Telephone $________
5. Transportation and gasoline $________
6. Food $________
7. Medical or health insurance $________
8. Clothing $________
9. Credit cards or charge accounts $________
10. Personal loans $________
11. Taxes state, local, etc. $________
12. Life insurance $________
13. Entertainment $________
14. Other $________

Total expenses $________

Add 5 percent for contingency $________

Total income $__________

Subtract total expenses $__________

Monthly operating budget $__________

Later, you will learn how to estimate the cost of your dream home. Divide that estimate by the monthly operating budget to determine the estimated number of months your project should take.

Solvency

Although site selection is important, having your property paid for is essential. Decide where your family really wants to live first. Check out school districts, crime rate, neighborhood standards, restrictive covenants, zoning regulations, proximity to shopping, health care, and other family concerns. Perhaps you'll need a year or two to pay off the debt on the land, or you might have equity in your current home that you will sell to get started on your dream home. Unless you have a clear plan for payment, which includes construction funds, try not to build on a mortgaged property. You could stand to lose all your investment and your labor. You'll also become accustomed to being in debt. Keep your commitment. Do not swerve down the road of indebtedness. By buying this book, you have demonstrated your goal of financial independence. Keep the faith. Keep your eye on the goal of home ownership and financial freedom.

Outside Help

I will concentrate on a simple, light frame, wood construction design. There is no substitute for in-depth consideration of a specific trade. If you plan to build a house other than wood frame, I strongly suggest additional research in the specific material or discipline. When you read a manual about a trade, keep in mind that local workmen may have a distinctive expression for a particular material or method. Ask a local tradesman to interpret when necessary. Remember how you acquired your current career. You had to study, learn, practice, and work at the position for a while. It will be the same for learning the building trades. Remember though, you don't have to learn enough to build a skyscraper, just enough to build your own home. Don't get bogged down learning about things that you'll never encounter. Concentrate on learning those things useful to your particular project. The friends you make in the construction field will help you distinguish between these matters.

Learning enough to build a home yourself may all sound a bit overwhelming, but it will be worth it. Keep in mind that each trade working to build a house makes a fee on their work. Their fee represents materials, labor costs, and profit. Profit on a house may represent around 10 percent of the cost of a new house. Labor may represent as much as 40 percent of the cost of a new home. You are attempting to save between 40 percent to 50 percent of the cost of a new home in exchange for a few short years of your labor and patience. However, the talent and experience of the tradesmen is valuable and cannot be discounted quite so easily. You must ascend to the journeyman level before you begin each step of your project. This is not an easy task.

You may seek help from friends, neighbors, and relatives. Be sure to be discriminating about accepting help. Before you agree to let your brother-in-law help you with framing, look at some of the framing jobs that he has completed. Remember that a professional's work is easy to spot, and the world is filled with amateurs.

If there is some aspect of construction that you are convinced you cannot do, even with help, you may be forced to employ a professional. Contractors are regulated by each state. They must validate their competence by passing examinations and have considerable experience. In most states they must also pledge a monetary bond to ensure their competence, honesty, and quality. Be sure to avoid any unlicensed contractor or tradesman. Usually, unlicensed contractors lack the skills or experience to become licensed. Also, you will have almost no recourse if they leave your job unfinished, with your money. Licensed contractors will ask you to sign a contract. Be sure to read it thoroughly and carefully. Watch for vague language. Do not sign anything you don't understand. Make sure that everything you agree upon is in writing and that you retain a copy of the documents. Verbal agreements are difficult to prove in court. When getting multiple bids, do not tell one contractor the amount of the bid offered by another. Remember, the lowest bid is not always the best. Look at previous work. Ask for references. Contact the state office charged with contractor regulation and ask about complaints. Consult the Better Business Bureau for any complaints. Ask for a copy of the contractor's performance bond. Ask about insurance, and review the policy. Verify liability for any accidents that might occur.

If you are compelled to hire some work, try to avoid paying 100 percent of the contract price up front. In fact, it would be ideal if you can agree to limit your down payment to just the cost of the materials. This would tend to guarantee the commitment and interest of the contractor. If there is any dispute between you and the contractor, you have limited your loss to the value of the materials, which are at least still on your property.

Advice

The best way to build a house, debt-free, is to build it totally by yourself. Nevertheless, you will need help, both technical and physical. You will invariably need to hire some work. Your goal should be to keep it to a minimum. Seek as much advice as possible—it is usually given free. The old adage, "free advice is worth what you pay for it," may be true, but free advice that is consistent and is from many different sources is probably reliable advice.

In order to reach your goal, you must become a limited expert at many things, both technical and nontechnical. Lots of first-time homebuilders build a second and third time out of pure joy in applying what they have learned the first time. Try to learn from others' mistakes by asking lots of questions and listening to advice. Brush up on your math skills. Read lots of product and material brochures. Ask questions and take notes. Keep a journal or scrapbook with collected advice, along with names, addresses, phone numbers, web sites, and other contact information.

A relatively new profession has developed within the building construction field during the last several years. Owner-builders who are complete novices and need a significant amount of help can still maintain control of their project by hiring the services of a project superintendent. The role of the superintendent will be to guide the owner-builder in the right direction and away from problems. Another role the superintendent sometimes offers is the supervision of any subcontractors needed by the owner-builder. Usually the project superintendent is a licensed contractor to satisfy the laws relating to construction contracting within your particular state.

Seek guidance from appropriate sources. Ask a plumbing tradesman to explain drainage and venting concepts. Ask a project superintendent to help you understand the vagaries of scheduling different aspects of construction such as plumbing and foundation work. Ask a framing tradesman to help with basic framing layout and design.

Another valuable source of help is your building inspector. Most often, the inspector has extensive experience in one or more particular trades, but is quite competent in all aspects of home construction. The inspector's job will not be to teach you how to build your house, but to ensure a minimum code compliant project. However, if you have some specific questions saved up, most inspectors would be very happy to work with you to achieve a high-quality result. To make your project go smoothly, I offer some more recommendations.

Suggestions

- Visit a nearby subdivision under construction on a regular daily basis, watch and take pictures and write down advice and notes. Greet everyone you see and ask questions of everyone who will talk with you. Watch various tradesmen. Follow a project from site work to carpet laying. These people will be your professors during the next few years. Treat them with the respect you would grant to a coach or an advocate. You will develop rewarding friendships.
- If you need skills in a trade that seems particularly complex, make a deal with a subcontractor to teach you the trade in exchange for your labor. Work several weekends with qualified professionals at this trade to see how specialists perform the task. You may only be doing grunt labor at first, but you will be near a professional. You may discover a skill you never knew you possessed. You will also build a network of contacts whom you can trust when you need to hire some professional help.
- Find others like yourself who want to be owner-builders, and establish a work co-op. Trade skills and labor with each other. You'll need this kind of help and make great friendships. Obviously, this concept is neither new nor unique. Our ancestors helped each other as a group and built entire communities.
- Seek permission to review the building plans of all the houses you see in construction stage and consider what design your family needs. Take your family on trips to construction sites and get to know their needs and desires. Gradually develop a design. Draft a floor plan and simple elevation views. Have all your contacts in the subcontracting trades review the basic design and critique it for you. Do the same thing with the local building department. Look for problems. This process will save you money and time. This will also serve as a brainstorming session for the framing design you select.
- Visit all of the building material supply stores in your area and get to know the manager or someone in authority. Ask questions about the quality and price of their products. Most building supply stores have a contractor sales division. This division will typically sell to large and small contractors who do business on a regular basis. Try to gain the privilege of conducting business through this division; the price will be reduced and the service will be better. Make sure that you select a store that you trust. Ask other contractors for recommendations. Employees should be knowledgeable enough to help you with answers to your questions about materials suited to your project. Pick the store with the widest variety of building materials and name brand tools and ask them to commit to a guarantee of the lowest price you can find during the course of your construction. If they agree, then agree to buy a significant amount of your materials from them.

Your Legal Rights

Even though your legal rights and obligations vary from state to state, there are some consistencies. There are myriad laws that you must abide by. Check with your local jurisdiction for a thorough list of these laws. Here are a few common points of law that affect you.

1. You have the right to submit an application for a building permit and receive a plan review. If you meet all of the requirements, you will receive a building permit and be entitled to call for inspections of your work. A building permit cannot be denied on any arbitrary or political basis. However, if you have outstanding building code violations on your property or another construction job, or if you owe money to a jurisdiction, a permit may be withheld.
2. You have the right to be secure in your home and property. The Fourth Amendment to the Constitution establishes your right to be secure in your home. It states: *"The right of the people to be secure in their persons, houses, papers and effects against unreasonable searches and seizures, shall not be violated, and no Warrants shall issue, but upon probable cause, supported by Oath or affirmation, and particularly describing the place to be searched, and the persons or things to be seized."*
3. The constitution generally protects individuals in their homes and property and requires that government action affecting those protections are protected by due process considerations. You have the right to use your property within the limitations established by law. These laws extend and include zoning codes, building safety codes, and sanitation laws. If you are in violation of a state or local law, you have the right to due process under the Fourteenth Amendment to the Constitution. It states in part "*... No State shall make or enforce any law which shall abridge the privileges or immunities of citizens of the United States; nor shall any State deprive any person of life, liberty, or property, without due process of law; nor deny any person within its jurisdiction the equal protection of the laws." (Ratified July 9, 1868, this amendment upheld the Civil Rights act of 1871.)*
4. The Civil Rights Act of 1871 established liability for actions by state and local governments that impinge upon your civil rights. Section 1983 states in part "*...Every person who, under color of any statute, ordinance, regulation, custom or usage, of any state...subjects, or causes to be subjected, any citizen of the United States or any other person within the jurisdiction thereof to the deprivation of any rights, privileges, or immunities secured by the constitution and laws, shall be liable to the party injured in an action at law, suit in equity, or other proper proceeding for redress."*

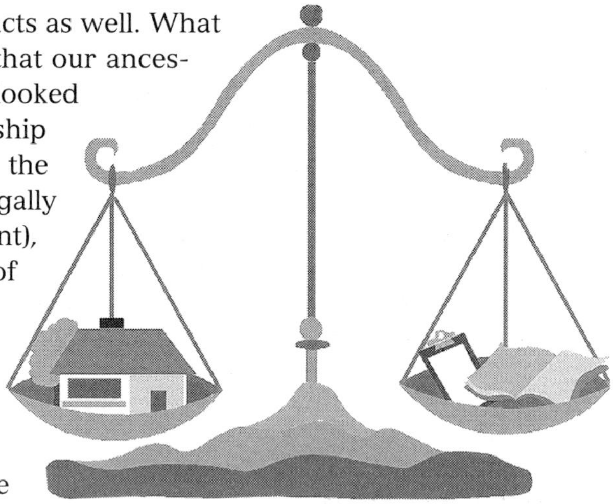

States may have their own versions of civil rights acts as well. What this means to you in your quest at homebuilding, is that our ancestors bestowed some legal protections to you. They looked forward and reserved certain rights for your ownership of property. These various civil rights laws preserve the claim of ownership and afford you the right to legally claim damages from parties (including government), who, without cause, attempt to restrict the legal use of your property. For instance, if you are denied a building permit by a building department on an arbitrary or capricious basis, these laws allow you to bring a claim or civil action against the jurisdiction.

It is important to cooperate as much as possible with your jurisdiction to achieve a common goal: a safe building. Make your inspector your advisor, not an adversary. While you may feel that some of your rights are at issue, the cost and ill will created by challenging or demanding those rights may exceed their value. The relationship between an owner-builder and the building inspector should be a partnership. You both really have the same goal. That common goal is sometimes obscured by edicts from the inspector or costs incurred from decisions made regarding your work. You will have an emotional tie to your project, and your work will be challenged from time to time by the inspector. Try to detach your emotional connection when discussing work with your inspector. Avoid feeling persecuted by suggestions that will really improve your project.

Your Legal Obligations

1. You must abide by all laws, ordinances, rules, and policies whether you know about them or not, so it's your responsibility to learn about them. Copies of all laws enforced in your community are usually available in the local public library. Some building departments have summarized these rules together in a booklet. Ask for a copy.
2. If a permit is required to alter your property, and you apply for one, you must submit all necessary legal documents and information that demonstrate your intent to comply with all laws and ordinances.
3. When you acquire a permit, you must still comply with all laws and ordinances, even if your plans have been approved, and even if they clearly show a code violation of some type. Approved plans are subject to an inspector's approval.
4. You must perform work that meets the minimum standard for code compliance. If your work is deemed substandard, you must bring it up to that minimum standard or your project may face abatement proceedings.
5. You must grant access for legitimate easements and a public right-of-way. These include electrical, water, sewage, or gas line right-of-way easements as well as access to repair or replace those lines.
6. You must abide by any legal subdivision restrictions or limitations on development.
7. You must abide by zoning regulations.
8. You must abide by ordinances such as flood plain, height regulations, slope restrictions, grading limitations, wastewater regulations, environmental regulations, and even certain planting regulations and animal protection acts.
9. Restrictive covenants within subdivisions sometimes supersede other laws and further limit your use of a property.

There are consequences to violating any of these legal obligations. You may be subject to potential fines, utility disconnections, or even an order to discontinue use of your structure. While most building code regulatory agencies will normally try to avoid such a serious legal sparring with an owner-builder, such actions may be deemed necessary to preserve public safety.

The Deed

A legal record of your ownership of your property may restrict certain uses you might want to make. These are identified as deed restrictions or covenants and are in addition to the laws and ordinances enforced by your local government. These deed restrictions usually increase requirements established by law. For instance, if the zoning code requires at least a 10 feet side yard setback, the deed restrictions may increase this to 20 feet, thus reducing the area permitted for building. Another aspect of deed restrictions may include the type of home permitted in the area. In order to create similar neighborhoods, deed restrictions may require a minimum size house or one built without a shiny metal roof. Other types of restrictions you could find in your deed may include public utility easements, a public right-of-way, or lender's rights if the property is secured with a mortgage. The deed is recorded in public record at your local recorder's office in the town hall or county office and is open to the public for review.

Staying Out of Court

The legal system will help you achieve justice, but nothing tarnishes the emotional tie and soils the attitude you have developed to a project more than prosecuting or defending a lawsuit. Also, nothing wastes as much time. There are good ways to stay out of trouble. Use the following as a guide to avoid the legal system.

1. Read everything you sign, thoroughly. Understand what you sign, or ask questions and get answers until you understand it. Then, keep copies of all of the documents you sign. This pertains to permit applications, purchase agreements, contract agreements, sales agreements, loans, disclosure statements, lien agreements, and everything relating to your effort to acquire a building permit.
2. Do not agree on anything substantial unless it is in writing. Written agreements are far more permanent than verbal agreements. If your neighbor agrees to cooperate by building a demising fence between your properties, get the terms and conditions in writing and sign with witnesses. If a contractor agrees to do masonry work in exchange for your used car, agree in writing, and there will be no dispute later.
3. Watch all work that you are hiring or paying for. Watch the quality and workmanship. If you don't like the way it is going, stop the work. Discuss your feelings with the representative you hired, not the laborers. Remember, it's your home! Don't be intimidated by a worker's threats. You have the right to a quality job. It's your money.
4. Try to settle differences early, as soon as you detect them. Do not let a problem fester in your mind and intensify until it can only be resolved in court. Begin with a casual dialogue. Then when an understanding is reached, get it in writing. Even a handwritten note is better than a verbal understanding. For instance, you may think that the subcontractor you hire will clean up the debris from his or her work. He or she thinks it is your responsibility. You get a citation from the zoning inspector, since it is your property. Think ahead and anticipate problems like this.

A Word About Building Codes

The state of building codes continues to be in transition, and it would be best to double check which code or codes are adopted locally, before you build. The rules you need to be most concerned with include those of your local building department. In the United States, prior to the year 2000, there were three model building codes. Generally, in the Western United States, the *Uniform Building Code*™ (UBC) was enforced. In the Southern United States, the *Standard Building Code©* (SBC) was applicable. In the North and Eastern United States the *BOCA National Building Code™* (BNBC) was administered.

These codes were developed, modified, and promulgated by a collective vote of those in the field of code enforcement along with input from the building industry. They primarily relate to life safety issues. They are not frivolous or capricious. They are the product of lifetimes of experience from consummate professionals in the field of building safety. Remember that if you are in violation of any of the codes, it's usually you and your family who are at risk, not the inspector. These model codes are usually only modified with local amendments in minor ways between state, city, town, parish, commonwealth or county jurisdictions across the country. Some jurisdictions, particularly large cities, or exclusive resort areas, write their own code.

There was a movement in the 1990s to unify the three model codes to establish a national building code. The *International Building Code* (IBC), which governs commercial construction, is promulgated by the *International Code Council®* (ICC); a body of code officials across the nation who vote for or against changes in code language proposed by any interested or affected parties. However, the IBC is not the only new code to be developed. It is part of a series of new code documents. This series of national codes represents more than 200 years of cumulative experience on the part of the three model code organizations and from the building industry. Over the last decade, each model code

Building Code publications.

group reformatted its own code in an agreed upon format, thus paving the way toward the *International Codes*. The first edition of the IBC was published in the year 2000. It is expected to be adopted by almost every jurisdiction nationwide in the near future.

The building trades such as plumbing, mechanical, and electrical installations are regulated in a mixed manner. The *National Electrical Code®* (NEC) previously was the only national code. It is developed and promulgated by the *National Fire Protection Association®* (NFPA). It is still accepted in most every state to enforce the installation of electrical systems. The NEC is currently the accepted code document for electrical installations by most states and local jurisdictions, due to its national imprimatur.

The mechanical code establishes the accepted methods for installing mechanical systems such as heating and cooling equipment within your building. Mechanical and plumbing installations were regulated by various model codes similar to the building code. The *International Mechanical Code®* (IMC) is a national model code publication that is adopted by numerous jurisdictions. It regulates the installation and maintenance of mechanical equipment such as heating and cooling apparatus for commercial construction. The plumbing code regulates all water supply and waste pipe systems that carry water to, and remove waste away from, your home. The materials permitted for the pipe, the pipe size, the installation standards, and assembly methods are all regulated by this code. Hot water heating systems, hose bibs, fixtures such as bathtubs, sinks, dishwashers, clothes washer, toilets, and bidets are all regulated as well. The *International Plumbing Code®* (IPC) is a national model code publication for commercial construction, which is adopted by numerous jurisdictions. Both of these code documents are developed and promulgated by the ICC and offered as model codes for jurisdictions to adopt.

Wouldn't it be nice if there was just one, smaller, easy-to-understand, building code that deals with every aspect of construction for dwellings? Because these codes deal with everything from skyscrapers to semiconductor fabrication facilities, they are complex. Most building code enforcement professionals know that. Because of this, a simpler code was developed by these professionals, just for residential construction. The IBC has a companion document entitled the *International Residential Code®* (IRC). This code book is significantly smaller and much more readable than any of its predecessor model codes. It deals exclusively with all aspects of construction including plumbing, mechanical, and electrical, as well as life safety and structural design, but only for residential buildings such as your project. If your jurisdiction is enforcing this document, I suggest that you invest a few dollars and buy this book. It will be worth it! It is filled with highly descriptive tables, charts, graphs, and other details, which explain code requirements in a clear, visual manner. Various code references in this manual, which relate to construction design requirements, refer to the *International Residential Code* 2000 edition.

There are basic aspects that pertain to the architectural and fire and life-safety requirements of the code for residential dwellings. Even with one national code, each state or town may still enact certain amendments or regional modifications that are uniquely suited to their local conditions such as seismic, wind design or flood hazard. But, generally, the following minimum requirements will apply to all dwellings.

Minimum Requirements for a Dwelling

- All habitable rooms such as bedrooms, living rooms, dining rooms, kitchens, hobby rooms, or even dens, need natural lighting and ventilation provided by an operable outside window or equivalent artificial light and mechanical ventilation. *Section R303*

- Rooms must be a minimum size of 70 square feet, and at least one room must be 120 square feet. Kitchens must be at least 50 square feet. Rooms must be at least 7 feet in any dimension. Ceilings must be at least 7 feet in height except for sloping ceilings. *Section R305*
- Sleeping rooms must be provided with an emergency egress window or door that opens directly to the outside. There is a minimum opening size of 5 square feet, a width of 20 inches, a height of 24 inches and a maximum sill height of 44 inches above the floor. *Section R310*
- Smoke detectors are required in every sleeping room, every hallway leading to a sleeping room, and on the ceiling of an upper story or basement. *Section R317*
- Any floor that adjoins a lower grade that is 30 inches below the floor must have a guard at least 36 inches in height and with vertical railing so that a 4 inch ball may not pass through. *Section R316*
- There must be at least one 3 feet wide, 6 feet 8 inches high, exterior door with a landing on both sides. Hallways must be 3 feet wide and at least 7 feet high. *Section R311*
- Unless exterior walls are rated as fire walls, houses must be at least 3 feet from property lines *(of course some zoning regulations may supersede this limitation). Section R302*
- Every house must be provided with a kitchen sink, and a bathroom with a toilet, tub or shower, and a lavatory. The kitchen must have cooking facilities. *Section R306*
- Every house must be provided with heating facilities that are capable of heating the house to a temperature of 68° at a point 3 feet above the floor and 2 feet from exterior walls. *Section R303*
- Stairs must be 36 inches wide, rise no more than 7¾ inches, tread at least 10 inches wide, headroom at least 6 feet 8 inches and a handrail between 34 inches and 38 inches above the nosing of the tread. *Section R314*
- All rooms, stairs and exterior exit doors must be provided with a switched light. *Sections R303.4 and R3803*
- Insulation must be installed and meet certain material standards. *Section R320*
- Foundation walls or finish floor level must be at least 6 inches above adjacent grade. *Section R404.1.6*
- Floor structural system must be capable of supporting 40 psf (pounds per square feet). *Table R301.4*
- Roof structural system must be capable of supporting 20 psf (pounds per square feet). *Table R301.4*
- Roof structural system may need to be stronger depending on local conditions such as snow load.
- Wood on concrete or near earth must be protected against decay and termites. *Sections R323 and R324*
- Glass must be safety type if within hazardous location. *Section R308*
- Some plastic pipe is permitted in some jurisdictions as plumbing water supply distribution. *Section P2608*
- Pipe and fittings must be able to withstand water system pressure without leaks. *Section P2503*
- Hot water is required to be provided to certain fixtures. *Section R306.4*
- Drainage pipe must slope at least ¼ inch per foot. *Section P3002.3*
- All plumbing fixtures and materials must have a listing by an approved testing agency. *Section P2608*
- Mechanical equipment must have a listing or evaluation from an approved testing agency or approved by the building department. *Section M1303*
- All code references are from the *International Residential Code* 2000 edition.

Remember that building codes only impose the very minimum standards that must be met. For example, all detached single family residences are permitted to be nonrated for fire resistance, unless they are too close to a property line. Residences do not need finish flooring, cabinets, painting, interior doors, trim work, food disposer or landscaping in most areas. They just have to meet the minimum life-safety aspects. So, in a sense, a man's home is still his castle as far as building code enforcement is concerned.

Inspections: What to Expect

The manner in which inspections are performed are unique across the United States. Some states enforce their own utility inspections, such as mechanical, plumbing, and electric and only allow local jurisdictions to enforce the building portions of the codes. You may need a separate permit from different jurisdictions for these installations. Some jurisdictions provide specialist inspectors for each trade such as electrical, plumbing or mechanical work. Some provide a combination inspector who performs all inspections at one time. The specialist inspector will be of the greatest value to you due to their experience in a particular field. Pick their brains. Most of them will respect what you're doing and be willing to assist you. The combination inspector will usually have a particular field of expertise. Find it and do the same.

Despite somewhat of a menacing reputation, your inspector can be your best ally against difficulties and your wisest coach. Remember, he or she has seen projects like yours more than a few times. They know how to guide you. Make friends and establish a working, professional relationship. Seek their guidance and their counsel and then follow it!

The code official or administrator is the plans examiner's and inspector's supervisor and has the authority to interpret the code and to accept alternate materials or methods of construction that meet or exceed the performance criteria set forth in the code. If your building inspector is not able to allow certain techniques and can find no remedy to your proposed installation and if you can provide for the basic safety and suitability of your proposal, ask the inspector for consent to appeal their decision to the code official. The code official can grant a special consideration. In some areas, a building code board of appeals is a body of experienced men and women selected to perform this appeal function.

Before an inspection, try to be prepared with specific questions that may affect inspections of your project. Inform the inspector of your plans to build by yourself. Ask about others he might know who can help with advice or trade in mutual labor. The inspector has a wealth of information about construction in general, not just about the code.

Ask your inspector for some details about your inspections. Try to learn what to expect. You can anticipate inspections at regular intervals. Some inspections may be clustered together. For example, a zoning check, footing, and foundation inspection may occur simultaneously. Termite treatment inspection will occur at this time. Groundwork plumbing will normally occur after the foundation is placed. Concrete slab floor is placed after backfilling around ground work plumbing. The first framing inspection may be either the first-floor framing or roof decking. Then, usually all the trades are inspected prior to the final frame inspection (just prior to insulation and drywall). This is to ensure that the tradesmen do not destroy critical framing elements while placing their materials and equipment.

Many jurisdictions are inspecting according to the energy code to ensure that proper insulation and windows are installed. Stucco lath or stucco siding is inspected next. The lath is usually installed before drywall is installed, and the stucco is usually installed after drywall is installed to avoid cracking from the pounding from within. The final inspection is last.

At the end of some strategic chapters, there are checklists of inspection details for which most building inspectors will be looking. At the end of Chapter 14, a through checklist is provided for the pre-insulation inspection. This includes a more exhaustive list of all utility installations, as well as framing construction. This is usually the most formidable inspection to pass and usually includes all of the rough-in utility work as well as the wood framing after all of the cutting, notching, and boring is complete. The complications between plumbing, electrical, mechanical, and structural aspects of the construction cause this inspection to be problematic. The checklist for the final inspection is at the end of Chapter 17. Code sections from the IRC are referenced at the end of each item in the checklist. This will help you more thoroughly research the requirements for that particular aspect of construction.

Do not call for an inspection until you are ready. You'll just be turned down and waste the inspector's time. Don't rush to complete work in a haphazard manner to beat the inspector. Become knowledgeable about the building code in your area. Impress your inspector with your knowledge of the building code. Buy a copy, or borrow it from the library, and read it. Ask for explanations where you need them.

Zoning

Most jurisdictions are divided into specific areas or regions that describe and limit the type of use permitted within that zone. These zoning districts usually have very tightly defined limitations for use within the zone. For instance, site-built, detached single family housing may be the only type of use permitted in a specific zone. In this case, mobile homes would be forbidden, except in some cases for temporary use while building a home. Since the value and appeal for any house is partly based on its neighborhood, the zoning code is intended to prevent the value of property from diminishing. This prevents you from gaining a factory or hog farm as a neighbor three years after you build your dream home. It also helps maintain a stable property tax base for the jurisdiction. But zoning codes also clarify where you can build on your property, including proximity to the property line, how big or tall your house may be, or even in some cases, what exterior color may be used to paint your home. All of this is intended to establish and maintain the overall appearance and character of your neighborhood.

Construction Safety

In the course of building your dream home, it is very likely that you will suffer some injuries. Since most building departments do not enforce job-site safety, you will need to learn some. While the Occupational Safety and Health Administration (OSHA) has jurisdiction over working conditions at construction job sites, it has little control over your home building project. However, there is nothing preventing you from using their rules to improve your odds against having a calamity. Contact the local OSHA office and request to speak to an inspector. Discuss your project and request advice or tips for job safety. In addition, ask for a list of the most frequent safety infractions in your area. If there is a safety training program provided, I would reccomend you attend.

You would be smart to acquire home owner's insurance. Even though insurance agents frown on insuring owner-builder projects, they will do it for a higher premium. When you talk with your insurance agent, request advice on construction job-site safety. Their confidence in you will grow and you may learn something useful as well. Your agent will also be impressed if you attend any construction safety classes, and your rates may go down as well. If you cannot get homeowner's insurance without a licensed contractor, contract with one to supervise and oversee your job for a small fee. An insurance company may accept this arrangement and provide insurance coverage.

I can't possibly enumerate every safety issue that may come up in your project. What I can give you is a description of the climate for a safe work site. In a sense, construction job site safety is a lot like beauty; you can't really define it, but you know it when you see it. In almost every project I have inspected over the last fifteen years, I could tell if there was going to be a problem with the construction when I drove up to the job site. Why? Because of how clean the project was kept. If it was clean, neat and well organized, the inspection rarely failed. Cleanliness is inextricably linked to organization and safety. If you religiously clean up every day, you'll be better off. This may be a job for your family while you're away with your day job. The following is just a starting point for you to explore your personal job-site safety regimen. Add to the list as you need, based on personal characteristics and regional aspects such as soil conditions and weather.

- Maintain job cleanliness at all times. Remember—it's first on my list. Enough said!
- Wear gloves, back harness, hard hat, safety shoes, and heavy work clothes during the appropriate times. Do not let items of your clothing dangle loose. Wear all your clothes in a firm manner, including shoes, shoelaces, socks, pants, shirt, overalls, and hat. Wear any glasses with a secure strap. Try to look the part of a carpenter. Avoid working in shorts, tank tops, or tennis shoes, although it's tempting when it's hot. Wear a loose bandanna around your neck for sweat. Wear goggles or other eye-safety glasses when using certain tools.
- Warm up with isometrics 15 minutes prior to engaging in heavy work such as digging, lifting, reaching, hammering, etc. Do not work on a full stomach. In fact, try to avoid heavy meals before beginning work. Eat a light breakfast and achieve satiety at lunch time. Keep lots of water around your job site. If you get dizzy or light-headed, rest in the shade for a while before continuing.
- Plan projects that fit into the time frame available for your work day. Do not try to bite off more than you can chew. Be careful and methodical in your work routine. You may rush when you see the sun beginning to set if you began rolling trusses at 3 p.m. Do not get into a hurry! Plan work for available time.
- Analyze things before you engage them. You do this everyday in other ways. While driving, you scan for pedestrians and automobiles before you reach an intersection. Look very carefully at each stage of your project and ask yourself, how could this become a safety hazard? Notice where people are standing. See if anything could fall on them. If someone is working below you, a loose hammer could drop on his or her head.
- When you have a close call (and you will), keep calm and do not overreact. Many car wrecks are caused by overreaction. Just as many job accidents are caused from the same thing. If a 20 ounce hammer falls from 14 feet and hits your child, your 200 pound body falling down on her because you were agitated will not help matters.
- Don't become distracted by anyone while handling power equipment. A phone call can wait. A sticker in your daughter's foot can wait. Your losing a finger is a heavy price to pay for not keeping someone waiting. Tell your family not to interrupt you while you are using power tools, short of an emergency, then work out a hand signal or sign for the interruption.
- Know where the nearest hospital or urgent care facility is and what hours they are open. When you're bleeding is not the time to thumb through the telephone book.
- Speaking from experience, do not let your nine year old daughter learn to roller-skate on the roof. Construction projects are attractive entertainment for kids. They will want to use your project to climb, jump, lift, carry, roll, hide, search, pretend, or even roller-skate. While they should be an integral part of the construction of their home, there must be limits on their participation. Keep the kids safe—and watch out for them!

- Put warning signs, barricades or tape where necessary to prevent a calamity. A delivery truck almost backed into our basement because he couldn't see it from his cab. An open trench for water, electrical, or sewage can be an especially nasty surprise when your neighbor is chasing her cat at night.
- Be very careful with heavy equipment such as backhoes or tractors. Delivery drivers will not know your project as well as you. They may not know that you have an autistic child or a deaf cat. Have a talk with them in the street before they enter your property. If you do not have the skills to use heavy machinery, learn them before you rent one to use at your project.
- Install lights for early morning or nighttime work Don't try to work in dim, natural light.
- Be prepared for injuries with a first-aid kit which has everything from tweezers to large bandages, to smelling salts.
- Use ladders exactly as they were designed to be used. If you have to stand on the top rung, the ladder is too small. Get a larger one. A rickety ladder is a dangerous job hazard. Would you leave an abandoned refrigerator on the job site? Your ladder represents an even more attractive nuisance for children of all ages. Wooden or fiberglass ladders are safer when working on electricity, since you deny electricity a convenient path to ground.
- Don't try to do more or lift more than you safely can. Don't over exert yourself. When being helped with a lifting chore, coordinate before you grab and lift. Rehearse and pretend lift, so each person knows what to expect.
- Assume that every electrical outlet, wire, or device is live and energized every time. Do not work on a hot electrical outlet or switch box, ever!
- Guard against accidental ingestion of poisons applied for termite abatement. Children and animals are especially vulnerable.
- Keep electrical cords in good working condition and out of contact with heavy or sharp objects and away from water.
- Never do electrical wiring work when the electricity is turned on. Do not stand in water while near electrical installations.
- Replace a blown fuse with the same size and rating to protect against starting a fire.
- Use the safety guards installed on power equipment. Do not discard them because you can't see your work as well. The guards are there for a reason, like seat belts on a car. You wouldn't discard them because they're a nuisance, would you?
- Use rebar caps to cover the sharp ends of rebar protruding from hardened concrete. A very good friend of mine was impaled on one of these while trying to jump over a trench.
- Do not leave nails in wood so that the point is aiming upward toward the bottom of your foot. If the wood is trash, bend the nail over.
- Watch where you are going and where you are stepping at all times. Materials, equipment, and tools are constantly changing locations. Don't try to predict where something is or isn't. Look for it.
- When using flammable, caustic, or hazardous materials, read and follow the directions for use. Look for adverse interactions between materials.
- Do not leave construction materials hanging in a precarious manner during unattended periods. Use bracing or scaffolds to effectively secure what you've built until you can finish. Wind and gravity will spoil your plans every time.
- Never work under the influence of medication, alcohol, or drugs of any kind.
- When you are really tired, stop, and rest. Don't push yourself beyond your limits.

Tools for the Job

You will need tools to build a house. You will need many different kinds of tools for completely different aspects of construction. Remember the age-old admonition, "Get the right tool for the job."

The kind of tools you'll need will vary according to the material used in the construction. Wood framing will require lots of cutting tools. Masonry will require grout and mortar mixers and hand tools. Concrete placement will require finishing tools. You might consider buying the basic tools for woodworking and rent some of the larger equipment that you may only use during a brief period unique to the nature of project at hand. Selecting tools to buy presents a dilemma: should you buy top of the line or bargain basement? I recommend half-way in between. Don't get the cheap model. It may be cheap for a reason. Buy a middle line version of the tool you need. Also, buy from a store that will accept returns easily. Mail in your warranty card and keep receipts.

Maintain jobsite cleanliness at all times. This is NOT what your site should look like!

Quality workmanship is partly the result of good quality tools as well as sharp replacement blades and bits. Do not rely on tools made for household use. These will not last very long during a construction project. Chose well-known brands with a good reputation. Buy hand tools which feel right for you. A 32 ounce framing hammer may be too large for you to swing. Be comfortable with the tools you select. Mark your tools to prevent them leaving the job site. You will have helpers with their own tools. Some of these will be identical to yours. Use spray paint or some other convenient method to clearly tag yours.

Power Tools

Most wood cutting tools are electric powered. No matter how silly you think it is, please read and follow all the safety precautions for each tool you use. Portable power saws are able to make short to medium rip cuts and cross cuts up to two inches in thickness. They include circular saws and worm drive saws. They are intended to be used to make a complete cut, separating two pieces of wood. The disadvantage to these saws is they cannot make a full-height cut in a piece of wood. They will always leave a semicircular shape when stopping inside the cutting pattern. Plywood sheets are usually cut in field situations with a circular saw. A power miter saw is a circular saw mounted on a frame that allows the saw blade to cut a piece of wood at specific angles. A compound miter saw will provide the saw with an extra pivoting angle to make compound angle cuts. It is good for cutting up to six inch wide and two inch thick wood. A jig saw, saber saw or reciprocating saw is used to make an intricate, full-depth cut or partial cut without requiring an over-cut or leaving the tell tale semicircular pattern from a circular saw.

Larger power tools such as a table saw or radial arm saw are power circular saws mounted onto a platform, which provides a framework for exact positioning of wood to be cut along the saw blade's path. A jointer is a power tool which will remove uneven surfaces from a side of a piece of wood. If this is done to two pieces of wood, they should fit together perfectly and be *joined* together. A power drill is used to bore holes into wood. The different size holes are made from different size drill bits. The size of most power drills is either ⅜ inch or ½ inch chuck. That means that the stem of the drill bit is either ½ inch or ⅜ inch. My advice is to get the ½ inch drill with enough horsepower to drill a hole into 4 inch thick Douglas fir board without hesitation or strain. Be sure to get a drill with a reversing switch. Also remember that drills work only as good as the bit. Keep rotating your drill bits and buy new ones or regularly sharpen old ones.

Electric band saws cut in the same fashion as jigsaws; they are just larger and positioned within a framework that allows for control of the wood to be cut. Routers will gouge out a running section of wood with a uniform pattern or shape, such as a bevel or rounded corner or even inside cove. There are hundreds of router bit templates from which to select. Router tables use routers and control wood better to ensure a more even, consistent pattern.

Hand Tools

Handsaws are necessary when in awkward situations where a power saw will not work. There are rip blade and crosscut blade handsaws. A ripping blade will make smooth cuts along the grain of the wood. A crosscut blade will make similar cuts across the grain. They have different number and size of teeth. A hacksaw is designed to cut metal. A coping saw is like a hand tool version of a jigsaw and is used to make fine turns in wood. Lots of ornate older gingerbread ornamental wood was cut with a coping saw. Wood chisels are used to gouge a section of wood out. For instance, if you are making recessed door hinges, the surface of the hinges must be flush with the surface of the wood. So, the thickness of wood must be removed. Chisels can do that.

Other hand tools include hammers, screwdrivers, pliers, wrenches, and pry bars. Hammers are available for framing, trim, or finishing and drywall uses. Screwdrivers are usually either Phillips head or flat head type. A utility knife is essential throughout your project. Get plenty of replacement blades.

Measuring Tools

Tape measures are used to measure lengths for cutting or installing framing members. They are usually very portable and even have a clip to attach to a tool belt. Do not use a broken tape measure or one that is damaged. Buy a new one. Get a wide, 25 foot long retractable tape measure. A framer's square is used to establish a perfect 90 degree angle from a reference line. A smaller type of square has an alignment guide on one edge, which comes in handy when making reference lines for quick cuts. A sharp carpenter's pencil is essential to doing good work in framing. The pencil is flat to prevent it from rolling off your project's working surface. Get several. A chalk line uses the concept of a string laden with colored chalk pulled taut to establish a straight line for cutting. It has a catch on the end, much like a tape measure to hold it in place. Pull tight and snap the string line, leaving a clear mark to use when cutting.

True and Plumb

A plumb bob can be used to establish a perfect *up and down* from a specific location. This will establish the true plumb from a reference point, based on gravity. A framing or masonry level is used to achieve straight and level walls. This is an essential tool! A builder's level is mounted on a tripod and looks like a survey instrument. It is important to have in order to establish a level surface for floor and foundation work.

Concrete Tools

Finishing concrete is performed with trowels, jitterbugs, screeds, and sometimes, power equipment. The power trowel should not be used by the novice without some experience. It takes a certain *feel* or control to keep from losing the machine, which could result in injury, damaged equipment, and a poor finishing job. Jitterbugs are used to force gravel down and bring cement up. Concrete finishing floats and hand trowels are used to achieve a smooth surface.

Pneumatic Tools

An air compressor will allow you to use pneumatic or air-driven power tools such as paint sprayers, power chisels, all-purpose texture sprayers, nail and staple guns, and even general clean-up. A pneumatic nail gun will make your life much easier, especially when you are nailing upside down. A pneumatic staple gun will provide similar increased production for tacking down floor and roof decking, roof shingles, as well as exterior siding. Follow the manufacturer's instructions for pressure settings, and use care with these tools. The added air pressure can cause injury if misused. When you have finished framing and have applied drywall, you can use the compressor again by spraying texture with a hopper gun. Later, you can even paint both inside and outside with the compressor. The compressor always comes in handy during cleanup of fine dust particles.

Inventory your tool crib. There are some essentials you will need for the entire project. You must have the following:

- Worm drive or circular saw with replacement blades
- A ½ inch drill with all sizes of bits including countersink bit
- Orbital and belt sander
- Reciprocating saw with all sizes of blades
- Miter chop saw (on long work surface)
- Shop vacuum
- Handsaws (crosscut and rip)
- Coping saw
- Hacksaw
- Wood chisels
- Pliers
- 25 foot tape measures
- 2 foot framing level
- Shovel

- Rake
- Nail claw
- Pipe wrenches
- Jigsaw
- Hoe
- Wrenches
- Adjustable wrenches
- Metal files (flat and round)
- Pliers
- Wire cutters
- Screwdrivers (Phillips and flat end)
- Nail belt
- Pencils
- Hammers (framing, sledge and trim)
- Hand winch (come-along)
- Belt sander
- Bench grinder
- Knives
- Tweezers (splinters)
- Chalk line
- Ladders (16 foot extension and 6 foot step)
- Clamps (all sizes)
- Calculator
- Pry bars
- Plumb bob
- Pocket knife
- Carpenter's pencil
- Carpenter's square
- Combination square
- Nail set
- Caulking guns
- Hand float
- Trowels
- Putty knife
- Staplers
- Wire strippers
- Extension cords (14 gauge wire)

Of course, there will be many tools that are very costly to purchase, but that you can rent for a brief period. Sometimes, if you can cluster the jobs together, renting larger equipment for a brief time makes perfect sense. Make plans for a week or weekend, and cluster those functions instead of laying out hundreds of dollars for a tool that you will not use again for years. You will need to make the determination as to which tools these are during your project, but some of them will include the following:

- Scaffolds
- Concrete float
- Survey instrument
- Pneumatic nail gun
- Pneumatic stapler
- Drywall jack
- Drywall texture gun
- Airless paint sprayer
- Concrete mixer
- Powder-actuated fastener
- Hardwood floor nailer
- A ½ inch right-angle drill
- Hammer drill
- Cutting torch
- Tile cutter
- Masonry wet saw
- Builder's level
- Steel bender
- Concrete power trowel
- Wench
- Survey transit
- Cutting or welding torch
- Mortar mixer
- Router and all types of bits
- Compressor (1 HP or over) with various quick-release fittings
- In the absence of electricity, a generator is essential

Scheduling

Scheduling various aspects of construction for a home is an art in that there must be some flexibility for delays, setbacks, problems, deliveries, and coordination. The most important thing to remember is to never get ahead of any phase of construction without completing the previous phase. For instance, don't install drywall before insulation. A perfectly scheduled project is the result of knowing what comes next. The scheduling for your project will depend on the type of construction. If you are building a basement home or are using post and beam construction with perimeter foundation walls, with plumbing only in the upper floor level, you will not need to install drain waste and vent before pouring a concrete slab foundation. If you are building a single story, slab on grade foundation, you must install the plumbing drainage pipes before forming for concrete pour. Study your plan and develop a schedule most appropriate for your particular project. When you plan your schedule, be sure and account for the time allotted for the inspector. You may be turned down and need to make some changes. These changes will cost you in extra time. Schedules can go awry, and time commitments for others may suffer if you don't anticipate the lost time.

SAMPLE SCHEDULE

STAGE	TIME PERIOD (ESTIMATED)
Read this book from cover to cover	2 Weeks
Site selection	1 Month
Negotiations with utility company	1 Week
Plan development	1 Month
Permit acquisition	2 Months
Excavation and grading work	1 Month
Utility connections*	1 Month
Septic tank installation*	1 Month
Basic habitation (RV or Mobile home)	1 Week
Footing form work and pour*	1 Month
Foundation form work and erection*	1 Month
Underground plumbing (drain, waste, and vent)*	1 Month
Slab floor form work and pour*	1 Week
Building layout	1 Day
Wall construction and siding installation*	1 Month
Roof truss installation	1 Month
Roof decking*	1 Month
Temporary roof *dry-in* installation	1 Week
Plumbing top-out*	1 Month
Electrical rough-in*	1 Month
Mechanical rough-in*	1 Month
Final interior carpentry work	1 Week
Insulation*	1 Month
Drywall*	1 Month
Tape, texture, paint	1 Month
Final trim; electrical, plumbing and mechanical*	2 Months
Flooring	1 Month
Moving time	2 Weeks

*Denotes work that normally requires a building inspector's approval prior to proceeding to next stage.

Organization and Management

The reality of the time needed to build your own home is that it may be longer depending upon several things. Weather may cause time delays. Deliveries may be delayed. Others may not be able to help when they promised. You must develop the skills to coordinate between many different groups of people. You will have to learn effective communication skills within and between each of these diverse groups. When you seek acknowledgment from a tradesman of his agreement to do a certain task on a specific date for a defined fee, ask yourself if his verbal okay means the same thing to each of you. Maybe he is saying okay to the date and task but thinks that he will negotiate the payment upward. When you are ordering your lumber from a clerk at the lumber yard, does her nodding her head to your query about return policy mean that the company will pick up defective lumber? Or did she bother to tell you that you must bring it back yourself to get credit? In every case of interaction between you and others, realize that this is your project, not theirs. Others will only have a passing interest in seeing you succeed in your goal of home building. It is up to you to organize the project and effectively manage it.

Work to make sure that all parties involved in your project know exactly what is expected of them and how they must interact between each other in order to earn their compensation. Plan for contingencies such as materials being delayed without advance notice. Do not be too optimistic about schedules or others' commitment of time. Be sure not to conceal work that is not really complete or missing inspection approvals.

Chapter 2

ARCHITECTURAL DESIGN

Site Selection Considerations

The first step in building your home will be to determine where it should be. The importance of proper site selection cannot be overstated. Where you live will affect every aspect of your life. Your neighbors will affect your life for better or worse. The school your children attend will shape their lives. Your commute to and from work will affect your attitude. The weather will affect your mood. The wind and dust may cause cleaning to be a nightmare. The scenery around your home will become your daily view. How your neighborhood is kept will color your feelings. The personality of the neighborhood will either agree or disagree with you. Your entire life will be affected by your site selection. A bad decision about site selection will haunt you every day. Your rewards in finding just the right site are worth the effort it takes. Use the following checklist of questions to begin a dialog with your family about their preferences in their ideal homesite.

- ❑ Look for problems now, not later.
- ❑ Is the size of the lot large enough for your home? Don't forget to take into consideration setbacks and easements. Will your neighbor wake you up when they open their garage door at 5 am? Will you bother them in some way?
- ❑ Is there a sound or traffic problem from a nearby airport, railroad or highway?
- ❑ Were there ever any toxic wastes dumped on site? Research land records. Ask neighbors.
- ❑ Is a legal access available? Are the roads too rough for your vehicle?
- ❑ Who owns the land adjacent to you on all four sides? What is their intent for future development?
- ❑ Is the land large enough for your plans? What are the distances between well and septic tank?
- ❑ What is planned for land in the immediate vicinity near your home?
- ❑ Are any major roads planned? Power plants? Subdivisions? Retail or industrial centers?

Utilities, Services, and Conveniences

- ❑ Is there water, sewage, power, fuel, gas, telephone, cable television, and garbage disposal available for your property? Are there any existing improvements? If so, are they functional? If not, how will you acquire them?
- ❑ What is the quality of the water? What is the depth of the water table?
- ❑ Is police or fire protection available? What is the cost? Taxes? Is fire protection subscription required?
- ❑ Are fire hydrants nearby? If not, will your fire district require an automatic fire sprinkler in your house?
- ❑ Does the school district suit your needs? At what cost? What are the school taxes? What is the travel distance?
- ❑ Is there a sanitary/storm sewer near your property? What is the cost? Initial fees? User fees?
- ❑ How far is the nearest grocery store, hospital or medical facility?
- ❑ How long would it take for an ambulance, police or fire protection to reach your site?
- ❑ Will a concrete delivery truck be able to access your site? Is on-site storage of construction materials permitted? Is there a time limit for construction activity? Do the materials have to be concealed?
- ❑ Is your lot safe from vandalism or theft during construction? Do neighbors lock their cars? Is fire insurance available? At what cost? Is there a volunteer fire department nearby?

Physical Condition of the Property

- ❑ What are the soil conditions? Was it ever a waste disposal site? Is a soil report necessary?
- ❑ Is there a percolation test for the soil on file?
- ❑ Who maintains the access roads and streets nearby? Do the roads carry heavy traffic?
- ❑ Has fill material or rock been dumped on property? It may need to be removed prior to construction.
- ❑ Is the land in a special flood hazard zone? Is a dam upstream? If near a river, a soil report may be cheap insurance.
- ❑ Does the lot provide shelter against wind, dust, rain, snow, and noise?
- ❑ How much excavation or earthwork will be needed just to start the building process? Will blasting be necessary to remove rock just to begin the foundation? Is that permitted?
- ❑ What is the water table? How good is the soil for construction? For drainage? To avoid wind erosion?
- ❑ Is this land subject to erosion from heavy wind, rain or dust?
- ❑ Is there an insect or vermin problem evident? Can you see termites present?
- ❑ Does the weather and temperature meet with your expectations for a homesite?
- ❑ Does travel to and from this site pose any problems throughout the year?
- ❑ Is the land appropriate for gardening?

Legal Questions to Ask

- ❑ Are there restrictive covenants that will affect your plans? Deed restrictions? Zoning restrictions? Are there any special laws, ordinances, rules or provisions that would prevent you from building the kind of dwelling you want?
- ❑ What is the zoning for the lot? For surrounding areas? Will industrial neighbors, such as a hog farm or a quarry be moving into your neighborhood in ten years?
- ❑ Are endangered species on or near your property that may affect your intended use?
- ❑ Can the property be subdivided any further? Can you sell a piece to pay for your portion?
- ❑ What are the required setbacks? Are there any existing easements on the property?
- ❑ Can you afford the property taxes, even while building? Will previously passed bonds increase your taxes in the near future?
- ❑ Is the land free and clear of all liens and encumbrances? Who verified this? What does a title report say? Is the title research insured?
- ❑ Will you need any easements to access the property? Who would provide them? At what cost? Are there any existing easements on your property? For what purpose?
- ❑ Are there any homeowners associations or architectural committees that have placed restrictions on the property to which you must abide? Would these significantly alter your lifestyle?
- ❑ Are there any moratoriums on development in this area?
- ❑ Are there any liens on the property? Any second mortgages? Any nonrecorded agreements? Will seller provide a warranty deed?
- ❑ Will an attorney be needed? Who will pay for the attorney—seller, buyer or realtor?
- ❑ Can you find the property boundaries without a surveyor? Is a survey necessary?

Do You Belong Here?

- ❑ Is the neighborhood appropriate to your family's needs and lifestyle?
- ❑ Will your preferred house design look appropriate on the lot? What do neighboring houses look like?
- ❑ Is the outdoor terrain conducive to games, parties or barbecues?
- ❑ Will your house be able to face the proper direction to suit your needs?
- ❑ Will your house be the biggest in the neighborhood? The smallest?
- ❑ Will your neighbors tolerate your living on the property in a RV or mobile home during construction?
- ❑ Think about how close your neighbors' interests are to yours and your family's.
- ❑ If you have children, will they have playmates?
- ❑ What will your neighbors think about you building a house at 7 am on a Sunday morning?

- ❑ Walk through the neighborhood and ask this question: Will your neighbors have the same values and expectations that you have about a neighborhood? What kind of cars are parked at neighborhood homes? What kind of house and landscape maintenance is evident?
- ❑ Would anyone complain about your attempt to improve your property in the manner you have chosen?
- ❑ Have you visited the site during all hours of the day and night? Is there any time when it causes you problems?

Architectural Planning and Cost Estimating and Design

The first step toward designing your new home is to match your lifestyle characteristics and those of your family with your new home. If the home is to be an expression of yourself, it should reflect your character. Be honest with yourself and every member of your family about your design needs. Accept compromises and trade-offs where you can.

Begin by talking to each other about your vision for your home. Then use the following questions as springboards for further discussion. Be respectful and acknowledge everyone's right to participate. The rules should be the same as for brainstorming; every idea is worth hearing! Someone should keep notes on all suggestions as they are proposed. Don't discourage any suggestion or attempt to find fault at the beginning. Great ideas come from a completely free flow of mental gymnastics. Discussion and a "devil's advocate" approach to suggestions should take place after all of the brainstorming is finished. Each person should prioritize their significant goals and objectives. Then, later, integrate all the best concepts within budget into a design.

Keep It Small

Your first house is a learning experience. You and your family will want to have all you have ever wanted in a home. Because of this, your expectations will probably be beyond your grasp. Your learning project can be a success if it is within your reach. Most first-time owner-builders will be attracted toward a larger house. If you do this, you will be risking the success you are seeking. Keep your project to a size and cost that you can effectively manage. The size will be up to you and will be unique to the minimum needs of you and your family. It will probably range somewhere between 1200 square feet and 1800 square feet. Do not overbuild for the neighborhood. Consider the aspect of resale.

Sample Room Sizes for Different Size Dwellings

Home Size	Kitchen	Bathroom	Master Bedroom	Living Room	Bedrooms	Dining Room
1200 sq. ft.	10′ x 10′	5′ x 8′	11′ x 14′	14′ x 14′	9′ x 9′	10′ x 11′
1800 sq. ft.	10′ x 14′	5′ x 10′	14′ x 16′	16′ x 16′	12′ x 12′	11′ x 14′

Keep It Simple

The shape you select will also have a lot to do with the ease of construction. You will be better off by selecting a small, rectilinear design, such as a rectangle or square. Avoid fancy designs, complicated architectural curves and offsets for your first home. Keep the span of roof or floor trusses at a manageable length. A 1200 square foot home that is rectangular might be around 30 feet by 40 feet. An 1800 square foot house may be 35 feet by 50 feet. The less complicated you make the design, the easier it will be to successfully complete.

Another aspect of simplicity is the number of stories designed. The fewer stories or floors, the easier a building is to design and build. Single story homes are much easier to build that a split-level, three story building or even a basement, which may require a retaining wall. Try to simplify the design in every way you can.

The simple house does not need to cramp your lifestyle or be confined or too compact. In fact, the design process could help turn a small, simple design into a more aesthetic home. You could begin by combining rooms such as dining room/kitchen or living room/study. Define your basic needs and then design the home around those basic needs. What size bedrooms do you really need? Look at your furniture and consider its location within each room.

Plan for the Future

Consider the future and your age. Will aging parents come to live with you someday? Will your children grow up and sprout wings? Will accessibility be an issue you should consider? Ramped access into a house will allow the disabled to gain access to your home. Would an upstairs kitchen be an inconvenience? Will you start a business in the future and need an office in your home?

Phased Construction

If you really want to build a mansion, and your family is committed to this endeavor, there is a simple means of achieving this. Build your project in stages. You might build a series of rectilinear shaped buildings with an end result in mind. The first project might be the basic habitation dwelling with a living room, bedroom, bathroom, and kitchen. Your second stage could adjoin the first and include a family room. Subsequent stages could include an array of bedrooms and bathroom clusters. This layout might look like this:

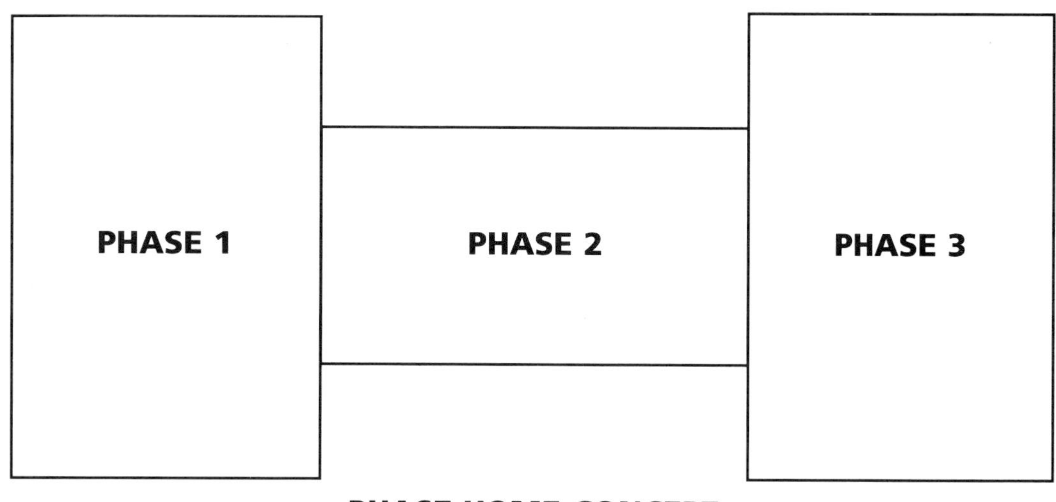

PHASE HOME CONCEPT

Interior Residential Design

In order to provide an efficient layout for a home which is aesthetically pleasing, you must learn the role of the interior designer. Interior home designers do more than just room layout, transition between rooms, and effective use of natural light. They provide a whole environment in which to engage the home. Their contribution is almost esoteric, in that it is more felt than seen. If their work is not successful, you will experience a general lack of comfort on a daily basis. When an interior designer is successful, the results often elicit a verbal response, "Wow!" or "Ooh."

While it is impossible to impart the skills of a carpenter in one book, it is equally impossible to impart all of the skills and secrets of an interior home designer in a brief section of a chapter. The most significant aspect of this role comes from the ability to "see" the desired result, before a house is complete. Years of experience and viewing countless other successful designs help designers in this visualization process. In fact, most interior designers increase their skills over a lifetime and become even more creative. But, the good news is that you can learn enough to create the design you want for your home by simply observing the successes they have achieved in other houses.

Effective use of space should be functional and artistic. The definition to ideal space planning is as enigmatic as that of beauty. You know it when you see it. Getting from here to there is the stock and trade of the interior home design professional. But remember, you only have to do this for one home, and it's yours. If you are resourceful, you can partake in some of the knowledge and experience of the professional interior home designer without much expense, enough to help design your home. There are several venues where you can tap into these resources. Since it is very hard to copyright design concepts, they are usually available for the taking. You can find ideas for design literally everywhere. Home shows, trade journals, home improvement magazines, real estate open houses, and home tours can help teach you a lot about interior design, but there are some specific concepts to notice. Look for professional designers' answers to the following considerations.

Area Planning

Your family did some brainstorming to determine a particular focus for each room. Then, considering costs, you collectively narrowed your desires to some indispensable components and wish lists. Now, you must integrate these individual aspirations into an efficient, aesthetic harmony. How do you do that? Use others' experience as a guide.

Your family's needs for individual room size are probably not all that unique. Builders respond to social trends. Even most custom homes bear some common space-planning attributes with tract housing. For instance, the entry room usually opens into a "great room" or living room. Bathrooms are usually near or within sleeping rooms. The kitchen is near the dining room and usually adjacent or near the garage or alternate entry door. The hallway defines the path toward the sleeping rooms. The master bedroom has a walk-in closet. The linen closet is nestled in the hallway. The laundry room is near utility equipment such as the water heater. Bathrooms and clothes washers are positioned close to the water heater. Private rooms such as bedrooms are near the rear of the home. The garage is near the front of the home. The front door is protected from weather with an entry porch, or it does not open directly to the street. The size of alternate bedrooms are a nominal size. The back door opens onto a covered porch. The back yard usually has a gate.

Circulation

The successful flow between rooms or spaces results from planning for efficient circulation. Some basic principals for efficient and aesthetic circulation spring from answers to the following questions:

Do you want to place your entry near the living room? Or do you want to use the living room as a family room and locate it closer to the bedroom? Will the kitchen successfully integrate with the dining area? Are bedroom doors visible from the main entry or living room? Will an office off the bedroom be conducive to its use or be a detraction? Will you want to get dressed in a large dressing room/closet or in the bedroom? Will media (TV or radio) play a part in your morning dressing routine? Should you locate laundry facilities near the bathrooms, bedrooms, or kitchen? Will the utilities (water heater, space heater, etc.) be located away from sleeping rooms?

Begin area or space planning design by considering that there are three aspects of space: private, semiprivate, and common space. Each of these aspects interacts with another in a home. For instance, the master bedroom and bathroom will be considered as private space, but will open into a hallway that leads to either another private spaces or common spaces like the living room or kitchen. The hallway leads to other bedrooms, but also may lead to a common bathroom, which may serve your guests as well. The kitchen may be a gathering space for friends as well as a space for dining. The family room may be used for private entertainment and will be off-limits to guests. A formal dining room might be the center for entertaining. You must decide which rooms will be relegated to these general use categories, then coordinate them into a design.

If you entertain regularly, dining rooms can open into a great room or near the entry, but function best when they are near the kitchen. If you want the separation, try a serving pantry as a buffer between the two rooms. If you intend to provide a bathroom for guests, do not isolate it inside a room considered private. Laundry facilities are most functional when located near the bathrooms, since that is where dirty clothes collect. Garages must never open into a sleeping room. That is a building code violation. They work best when opening into a buffer space, such as storage or laundry room, adjacent to other living areas.

Design and Drafting Tools

At this point, you may want to use graph paper and do simple sketches with a scale to see if your concepts will work. Measure the grids on the graph paper and select an appropriate scale, such as one line per foot, and draw each room. Then use the same graph paper to cut out little pieces of paper representing the size of your furniture.

Place these simulations of furniture within the rooms and see if your concepts will work. There are very good computer programs that do this as well. After a while, your level of sophistication for the sketches will evolve into the actual plan drawing.

Lighting Sets the Mood

While it is usually the last thing most owner-builders think about, artificial lighting is an important part of interior design. General lighting will be required by building codes in each habitable room. However, the basic lighting does not necessarily enhance the mood or look desired within a particular room. Light fixtures can be either surface mounted on a wall or ceiling, or hung from a ceiling. A chandelier in a dining room may be a means of providing the required light and provide a theme for the room as well.

A ceiling fan with a light can provide illumination as well as ventilation. You may want to accent certain aspects of your furnishings or artwork. In this case, track lighting may be appropriate. The intensity of the light will modify the mood of the room. Dimmer switches can achieve this effect and can be increased or decreased as needed.

Use of Natural Light

How does the plan for natural lighting affect the colors in each room? How do wall colors affect the personality of each room? Does light (or lack of it) help define the character of each room? Will you gain morning sun in the entryway? Living room? Kitchen? How does the use of light in each room match your family's personality? Visually imagine each room throughout the day. How does the sun provide light in each corner? Is there a shadow cast?

A scenic view may be paramount to your natural lighting plan. You may have bought the lot for its setting. You can just imagine sitting in your great room and gazing at the mountains. In that regard, an expansive picture window may be appropriate. Do you want to avoid light in certain rooms such as darkrooms or media rooms, or even exercise rooms? If so, you will have to contend with the diminished natural light by providing artificial light. There are both economic and aesthetic aspects of natural light. Solar energy is free. Natural light improves our mood. We become more productive when our mood is better. Take advantage of this in your design.

Noise and Sound

Locate sleeping, reading or meditation rooms away from rooms from which sound is inevitable. Kitchens, media rooms, and game rooms should be isolated from areas in need of privacy and quiet. Consider outside noises as well. Locate these quiet rooms away from the street or the neighbor's garage entry or basketball hoop. Some sound will breech even exterior walls. You can alleviate certain exterior noise with sound attenuation techniques such as insulation, wall veneer, acoustic walls, insulated windows and doors, shutters, screen doors, as well as simple behavior modification such as shutting the doors and windows. Outside patio walls can also serve as a buffer for noise abatement.

Practical Considerations

While aesthetic and environmental considerations are what makes a home enjoyable, you must not forget certain pragmatic factors. Plumbing layout or the location of electrical or mechanical heating and cooling equipment may not be flexible. Where you install the HVAC system may cause sleepless nights from noise. Walls that have plumbing vents may prevent the installation of a window at a certain location. Plumbing walls may also prevent the placement of a medicine cabinet at a strategic location. A large mirror above a lavatory may accidently cover a required electric outlet box. The desired location of a view from the kitchen sink may be altered by the standard sizes of kitchen cabinets. The distance between a door and window may affect the size of the kitchen cabinetry. The location for an attic access panel may conflict with the desire for elegance. The location for a return air grill near the kitchen may cause unnecessary smells to be spread over the entire house. The need to locate a stair to enhance one floor level may cause a loss of adequate ceiling height in another.

There are too many possible permutations of adverse effects to list. It would be a good idea to discuss your preliminary plans with a housing project superintendent for these kinds of issues. Ask for help in spotting any problems of this kind.

Structural Considerations

Interior bearing walls cause a house to be broken up into rooms that affect visual appeal and circulation. Removing those interior walls will create the need for beams, posts, and point-load footings. You cannot postpone decisions like this. How open do you want your home to appear? If your design will allow for interior walls, then solid wood joists such as 2 x 12s or composite wood joists can be used as a floor or roof joists due to their allowable span. On the other hand, if you wish a totally open house, you must consider the use of clear-span, engineered trusses, which will allow for a significant span, eliminating the need for any interior bearing walls. Although you can achieve similar results with posts supporting beams in interior locations, engineered trusses eliminate the need for these elements.

There are other considerations related to economical use of structural members. First, dimensional lumber is sold in increments of 2 foot lengths. For instance, 2 x 12s are available in 10, 12, and 14 foot lengths, and so on. You can design your home to accommodate this fact and plan interior bearing walls by spacing the walls in increments of 2 feet in width or length. For instance, instead of making your daughter's bedroom 11 feet by 11 feet, 6 inches, just make it 12 feet by 12 feet. This conserves the lumber purchased as well as the time spent in cutting the lumber to fit the smaller room. If you want a wall that was at least 9 feet high, you would have to purchase 10 foot long studs. In order to sheath the walls, you would need at least two sections of plywood anyway. Why not add an extra foot to the building's nine foot height and have the spaciousness of your 10 foot ceiling height? You will avoid having to cut all of the studs and also enjoy a higher ceiling.

Harmony of Elements

Successful design is the result of blending all of these elements into a plan for a house. Functionality, circulation, natural, and artificial lighting, ease of use, practical considerations, structural design, noise and sound, space use and appearance must all be incorporated into the exact result you are after. This planning process is not easy and will be filled with many compromises and trade-offs along the way. Exercise your "compromise gland" and work together to achieve harmony in your design.

Family and Use Considerations

- ❑ How long do you expect to live in this house? Do you expect to pass this house on to your children?
- ❑ Does your family entertain together? How much entertaining does your family do, and how large are the parties?
- ❑ Do you ever have overnight guests? Weekend guests? Out-of-town family on vacation?
- ❑ Do you desire a view between rooms? For example: from dining room to kitchen?
- ❑ Do you like the sun shining through windows in certain rooms at certain times?
- ❑ Do you like indoor plants? What kinds? What are their solar needs? Outdoor plants viewed from inside the home?
- ❑ Do you anticipate future expansion of the house? Consider the needs of the following:
    - ❑ Electrical equipment size
    - ❑ Sewer/septic size
    - ❑ Foundation strength/size
    - ❑ Water service pipe size

- ❑ How many television sets and radios will you have? Where?
- ❑ Where would you like to have telephone outlets? How many phone lines? Computer terminals? Fax terminals?
- ❑ What types of doors (if any) would you like on closets?
- ❑ Do you have any art work or family photographs to display? Where will they be most appreciated?
- ❑ What type of interior ceilings appeal to you? Low? Flat? Vaulted? High flat? Skylights?
- ❑ What type of stairs do you want? Straight? Spiral? Circular? Open railings? Open tread? Carpeted? Hardwood?
- ❑ Is there any allergy concern in your family? Dust? Wool? Feathers? Pollen? Chemicals? Wood? Fibers? House pets? Will you need a humidifier or vaporizer?
- ❑ Will you display holiday lights? Outdoor outlets?
- ❑ How often do you throw out the newspaper? Empty bottles? Do you recycle/store them? Do you subscribe to magazines? Newspapers?
- ❑ How many books do you have? Will you display them?
- ❑ Will you have pets inside or outside the home? Consider the need for a fenced yard.
- ❑ Describe the one-word feeling you wish to have for the following portions of your home:

Building exterior ____________________

Living room ____________________

Kitchen ____________________

Master bedroom ____________________

Entry ____________________

Dining room ____________________

Family room ____________________

Bathroom(s) ____________________

Porch ____________________

Structural Considerations

- ❑ What architectural or structural style do you like the best?
- ❑ What architectural or structural design is most appropriate for the site conditions?
- ❑ What architectural or structural design can be built with the greatest economy?
- ❑ What architectural or structural design is most common where you live?

What type of foundation would best suit this site:

- ❑ Full basement?
- ❑ Partial basement?
- ❑ No basement?
- ❑ Exposed wall? (unfinished)
- ❑ Monolithic slab on grade?
- ❑ Combination footing/foundation with slab infill?

Describe the desired exterior finish:

- ❑ Wood siding (vertical or horizontal?)
- ❑ Panel siding?
- ❑ Metal siding?
- ❑ Stucco? Siding which looks like stucco?
- ❑ Brick veneer?
- ❑ Vinyl siding?

What type of ventilation, heating and cooling equipment do you want:

- ❑ Heat pump?
- ❑ Baseboard heaters?
- ❑ Wall heaters?
- ❑ Passive solar?
- ❑ Wood stove?
- ❑ Fireplace?
- ❑ Air conditioner?
- ❑ Evaporative cooler?
- ❑ Ceiling fans?
- ❑ Natural ventilation using windows?

Access and Exits

- ❑ How many and what type of entrances do you want and where will they be?
- ❑ Should the main door be single or double?
- ❑ Do you want a sidelight by the door?

Consider the following locations for access:

- ❑ Garage to house? Carrying groceries?
- ❑ Garage to basement?
- ❑ Garden to basement?
- ❑ Garden to utility area?
- ❑ Bathroom to swimming pool area?
- ❑ Master bedroom to garden area?

Circulation and Flow

Consider access between the following locations or rooms:

- ❑ Living room to family room?
- ❑ Family room to kitchen?
- ❑ Kitchen to dining room? Private doorway?
- ❑ Kitchen to sundeck?
- ❑ Kitchen to garage?

Do you want a balcony? Off which room(s)?

- ❑ Master bedroom?
- ❑ Living room?
- ❑ Family room?
- ❑ Kitchen?
- ❑ Library?
- ❑ Entryway?

Do you want an enclosed garden or atrium? Off which room(s)?

- ❑ Entry?
- ❑ Kitchen?
- ❑ Library?
- ❑ Master bedroom?

Individual Room Considerations

Go through each room and discuss the following general questions about the use of that space:

1. What use is anticipated in this room? Will the use change over time? Can both uses be accommodated?
2. What size should the room be to achieve usability?
3. What is the theme of expression your family wants in this room? What outdoor view do you want from this room?
4. How important is natural or artificial light in this room? How important is the color of walls and ceiling?
5. What is the kind of window your family wants in this room? What sort of window treatment do you plan?
6. What type of wood trim finish do you want in the room?
7. What kind of light fixtures would best work in this room?
8. Will music or media be accommodated in this room?

Then discuss the following specific questions about each room.

Living Room/Family Room

- Approximate minimum size?
- Direction of windows?
- Ceiling height?
- Sunken floor? Raised floor?
- Do you want a fireplace?
- Do you plan slide or movie projection facilities?
- Large screen or multimedia television system?
- Will you have a piano? Bookshelves? Special lighting?
- Will you display books? Bookcases? How many? What size will you need?
- Will you serve alcohol? What facilities will you need for this?
- Will you have a liquor cabinet?
- Will you have house plants?
- What type of chairs will be in each room?
- What size tables will be in each room?
- How large is your sofa? Do you have more than one? Where will they set?

Dining Room

- Is this room really necessary? Do you entertain regularly?
- How many seats around the table are needed?

- What other furniture will be in this room? Buffet? China cabinet?
- Will the room be combined with kitchen? Living room? Family room?

Kitchen

- What shape kitchen? U-shape? L-shape? Corridor? Work island?
- What size will your kitchen be?
- How many people will be working in the kitchen at one time?
- Will you entertain in the kitchen?
- Window direction? Morning sun?
- What facilities do you want: Range/oven? Range top? Built-in oven? Microwave? Barbecue grill? Warming oven?
- Will you have warming lights?
- Will you have a standard range?
- Cutting table?
- Built-in dishwasher? Trash compactor?
- Single, double, or triple sink? Separate sinks? Island sink?
- Disposer?
- Spray nozzle in sink?
- Pot hanger?
- Ice dispenser for refrigerator?
- Exhaust fan for grease or smoke over range/oven?
- Desk area?
- Area to store cookbooks?
- Refrigerator/freezer? Chest freezer? Upright freezer?
- Food storage and pantry facilities? Seasonal equipment storage?
- Pantry size? What must be stored? Food? Utensils? Supplies? Canning facilities?
- Linen storage? Dinnerware? Cleaning supplies?
- Broom closet? Garbage pail?
- Built-in spice rack?

Food Mixing Center and Serving Center

- Will it be close to range/oven? Will it be close to refrigerator?
- Large enough for coffee pot? Toaster? Waffle iron? Room for blender? Food processor?

Laundry

- Do you plan to have a washing machine? Clothes dryer?
- Do you plan to have a built-in ironing board?
- Do you plan to have a laundry tub? Spray nozzle?

- Do you want to have a laundry chute from the bathroom?
- Do you want a wall phone?
- Do you want a flat work area for folding clothes?
- Do you want a separate sewing area?

Consider the following vicinity questions about the laundry room:

- Close to bedrooms or bathrooms?
- Close to kitchen?
- Near back door and clothes line?

Bathroom(s)

- How many do you need?
- What kind of bathrooms do you want? Toilet? Bath? Tub? Hydro-massage tub? Vanity? Bidet?
- Do you want a shower? What will be the size? Do you want a multihead shower?
- Heatlamp?
- Do you want a half bath (mud room) off back entry?
- Do you want a medicine cabinet?
- Will you store towels and face cloths in bathroom?
- Do you want a full length mirror in bathroom?
- What size vanity would you like? How many sink basins?
- Would you like roughed-in plumbing in basement for future addition? Will you need to install a sump pump?
- Size of bathroom needed?
- Orientation of windows? Number of windows? Operable or fixed windows? Glass block?
- Exhaust fans? Steam generator?

Closet(s)

- How much storage do you need in closet?
- Do you want a walk-in closet?
- Do you want to store a security safe in closet?
- Should a closet be accessible from the bathroom?
- Should there be a dressing room in closet? Dressing bench?
- Should their be a full-length mirror?
- Should there be natural lighting in closet? A skylight?
- How will clothes storage space be divided between partners?
- Do you want a television or radio within the closet? Telephone?
- Is this a good space for a central security alarm?
- Is this a good place for secure gun storage?

Bedroom(s)

- Dressing room in bedroom?
- Do you want a makeup table?
- Will there be an armoire?
- Do you want a chest of drawers? Where will it be located?
- Will you have a dresser? Where will it be located?
- Do you want a full length mirror?
- Do you plan built-in drawers?
- Will there be built-in toy storage for kids rooms?
- Do you want ceiling fans?
- Do you want special lighting?
- What will be the window orientation to achieve desired natural lighting?
- Television? Stereo? Telephones? Computer terminal?
- Smoke detector location?
- Small adjoining office? Does office need independent outside access to conduct business?
- Adjoining sitting room?
- Family pictures on wall?
- Outside door? Opening into porch?

Porch

- Front porch? Rear porch? Side porch?
- Enclosed porch? Screened porch? Locking doors?
- Facing which direction?
- Pet's area?
- Door bell location?
- Hobby area?
- Extra sleeping area?
- Storage area?
- Will the porch be a formal entryway?

Garage/Carport

- Do you need one?
- Size? Single? Double?
- Purpose for garage? Cars? Workshop? Storage?
- Location with respect to house?
- Attached? Detached?
- Shape to match house?
- Which way will garage doors face? Does this direction pose a challenge for backing out?

Driveway Access

- Driveway location? Backing out? Circle driveway?
- Driveway materials? Rock? Gravel? Asphalt? Concrete?
- How will weather affect the driveway?
- How many drivers are there now? How many future drivers?
- Will cars be safely hidden from street?
- What are the noise considerations?
- What are the dust-control problems caused by driveway location?

Landscaping

- What type of landscaping will you have?
- Water needs? Maintenance needs?
- Will it match the house?
- Will it match surrounding area?
- Will it be productive? Fruit trees? Shade from summer sun?
- Will it be decorative? Roses? Flowers? Will you have a garden?
- Should any existing vegetation remain? Should any be transplanted elsewhere?
- Will any proposed landscaping invite termites or other vermin?

Heating and Cooling

- What do other houses in the area or region predominately use?
- Ease of installation?
- Availability of fuel source?
- Familiarity with equipment?
- Comfort level provided by equipment?
- Safety aspects?
- Most appropriate for first-time home builder?
- Efficiency?
- Costs of installation?
- Cost of service? Cost of maintenance?
- Effective life of equipment?
- How complicated is it to install or run? How many moving parts?

Other

- Will you have a central vacuum system?
- Will you have storm windows?
- Will you have security doors? Screen doors?

Passive Solar Design

There are simple design ideas you can add to your design that will decrease your heating and cooling costs during the life of your home. Consider that a home may cost you $75,000 in total costs (if you build it yourself). Now, consider what your heating and cooling costs will be during the life of the home. For a family of four, a reasonable annual cost of heating and cooling a 1500 square foot house will average around $2,000 annually or $165 per month. If you expect to live in your home for 40 years, the cost of utilities will be $80,000, assuming there is no increase in utility rates. You can see that the costs to heat and cool your home will exceed the cost of building. At this stage, you can add design aspects that lower the costs of providing conditioned space. You have made the choice to save money by building yourself. Don't stop there. Add some simple design techniques that will save you more in the long run.

For instance, if you orient your home properly, you can invite passive solar energy into your home during the heating season, when you need it. That means situating your home roughly along an east-west axis to gain solar energy. Windows have less ability to resist heat flow than insulated wood frame walls, but southern exposure windows are net gainers of energy, so the more south-facing windows on your home, the better. Conversely, northern windows are net losers of energy, so use as few of them as possible. East and west windows gain as much as they lose. You can increase energy efficiency of windows with a thermal barrier such as heavy drapes designed to resist heat flow out at night.

If you have a pitched roof design, use the roof overhang to your advantage. Prevent the summer sun from entering and allow the winter sun to heat your home. Study the difference in angle that the sun is at between the heating and cooling seasons. In the northern hemisphere, the sun is lower in the daytime sky in the winter and higher in the summer. Use that to your advantage. Orient your home to capitalize on solar energy.

Maintain deciduous trees around your home, which shed leaves to permit sunlight to enter in the winter and grow thick healthy leaves during the summer to provide a thermal shield from the sun. Masonry walls and concrete floors can absorb and maintain passive solar heat in front of windows that receive sunlight during the heating season. Use the following discussion list to consider the value of adding passive solar heating to your project.

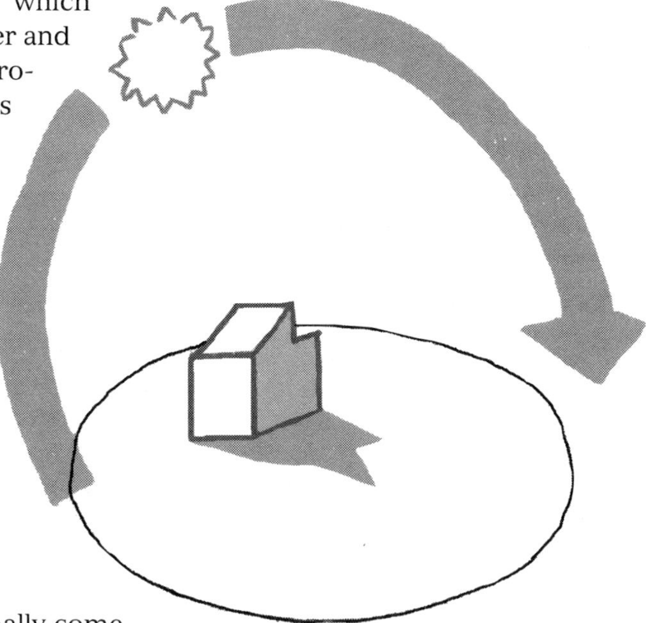

- Are there any natural obstructions to solar heat gain such as trees or mountains?
- Look true south from your building site. What is in that direction? Roads? Driveway? Houses?
- What type of trees are near your home site? Deciduous or evergreen?
- What direction do rain or snow storms normally come from?
- How many hours of winter sunshine do you normally have?
- Will using lots of south-facing glass destroy privacy interests?
- What latitude is your home located?

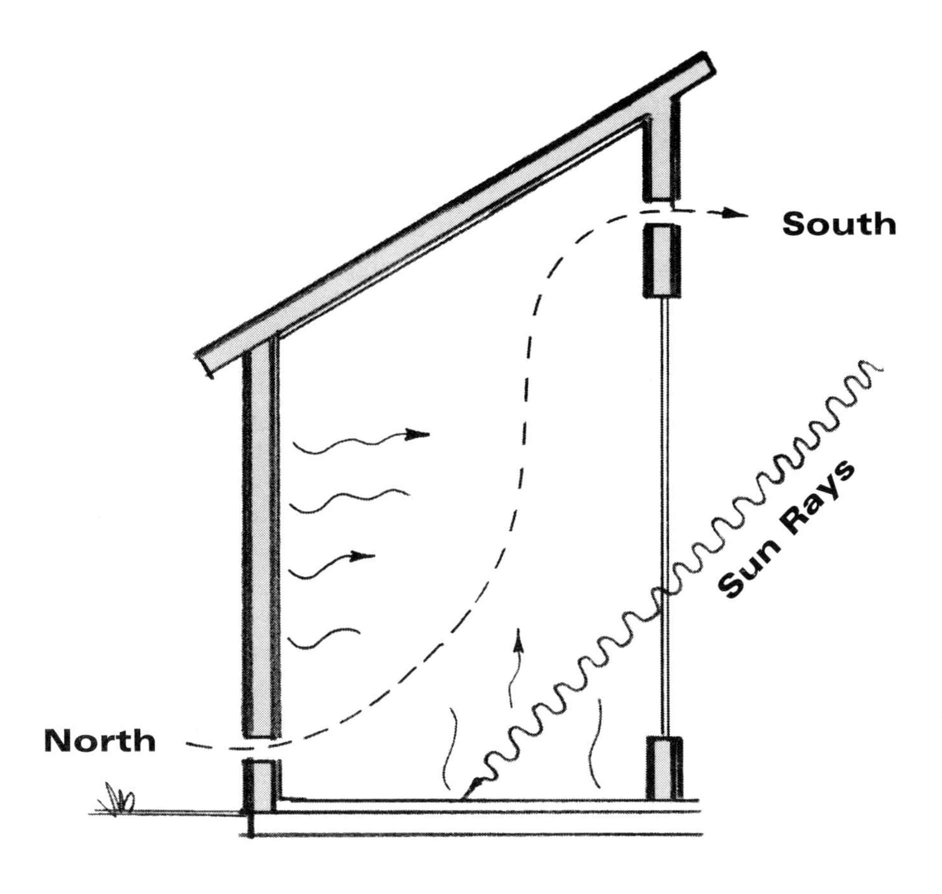

Active Solar Design

You may have an interest in using renewable energy systems. While passive solar heating can reduce your required heating costs with a minimum investment, active solar energy methods can reduce your energy consumption costs even further. The question is often asked, "Can I use renewable energy systems to provide some or all of my electrical energy requirements?" The answer is: It depends. At this time, renewable electrical energy generated by wind turbines or photovoltaic (PV) power systems costs considerably more to install and maintain than just connecting to the local utilities and paying the monthly bill. However, if some or all of the following conditions exist, then maybe you will want to investigate renewable energy systems further.

1. The electrical utility is a considerable distance from your building site, and the connection fees are tens of thousands of dollars.
2. Your location does not have weather extremes requiring large heating and cooling systems.
3. Your electrical requirements are modest, and you do not need large amounts of energy for loads like farm irrigation.
4. You have good unobstructed solar access from 9 am to 4 pm every day of the year.
5. You have steady winds above 10 mph all year around.

6. You are a do-it-yourselfer who can maintain battery systems (periodic watering) and generators (oil and filter changes) or does not mind the heights involved in servicing a wind generator.
7. You have disposable income that might be well spent generating some or all of your energy.

The life cycle costs for these systems varies depending upon the initial cost, the cost of energy in your area and the extremes in weather conditions as well as your personal comfort level. Some systems can pay for themselves within just a few years, while others will take as much as twenty years to recover your expenditures in energy savings.

In general, renewable energy systems come in two versions. The stand-alone, off-grid system is, as the name implies, a system that is self-contained and provides all the energy necessary for a given site. These systems have deep-cycle storage batteries for energy storage for nighttime and cloudy (or no wind) periods. Many of these systems have a gasoline, propane, or diesel-fueled generator for backup. Photovoltaic systems are more common than wind systems because of the wider availability of the solar resource. Utility-interactive or grid-connected systems are the second version of renewable energy systems. These renewable energy systems are used to supply energy to a site where the utility power is also available. The grid serves as the storage and backup medium, and there are no batteries or backup generators.

Energy produced by the renewable energy system is used on-site, or if in excess of site requirements, is fed into the utility grid. When renewable resources are not present, such as a cloudy day or at night, the utility supplies all of the energy for the site. The utility interactive renewable energy system is interlocked with the utility system. If for some reason the utility has an outage such as a blackout, brownout, or line failure, and the renewable energy system cannot produce any energy, the site will be without power.

Renewable energy systems, while representing several mature technologies, are used infrequently for residential applications because of the costs. Many electrical inspectors and electricians may be unfamiliar with these systems. The best way to get a safe, reliable, and durable renewable energy system installed is to follow a few simple steps.

1. Educate yourself through books, magazines, and internet sources on renewable energy systems.
2. Search out a competent renewable energy equipment dealer/installer who has a good track record with references.
3. Ensure that a licenced electrician is involved in the installation of this electrical power system.
4. Include the electrical inspector in your planning stage and throughout the project.
5. Ensure that everyone understands that you will only accept an electrical system that uses electrical components that are fully listed to the appropriate standards established by Underwriters Laboratories® (UL), and that the systems must be installed in full compliance with the NEC.

Cost Estimating

Where Wants and Needs Meet

At some point you will need to determine which you can afford: a two-car garage, a balcony or a third bathroom. The tradeoffs will be yours to make. As I pointed out earlier, I strongly suggest that you reach these decisions as a family. Communication and compromise can save hurt feelings later. I will show you how to estimate your costs. Collectively, your family must reach a consensus on the design.

The first consideration is how much of the work you are willing to do yourself. A follow-up question is how many of your friends will be willing to work for free to help you in your effort? You may not be able to do everything yourself. I will teach you how to calculate amounts of material. You must seek unit pricing in your area and determine the labor costs of that work for which you need help. Note that the following calculations assume a square house with sides that are 40 feet long. The building wall is 8 feet high.

Foundation/Concrete Floor Slab

For monolithic concrete slab foundations, the volume of concrete will include the slab thickness as well as the volume of the footing. For split foundation systems, the footing may be calculated separately.

1. Footing width (ft) x height (ft) x length (ft) = Volume (cubic feet)
2. Foundation wall width (ft) x height (ft) x length (ft) = Volume (cubic feet)
3. Area of slab floor (ft^2) x depth of floor (ft) = Volume (cubic feet)
4. Add the first three sums together and divide the results by 27 for volume of concrete in cubic yards

Example:

A square house with a floor area 1600 square feet is 40 feet by 40 feet. The footing is 12 inches wide, 12 inches deep around the perimeter of the building. The slab is 4 inches thick.

Solution:

Footing: 1 foot x 1 foot x 40 feet x 4 sides = 160 cubic feet
Slab floor: 1600 square feet x .333 (4 inch slab thickness divided by 12 inches per foot) = 533 cubic feet
160 + 533 = 693 cubic feet
693 cubic feet ÷ 27 = 25.67 cubic yards of concrete

Wood Wall Framing

Wall framing should really be the easiest of all estimates.

1. Calculate the number of studs by determining the perimeter of wall length in feet and dividing by 1.33 (for 16 inch o.c. stud spacing)*

Example:

Perimeter = 4 x 40 ft = 160 lineal feet. Then count the studs: 160 ÷ 1.33 = 120 studs within the wall frame. Add extra studs for corners and to account for differences in wall layout.

* **Special Note:** *Technically, you will not have a full-height stud every 16 inches due to windows or doors, but you will have trimmers, cripples and king studs that will make the difference. Another word to the wise is this: Order 10 percent more than you think you'll need, because you'll probably use it.*

It's a good idea to order about 5 percent to 10 percent more concrete than your calculations call for. You can always be prepared to use any extra for porches, door steps, or landscaping.

2. Calculate the amount of plate material (two top plates and one treated bottom plate)

   2 x 160 = 320 feet (using 16 foot plates) = 320 ÷ 16 = 20 top plates 16 feet long
   160 feet (using 16 foot pressure treated type) = 160 ÷ 16 = 10 bottom plates 16 feet long

3. Calculate header material (using 2 by 12s). Count the number of windows and doors. Check their widths and multiply two pieces per opening for a 2 x 4 wall frame and three pieces per opening for a 2 x 6 wall frame. Headers will be 3 inches longer than openings to allow for bearing support on either end.

Roof/Floor Framing

This cost is totally dependant on your design. I encourage your selection of engineered trusses for the roof structure. Floor joists may be solid wood joists or composite wood joists with bearing walls at appropriate spans. Use Chapter 3 to help determine the structural members, then calculate the number as follows:

Example:

Span of 40 feet, length of 40 feet. Trusses every 2 feet: 40 ÷ 2 = 20 trusses, using trusses 40 feet long between bearing points. If you build gable end walls (rake wall), you can save on the cost of two trusses.

Roof or Floor Decking

The use of plywood or oriented strand board (OSB) decking is appropriate for roof or floor decking. Calculate the projected area to be covered and divide by 32 (32 square feet per sheet of plywood).

Example:

A simple pitched roof, where the roof pitch is 6:12, the hypotenuse of a right triangle is equal to square root of sums of squares of two sides. Use half of roof to calculate height.
Span = 40 ÷ 2 = 20 feet
At a 6:12 pitch, height is = 6 ÷ 12 multiplied by 20 feet = 10 feet
Now that you know two legs of the triangle, use the principal of the right triangle to calculate the long side.
$(10^2 + 20^2) = (100 + 400) = 500$
Square root of 500 = 22.36 feet or 22 feet 4⅓ inches
This leg represents the roof line.
Actual roof area on one side = 22.36 x 40 = 894.4 square feet
Actual roof area on two sides = 894.4 x 2 = 1788.8 square feet (approximately 1,789)
The number of sheets of plywood is 1789 ÷ 32 = 56 sheets (always round up)
For more complicated roofs, do the same calculation and give a larger safety margin for waste.
Don't forget roof overhang on pitched roofs.

Siding

The amount of siding is the easiest to estimate if you plan on using tongue and groove or lap-type hardboard siding. For a wall height of eight feet, any multiple of four feet, in perimeter measurement, counts as one sheet on each wall section. Heights over 8 feet should be included in an estimate for additional siding. For example, our square house, 40 feet in length and width, would need 40 ÷ 4, or 10 sheets per side or 40 sheets for all four sides. Remember to account for the siding on the gable ends of walls. Calculate this area carefully. You may have some waste here.

Roofing

A *square* of roofing represents the coverage of 100 square feet of roof deck. This includes any required overlap in the roofing. Remember the area of the roof decking calculated above? Divide this area by 100 and order this many squares of roofing. For the sample house, with 1789 square feet of roof surface, you would order 1789 ÷ 100 or 18 squares of roofing. Be sure to add cost for metal flashing, nails, mastic, metal drip edge, gutters, down spouts, and underlayment that is required under shingles.

Doors and Windows

These will be easy for you to price. I recommend prehung type doors for the first time installer. Windows should be a mid-range quality and meet an adequate energy efficiency rating. Shop around for prices!

Plumbing

This is a tough area to estimate. You can purchase used fixtures such as toilets, bathtubs, sinks, or lavatories from a secondhand store or yard sales and lower your costs, or you can blow your budget and buy a $2,000 hydro-massage bathtub. I'll leave the fixtures up to you. Plastic drainage and vent pipe and fittings must be counted. The materials most widely used for water supply lines are copper, CPVC plastic, polyethylene, and polybutylene. Consider the costs of plumbing pipe fittings and solder or glue for the pipe connections.

If you are adept at sweating joints, copper is an excellent pipe to use and has had many years of assessment. CPVC is approved for use as indoor hot or cold water supply lines in most areas of the country. CPVC pipe is sold in ten foot lengths. You'll need many, as well as several boxes of joint fittings. They are glued together to create a water-delivery system with a special solvent. The product goes together very easily. If you have patience, this may be the easiest way for you. Cost estimates for any of these systems vary according to product. Call a plumbing supply store and inquire. Many stores will give you a complimentary estimate or "take off" from your plans. Don't forget the toilet, bathtubs, shower, sinks, lavatories, water heater, disposer, dishwasher, and other desired fixtures.

Of course, sewage disposal systems are included in this cost category. Be sure and estimate the cost for either a private septic tank or the cost to connect to a public sewer.

Electrical

Estimates for your electrical installation will include wire, electrical service equipment, outlets, switches, lights, and circuit breakers. The cost of wire is based on the distance of service equipment to each branch circuit location. Wire is sold in packages of 250 feet. Most branch circuits will use between 40 to 80 feet. For a small house, 4 or 5 packages of wire will complete all the general lighting and outlet load. A medium size house may use 8 to 12 boxes of wire.

Depending on whether you have electric cooking equipment, an electric water heater, electric space heating, a welder, or other equipment, you will need additional, heavy gauge wire. These are available in boxes or they can be cut to a prescribed length. Electrical outlet and light boxes cost very little. You may need 50 outlets and 10 lights for a standard size house. Lights, special fixtures, fans and other unique devices must be estimated. These are personal choice items. Your electrical service disconnecting equipment and circuit breakers will be the most expensive parts of your electrical package. An *All In One* unit that houses the meter base and service disconnecting equipment is the best buy. Circuit breakers must fit the particular panel and are sized according to the maximum overcurrent protection afforded the wire it serves. Do not forget the costs of bringing electrical power to your doorstep. Sometimes the electric company will not provide a service without these costs being paid for in advance by the customer.

Mechanical

If you can install your own system, price the necessary equipment and duct work from a reputable dealer. If you need help, ask around for a qualified licenced contractor. He'll give you an estimate for a turn key operation and may allow you to do some of the work, like duct or discharge box installation.

Insulation

Count the number of stud cavities and get a price on batt insulation per bag. The bags have a yield of a specific number of batts. Determine how many bags by simple arithmetic. The same for the ceiling. This should be an adequate estimate. If you use blown-in cellulose for walls and ceiling, use the yield estimated by the manufacturer for their product.

Drywall

Drywall comes in two basic sizes: 4 foot x 8 foot and 4 foot x 12 foot. Count joists spans and measure lengths and divide by the appropriate divisor (32 square feet for the 4 x 8 or 48 square feet for the 4 x 12).

Cabinets and Trim

Most large retail building material supply stores will do free *take off* estimates for cabinets. Counter top material is sold by the lineal foot. Trim is sold by lineal foot. Be sure to include bathroom, kitchen and laundry room cabinets.

Flooring

Most new homeowners want to put wall to wall carpet in to make it feel like a new home. My wife recommends installing an easier-to-clean, durable floor such as tile, hardwood, or even a painted concrete floor. Throw rugs will still make it cozy and warm, if that is the effect you're after. Carpet is sold by the square yard based on 12 foot widths. Tile or wood parquet is sold by the square foot. Hardwood or even softwood is sold by the board foot.

Homeowner's Insurance

Homeowner's Insurance, and even builder's insurance should be purchased as soon as possible. If you have friends, neighbors or even relatives working for you, it is good to know that you are covered from any injury suffered.

Permits

Permits for construction are essential and may be higher than you expected. Check before you decide for an estimate of permit fees.

Road Development

Road development or curb-cut street access may cost you more than you expected and you may not be able to perform the work yourself. The development or jurisdiction may require that a licensed, bonded contractor familiar with development standards perform the work.

Well Drilling

Well drilling costs if you are not connected to a water company system.

Electricity

Getting electric power to your site may be excessive and may be the reason for the lower cost of the property. If you can convince the utility company of your intent to use as much electricity as possible, they may not charge as much for the cost of bringing power to your site.

Other Utilities

Utilities such as cable television, telephone, mail delivery, or even newspaper delivery may have hidden costs. Now is the time to discover what they will be. Additional costs to consider are:

Portable toilet rental for construction workers

Tool purchases necessary to begin project

Books or classes that may improve your skills

Vehicle expenses and repairs

Drafting and design work and plan development, if you can't do it yourself

Legal fees for land purchase, closing costs or recording fees, etc.

Managing Your Project

One way in which production builders save money is through thorough, effective organization and maintaining scheduling commitments from subcontractors. The production builder hires job superintendents or foremen to oversee a job. Sometimes, these superintendents will be in charge of as many as a dozen projects and are stretched to their limits just directing the actions of subcontractors. A superintendent's function is to effectively organize labor within a project in the most efficient manner so as to save money and of time. These superintendents sometimes need to be of strong character and demeanor in order to get the results they want.

Use the following worksheet to begin to get an estimate for costs. Allow for whatever funds you can reasonably spend on a monthly basis. From that, you can extrapolate your anticipated construction period. For example, if your cost of construction to get a livable house is $35,000 and your available funds are $1,000 per month, you can expect to be building for 35 months. If you can't live with the length of time, reduce the size or cost. Be realistic. You don't want to get yourself into a project that rivals the Taj Majal and then lose interest. Remember, this is the first house you will have built all by yourself. There is time to experiment with other concepts later on. Don't bite off more than you can adequately chew. Keep your first home project small and manageable. There are intangible costs, which must be considered. Sometimes these can be forgotten, though they are very important or essential.

Cost Estimating Worksheet

Land

Initial cost ______

Taxes ______

Real estate fees ______

Legal fees ______

Utility connection fees ______

Water ______

Electricity ______

Phone ______

TV? Miscellaneous ______

Septic ______

Plan preparation ______

Building permit ______

Zoning permit ______

Mechanical permit ______

Plumbing permit ______

Electrical permit ______

Foundation/Site Work

Land clearing/leveling ______

Footing concrete ______

Footing steel ______

Labor ______

Slab concrete ______

Foundation block ______

Labor ______

Wood Framing Package

Wall frame ______________________________

Roof trusses ______________________________

Wood roof joists ______________________________

Floor joists ______________________________

Floor decking ______________________________

Roof decking ______________________________

Nails ______________________________

Labor ______________________________

Wall Siding

Hard board siding ______________________________

Windows/Doors

Windows ______________________________

Doors ______________________________

Roofing

Shingles ______________________________

Hot mop ______________________________

Roll roofing ______________________________

Labor ______________________________

Plumbing

Drainage fittings and pipe ______________________________

Supply fittings and pipe ______________________________

Fixtures toilets ______________________________

Lavatory basins/cabinets ______________________________

Bath tubs or showers ______________________________

Sinks ______________________________

Water heater ______________________________

Other ______________________________

Labor ______________________________

Mechanical

Equipment ______________________________

Duct work ______________________________

Supply boxes ______________________________

Electrical boxes ______________________________

Labor ______________________________

Electrical

Service equipment ______________________________

Breakers ______________________________

Wire ______________________________

Exhaust fans ______________________________

Smoke detectors ______________________________

Outlets/switches ______________________________

Light bases ______________________________

Fixtures/Lights

Heaters ______________________________

Disposer ______________________________

Range/oven ______________________________

Dishwasher ______________________________

Labor ______________________________

Insulation

Batt type roof ______________________________

Batt type wall ______________________________

Caulking ______________________________

Sheetrock

Wall ______________________________

Ceiling ______________________________

Tape ______________________________

Texture ______________________________

Paint ______________________________

Brushes______________________________

Labor______________________________

Trim

Floor base ______________________________

Door trim______________________________

Cabinets ______________________________

Labor______________________________

Flooring

Hardwood ______________________________

Tile ______________________________

Carpet ______________________________

Labor______________________________

Equipment

Purchases______________________________

Rental ______________________________

Insurance

Homeowners ____

Life insurance ____

Workman's compensation ____________________________________

Other

____

____

____

____

____

____

____

____

____

____

____

____

____

Remember there will be lots of "Other." Allow 10 percent over and above your final tally to include incidentals or accidents. You may cut the wrong side of the stud you were measuring. You may break the window while installing it, or you might have spilled the paint bucket or left the paint roller out overnight without cleaning it first, or any number of things. Actually, 10 percent would be an average to strive to attain. Expect losses while building a house, or you'll be disappointed! Just work to keep it at a minimum and seek uses for the spilled milk, which you will have. For instance, if you cut a stud or trimmer too short, don't throw it on the scrap pile. Use it as fire blocking or truss bracing, which you will need to cut later. Remember, waste not—want not!

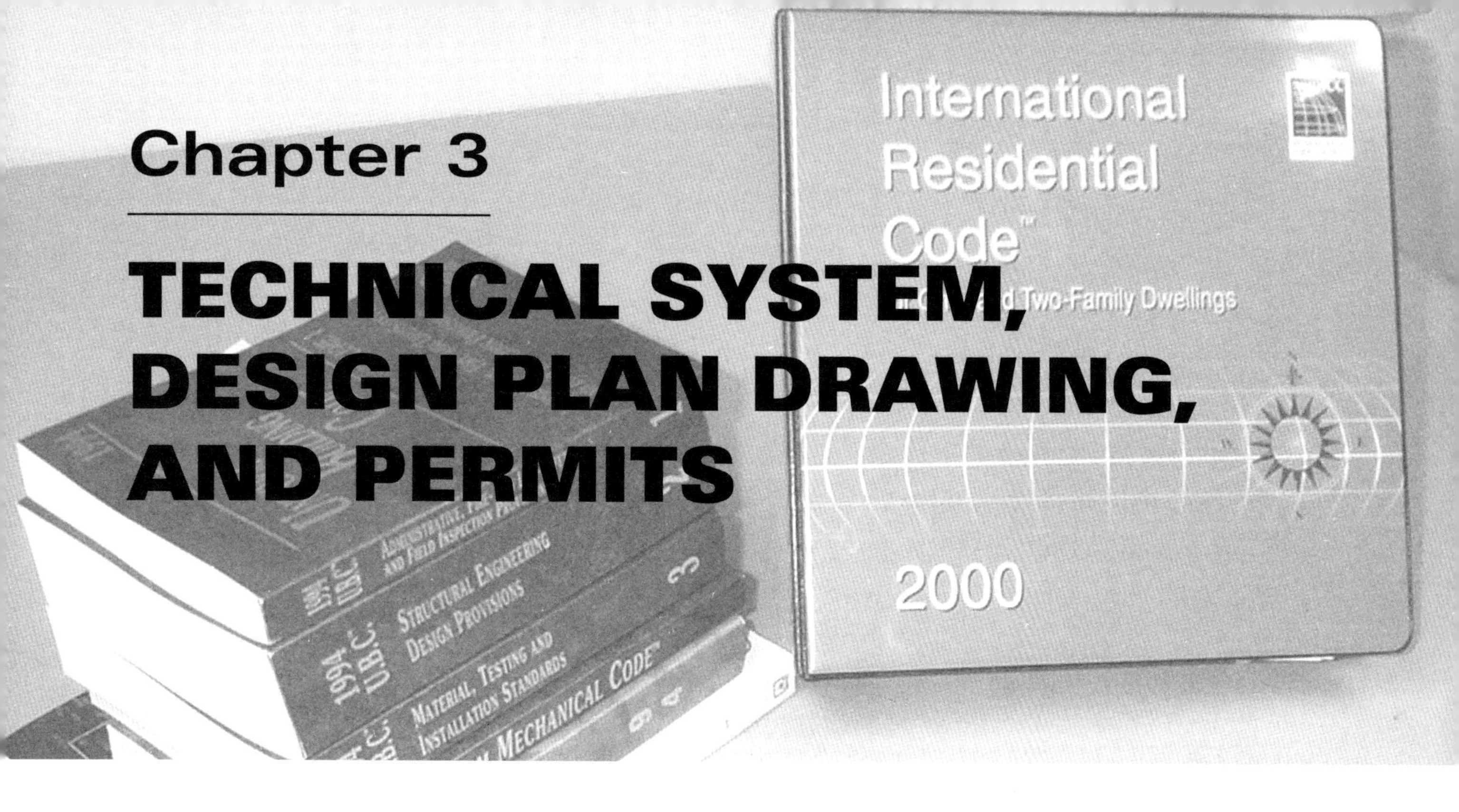

Chapter 3

TECHNICAL SYSTEM, DESIGN PLAN DRAWING, AND PERMITS

This chapter explains the methods of putting your concepts on paper. There are architectural standards that you must use to create a drawing. There are minimum standards for clarity that must be met in order to acquire a building permit. These plans must be reviewed and approved by plans examiners within the building department.

Preliminaries

Before you visit your local building department, sketch your concept on paper. Your first consultation should be informal and casual. Plan on a brief visit with the office personnel and ask to visit with the inspector and/or plans examiners if their time permits. Ask for a general handout that outlines the basic plan submittal criteria. Usually each jurisdiction has a checklist like this that answers the common question, "What kind of plans do I need to draw to get a permit?" Ask questions such as the following:

- ❑ Do the plans need to be a certain size or on special paper?
- ❑ How many copies of the plan are needed?
- ❑ At what scale do the plans need to be drawn?
- ❑ What particular plans must be drawn?
    - ❑ Site plan?
    - ❑ Foundation?
    - ❑ Floor?
    - ❑ Structural framing? Floor framing? Roof framing?
    - ❑ Plumbing plan? Plumbing isometric plan?
    - ❑ Electrical plan? One line diagram? Panel schedule?

- ❑ Mechanical plan? Mechanical equipment list?
- ❑ Elevation views?
- ❑ Structural cross sections?
- ❑ Details?

- ❑ What items must be on each sheet?
- ❑ How much detail must be included on each sheet?
- ❑ What will the permit cost?
- ❑ How long will it take to get a permit?

Asking these questions will give you the answers you want, and you'll gain insight into the personality of the department. This will help you decide how best to obtain your building permit. Get to know at least one person on a first-name basis. Show the professional at the building department staff your sketch and relate your concept. Ask about the department's position on owner-builder projects. Ask if any special assistance is available for customers such as yourself. Pick up a business card and ask permission to call later for other questions. A plan should be thorough enough to allow a qualified builder who is not familiar with you to build what you want, even at times you are inaccessible for questions. From your perspective, the purpose for construction plans is simply to provide a guide from which to build. From the perspective of the building department, the purpose is to ensure your compliance with other codes that they enforce. From their perspective, the more detail you have on the approved plans, the more likely you are to know about those details during an inspection.

In general, most jurisdictions require the same basic set of plans. A handout from the jurisdiction should illustrate exactly what is required. If none is available, the following checklist should suffice for most departments to verify compliance with the code.

Permit application process at building codes counter.

Minimum Building Plan Requirements

Two sets of plans drawn to ¼ inch scale including the following:

___ Site plan showing setbacks, adjoining streets, easements, etc.

___ Foundation plan with details as needed

___ Floor plan showing rooms, their uses, and dimensions

___ Stairway details (for two stories)

___ Electrical plan showing service location, lights, outlets, and panel schedule

___ Plumbing plan showing drain, waste, vent, and water lines

___ Mechanical plan showing equipment, supply outlets, and return air

___ Roof/floor framing plan with details as needed

___ Elevation views of all sides

___ Cross-sectional view of structure with various connection details

___ General architectural and general structural notes tied to plan

In addition, there are several other authorities which must approve your plans before you begin to develop your property. Sometimes they are located within the same building. Where this ensemble of agencies are collected in one building, it is known as a *one-stop shop*. It makes the building permit process very user-friendly. If your jurisdiction does not have a one-stop shop, you may need to seek out separate agencies for their approvals. Ask if there is any coordination between these agencies. Some of the following different permits may be necessary:

- ❑ Water connection or well drilling
- ❑ Endangered species protection
- ❑ Road permit or curb-cut into your property
- ❑ Zoning permit
- ❑ Grading permit
- ❑ Tree cutting permit
- ❑ Percolation test for septic tank design and approval
- ❑ Certificate of Occupancy
- ❑ Flood prevention permit
- ❑ Temporary electric permit
- ❑ Fire district approval
- ❑ Equipment moving permits
- ❑ Native plant protection

How to Draw Plans

Let's start with some simple observations. First, a plan is a conception and a proposed design used to achieve a tangible objective. A plan is drawn to create an agreement on all the design parameters of the project. In legal and construction circles, it represents a binding contract between parties. How complex and detailed does your plan need to be? Enough to satisfy two purposes. First, if you are doing most or all of the work yourself, it has to be adequate for you to build your house. Second, it must be sufficiently thorough to acquire a building permit. An unwritten rule I have heard used to express how detailed a plan must be is this:

> *Imagine that you want your dream home built by a contractor you have never met. Now imagine that you have hit the lottery and are going on a cruise for six months without being able to talk with the contractor about your house. When you return, the house is completed exactly as you imagined it because the plans left so little to the imagination.*

That may be more thorough than you need, but it may be the prerequisite for a permit. The following instructional material may seem overwhelming, but it's the basis for home design. This won't make you an engineer or architect, but it will allow you to communicate effectively with them. Have patience, ask questions, learn, and apply your education by designing your own home. There are methods for preparing a plan. The following is a step-by-step approach. Tools you need are:

- ❑ Drafting board
- ❑ T Square
- ❑ Triangles
- ❑ Mechanical pencil
- ❑ Architectural scale
- ❑ Erasers
- ❑ Calculator

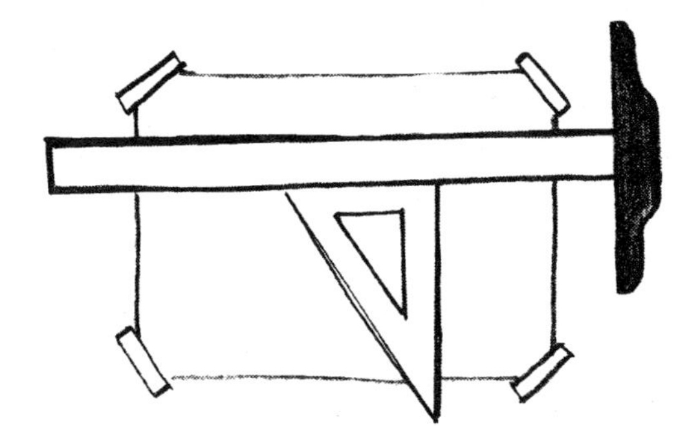

Computer-Aided Drafting

You may also consider using a computer aided drafting program to develop your plan. This approach is generally abbreviated CAD, an acronym for computer-aided drafting. If you have access to a computer and software with this capability, you will also need to plot or print your plan with a large enough scale to ensure legibility. Before you purchase a particular program, contact various printing businesses in your area and discuss the possibility of them printing your plan from your CAD file. Some print shops will be able to plot drawings from standard types of CAD programs for a reasonable fee. However, they may not have the necessary configurations for every CAD program.

CAD drafting has achieved a considerable level of sophistication. There are architectural drafting components which save you lots of tedious drafting or computer work. With some programs, a thorough floor plan can be drawn within a few hours. Plumbing, mechanical, and electrical symbols are included in some programs, which add to the professional look of these utility drawings. Scaling and dimensioning is automated with most CAD programs. Some even have standard general notes that apply to residential construction.

CAD drawing.

Another advantage to CAD drafting is that changes to the plan are much simpler and cleaner than manual drafting. You may decide to alter a bathroom layout or relocate a window along the wall frame. The building department may request that you alter your plan to clarify a code requirement. In some fashion, changes will invariably be made to your plans. These changes can be made on the file, and a re-plot will look as good as the original.

Drawing a plan using a CAD program has several distinct advantages. However, for the novice, hand drawing encourages learning the connection details, which may be somewhat generic in the library images of the CAD program. These methods of connection or specific details may not even be available in your region or appropriate to your design. Learn what the different symbols mean before you place them on your drawing, or you may need to change your plan later.

Site Plan

Let's start with the site plan, which is a layout of all buildings and structures on the property. An engineer's scale is used to draw large scale drawings such as site plans. Draw the site plan to a reasonable scale such as 1 inch = 10 feet. An acre of land is 43,560 square feet or roughly 200 feet by 200 feet square. A scale of 20 feet per inch will require a 200/20 or 10 inch square piece of paper for the plan.

You can either orient the top of the paper as north or as the back of your property and begin your drawing. Draw the boundaries of your property. Then use a compass to turn an angle for the adjacent side(s), and then draw the other sides and corners. Be sure to locate your house exactly where it is planned on the property, and reference the setbacks by dimension lines with distances specified. Add all the details below and you will have a site plan. Follow the example.

Site Plan

___ Name and address (including legal description). *Section R325*

___ North arrow and unit drawing scale.

___ Lot, block, subdivision name, and vicinity map. *Section R106.2*

___ All adjacent streets, right of ways, and easements. *Section R106.2*

___ All existing and proposed structures on the property must be shown. *Section R106.2*

___ Dimensions to nearest structures and property lines must be indicated. *Section R302.1*

___ All walls and fences must be shown. *Section R106.2*

___ Proposed driveways and parking areas. *Section R106.2*

___ Ponding areas for on-site retention of rain water or similar control of storm water. *Section R401.3*

___ Utility locations—water, electricity, gas, sewage, or septic tank. *Section R106.2*

___ Show all required setbacks, including zoning requirements. *Section R302.1*

___ Percent of land covered by development including any grading. Zoning Code

___ Survey of property, if requested by the Administrative Authority

___ Tax assessors parcel number for identification for taxing purposes.

___ Location of septic tank or sewer connection. *Section R106.2*

___ Elevation heights in topographic style to denote terrain features such as hills or valleys or drainage.

___ Any native plants targeted for protection if required by local ordinance.

___ Major landscaping or trees and other significant flora targeted for protection if required by local ordinance.

Typical Site Plan

1 inch = 20 feet
Tax Assessor's Number 23-23456789-00
Lot Coverage = 1859 square feet /10,890 = 17 percent coverage

Structural Design

Next, we have to get a little more involved in structural design. The practice of engineering is regarded as a professional career. This field of study is based on mathematics, statics, dynamics, physics, and other scientific methods. A registered professional has attended an approved curriculum in a college or university for at least four years and then passed a preliminary exam, worked for four more years under a structural engineer and then passed another more rigorous exam before he

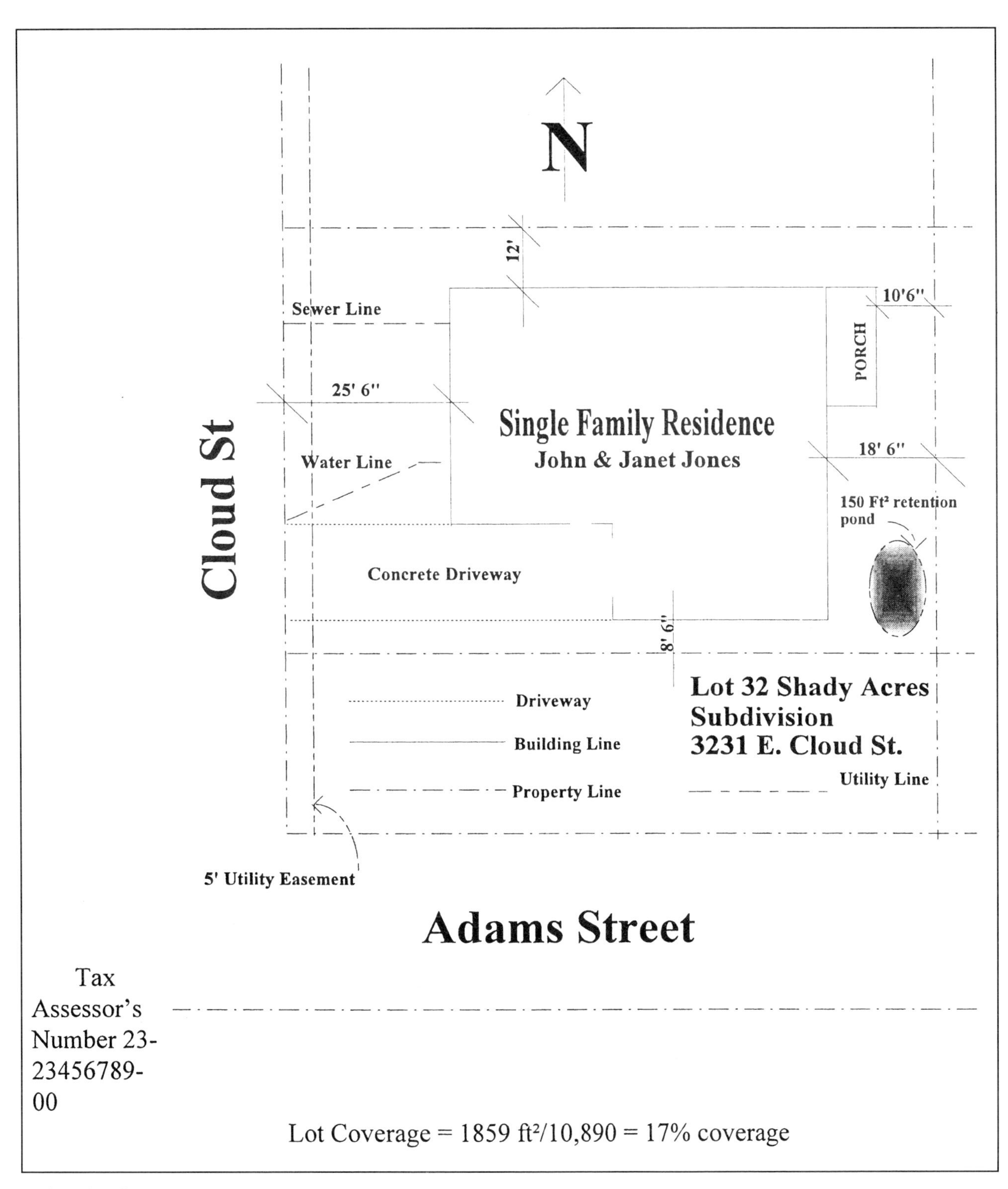

Typical site plan.

or she has the right to perform design work. But don't panic! Remember, you just have to know enough to build a house, not design a large multistory bank building.

First, think about what structural design really is all about. A building must resist forces, both natural and artificial. How does a building resist these forces? Gravity, wind, and earthquakes are the main natural forces that affect structural design for smaller structures. Footings and foundations resist gravity loads. Gravity wants to draw things closer to the surface. If the soil is compacted enough to resist the gravitational force, a concrete footing will absorb, then transfer that force into the soil. If the soil is not compacted enough or the load exceeds the soil's carrying capacity, the footing will sink into the soil causing failure in the structural frame in the building.

Wind and earthquake forces are referred to as *lateral* loads, in that they are applied laterally to a building or structure. Those forces are resisted by the roof and wall systems, which transfer the forces into the footing. Imagine a kite, which stops wind. The string holds it in place. You can feel the wind against the kite, even though it is a long way up in the air. The wind energy is being resisted by your fingers and string holding the kite in place. The wall and roof sheathing transfer that same wind (or seismic) energy along trusses, joists or studs to the bottom plate and anchor bolts into the footing, which anchors it to the ground with concrete. A structure is said to be static or stable when it is able to resist these forces without failure.

Soil Stability

In order to ensure a stable structure, an adequate footing and foundation system must anchor the superstructure to the earth. This is the job of the foundation. Every force that is applied to the house is transfered to the foundation. The foundation transmits these forces into the surrounding soil. If the soil shifts or settles in a manner that causes collapse of the foundation or unexpected movement of wall or roof systems, serious damage is inflicted upon the superstructure of the house above the ground.

Why Soil Stability Is Important

The weight of the structure (house) rests on the footing and therefore the soil below. Its ability to resist is quantified in pounds per square foot (psf). Soil bearing capacities are related to soil classification. Nature imposes many forces upon a structure such as a house. There are vertical loads such as gravity, the live and dead load of the building, as well as atmospheric pressure. There are also horizontal, or lateral loads, applied to a building. Wind and earthquakes can cause serious damage to even small structures. With proper design of soil and foundation, these forces can be managed.

Empirical Approach

The most common method of dealing with foundation design is the empirical approach. This approach is the process of designing new structures on the basis of past experience. The tables in the IRC are based on this approach. Look at houses near your site. Ask questions of your neighbors about the behavior of the local soil and their foundation's size, depth below grade, and width. Ask about underground structures on the site. Look around your property for anomalies such as surface boulders, dry washes, or standing water. Look for other danger signs such as dramatic hillsides, alluvial soil, expansive soil, or a poor mix of soil particles that contain too much organic matter.

Look at the soil yourself and make some judgements regarding your site's soil condition. Look around your site. What kind of vegetation is growing nearby? Does the flora suggest a high water table? Is there standing water on site? Ask neighbors about the water table depth. Do any streams flow through the property? Look for signs of erosion that may adversely affect a structure. Dig about 1 or 2 feet down and look at the composite materials. Is there a good mixture of clays, sand, fines, and loamy soil? Is it hard to dig? Is there any strange discoloration of the soil below grade? Soil consolidation suggests that it has compressive strength and can withstand loads from footings. Dig a hole with a shovel and look at the soil at the proposed footing depth. The soil at a depth of 18 inches should be slightly damp. Make a clod by squeezing with your fist and then break it apart. I have found that a sturdy building soil will consolidate together to make a fist size clod and then break apart in two or three major parts that remain more or less intact.

Soil Classification

A soils engineer classifies soil by texture and size. There are four major categories of texture: gravel, sand, silt, and clay. These are classified by their grain size established by their propensity to pass through different sieve sizes. Table R405.1 from the IRC describes these different soil types and characteristics.

TABLE R405.1
PROPERTIES OF SOILS CLASSIFIED ACCORDING TO THE UNIFIED SOIL CLASSIFICATION SYSTEM

SOIL GROUP	UNIFIED SOIL CLASSIFICATION SYSTEM SYMBOL	SOIL DESCRIPTION	DRAINAGE CHARACTERISTICS[a]	FROST HEAVE POTENTIAL	VOLUME CHANGE POTENTIAL EXPANSION[b]
Group I	GW	Well-graded gravels, gravel sand mixtures, little or no fines.	Good	Low	Low
	GP	Poorly graded gravels or gravel sand mixtures, little or no fines.	Good	Low	Low
	SW	Well-graded sands, gravelly sands, little or no fines.	Good	Low	Low
	SP	Poorly graded sands or gravelly sands, little or no fines.	Good	Low	Low
	GM	Silty gravels, gravel-sand-silt mixtures.	Good	Medium	Low
	SM	Silty sand, sand-silt mixtures.	Good	Medium	Low
Group II	GC	Clayey gravels, gravel-sand-clay mixtures.	Medium	Medium	Low
	SC	Clayey sands, sand-clay mixture.	Medium	Medium	Low
	ML	Inorganic silts and very fine sands, rock flour, silty or clayey fine sands or clayey silts with slight plasticity.	Medium	High	Low
	CL	Inorganic clays of low to medium plasticity, gravelly clays, sandy clays, silty clays, lean clays.	Medium	Medium	Medium to Low
Group III	CH	Inorganic clays of high plasticity, fat clays.	Poor	Medium	High
	MH	Inorganic silts, micaceous or diatomaceous fine sandy or silty soils, elastic silts.	Poor	High	High
Group IV	OL	Organic silts and organic silty clays of low plasticity.	Poor	Medium	Medium
	OH	Organic clays of medium to high plasticity, organic silts.	Unsatisfactory	Medium	High
	Pt	Peat and other highly organic soils.	Unsatisfactory	Medium	High

For SI: 1 inch = 25.4 mm.

a. The percolation rate for good drainage is over 4 inches per hour, medium drainage is 2 inches to 4 inches per hour, and poor is less than 2 inches per hour.

b. Soils with a low potential expansion typically have a plasticity index (PI) of 0 to 15, soils with a medium potential expansion have a PI of 10 to 35 and soils with a high potential expansion have a PI greater than 20.

The primary unit of soil is the *soil series*. Most soil profiles within a soil series tend to be nearly alike. Nomenclature for soil series usually bear location names such as Dona Ana, Vail, Luna, or Mesa. A local soil survey map can indicate what type of soil is in a particular location by referring to a name such as these. The soil survey contains a list of the characteristics and properties of the particular soil at that site. These unique soil mixtures feature specific structural properties. Determination of those properties is the job of the soil engineer. If you can identify a problem soil, consult with a soil engineer to help you in your design. Your soil will be graded by size of its constituent materials such as sand, clay gravel, etc. This is important because it is the first step in the classification process. Size of constituent parts may be identified as follows:

SOIL SIZES

Component	Size Range*
Boulders	Above 12 inches
Cobbles	3 inches to 12 inches
Gravel	No. 4 to 3 inches
Course	¾ inch to 3 inches
Fine	No. 4 to ¾ inch
Sand	No. 200 to No. 4
Course	No. 10 to No. 4
Medium	No. 40 to No. 10
Fine	No. 200 to No. 40
Fines (silt or clay)	Below No. 200

* *No. Ranges refer to standard soil sieve analysis testing sizes.*

Soil Report

The code official has the authority to require that you provide an analysis of soil conditions by a qualified engineer or architect. The code official (or other designee) will make that request when they suspect that you are in an area which is prone to poor soil conditions. Compaction of soil is required if fill dirt is needed to raise the building pad above surrounding terrain to allow for water runoff. A soil report is required if the footing is placed on fill dirt. In certain seismic zones, an additional evaluation pertaining to soil liquification and soil instability may be required. A report may also be required if you plan a multistory structure or one that may be subjected to extreme natural forces such as high winds or a seismic event. If you believe that your soil may have adverse structural qualities, I strongly recommend that you acquire the services of a soils engineer for analysis and design.

A soil report will provide the classification and design bearing capacity of the soil. The elevation of the water table may be specified. Further recommendations for foundation types and various design criteria may need to be included in the report. If a soil report indicates that expansive or collapsible soils are present, a specific design to mitigate the conditions must be developed and submitted for review.

If a soil report indicates that poor soils are present, don't despair. There is always a mitigation that will improve your soil stability. Bearing capacity depends on the density and consistency of the soil. This can be altered, either intentionally or accidentally; favorably or unfavorably. The answers will be in the soil report. That is why you hire a soils engineer in the first place. Ask for a variety of structural recommendations and select the best for your project.

Foundation Plan

A foundation plan shows the footing design and any foundation walls that will support bearing walls, such as basement walls. In order to design a foundation, it is necessary to understand soil stability, wind load, earthquakes, and the weight of structure such as a house. Several conditions will adversely affect a soil's ability to withstand continuous loads imposed by bearing walls onto footings. Your foundation design must begin by knowing the soil's condition.

Soil is classified according to its ability to withstand an imposed load based upon its soil classification. There are stable soils, expansive soils, collapsible soils, alluvial soils, and a variety of other types of soil conditions. A soil survey for your area will approximate the type of soil in your proposed building site. A soil survey is available at either the state or county engineer's offices. Site inspection and soil investigation will exactly determine the qualities of your soil. Normally, this evaluation will be made by a soil scientist or geotechnical engineer specializing in soil analysis of building sites.

Conventional Design

The IRC establishes the minimum parameters for an acceptable base for a foundation. There are certain minimum standards for compaction and stability. Without a specific design, a soil is determined to have a soil bearing pressure dependent upon its class of materials. Each major soil group has assumed soil bearing pressures. They are outlined in Table R401.4.1 from the IRC as follows:

TABLE R401.4.1
PRESUMPTIVE LOAD-BEARING VALUES OF FOUNDATION MATERIALS[a]

CLASS OF MATERIAL	LOAD-BEARING PRESSURE (pounds per square foot)
Crystalline bedrock	12,000
Sedimentary and foliated rock	4,000
Sandy gravel and/or gravel (GW and GP)	3,000
Sand, silty sand, clayey sand, silty gravel and clayey gravel (SW, SP, SM, SC, GM and GC)	2,000
Clay, sandy clay, silty clay, clayey silt, silt and sandy silt (CI, ML, MH and CH)	1,500[b]

For SI: 1 pound per square foot = 0.0479 kN/m^2.

a. When soil tests are required by Section R401.4, the allowable bearing capacities of the soil shall be part of the recommendations.

b. Where the building official determines that in-place soils with an allowable bearing capacity of less than 1,500 psf are likely to be present at the site, the allowable bearing capacity shall be determined by a soils investigation.

Concrete has strength based on the proportions of its mixture of constituent materials such as sand, gravel, cement, and water. The plant which makes the ready mix concrete for your project can design and prepare the strength you desire. There are minimum standards for different uses based on weathering potential. Table R402.2 shows the required strength for each use.

Based on Table R401.4.1, an unclassified weak soil is assumed to be able to withstand 1500 pounds per square foot (psf) of allowable foundation pressure. The useful translation is that a footing 1 foot wide can only support 1500 pounds per lineal foot. All loads on walls supported by that foundation may not exceed 1500 psf. The way to increase the allowable load on a bearing wall is to widen the footing, displacing the load over more surface area of soil. Therein lies the job of the structural designer. To gain a more clear understanding of how footings are designed, the following example will demonstrate the methods used by a structural engineer.

Calculate all the loads on a bearing wall. A bearing wall is assumed to support the floor or roof load that is exactly half the distance to the next bearing wall. That distance is referred to as *tributary width*. The assumption is that the adjacent bearing wall or beam supports the other half of all loads.

TABLE R402.2
MINIMUM SPECIFIED COMPRESSIVE STRENGTH OF CONCRETE

TYPE OR LOCATIONS OF CONCRETE CONSTRUCTION	MINIMUM SPECIFIED COMPRESSIVE STRENGTH[a] (f'_c) Weathering potential[b]		
	Negligible	Moderate	Severe
Basement walls, foundations and other concrete not exposed to the weather	2,500	2,500	2,500[c]
Basement slabs and interior slabs on grade, except garage floor slabs	2,500	2,500	2,500[c]
Basement walls, foundation walls, exterior walls and other vertical concrete work exposed to the weather	2,500	3,000[d]	3,000[d]
Porches, carport slabs and steps exposed to the weather, and garage floor slabs	2,500	3,000[d,e]	3,500[d,e]

For SI: 1 pound per square inch = 6.895 kPa.

a. At 28 days psi.

b. See Table R301.2(1) for weathering potential.

c. Concrete in these locations that may be subject to freezing and thawing during construction shall be air-entrained concrete in accordance with Footnote d.

d. Concrete shall be air entrained. Total air content (percent by volume of concrete) shall not be less than 5 percent or more than 7 percent.

e. See Section R402.2 for minimum cement content.

The weight of all structural elements including wall frame, drywall, siding, trusses, rafters, floor joists, floor decking, insulation, roofing material, plumbing, mechanical, electrical, and other associated building materials are referred to as *dead load.* This dead load is commonly ascribed a weight value of between 10 and 20 pounds per square feet (10 psf to 20 psf). The remaining vertical loads, which must be considered, are atmospheric pressure, rain and snow loads or other implied loads. These are referred to as *live load.* Live load for a floor joist or truss includes furniture, and people. Live load for roof structures is a function of the weather where you live. Ask your local building department what live load should be used for design purposes. Usually the design live load is between 20 psf and 40 psf for roof load. Live load for floors used for residential purposes vary between 30 to 40 psf.

Example:

Let's go through a sample calculation for foundation design. Follow the illustration for our calculation. The factors are as follows:

Soil bearing pressure	1500 psf
Live load for roof	20 psf
Live load for floor	40 psf
Dead load calculated at	17 psf
Tributary width for floor	16 feet
Tributary width for roof	24 feet

Solution:

Step 1: Calculate weight of roof on bearing wall: (20 psf + 17 psf) x (24 feet) = 888 #/foot

Step 2: Calculate weight of floor on bearing wall: (40psf + 17 psf) x (16 feet) = 912 #/foot

Step 3: Add loads: 888 + 912 = 1800 #/ft

Step 4: Calculate required footing size: The required width = (1800 #/ft) ÷ (1500 #/ft) = 1.2 feet, or 14.4 inches.

More than likely, you will dig at least a 16 inch wide footing.

Remember you can vary the load imposed upon the bearing wall by adding or subtracting bearing walls under the roof or floor joists or trusses. You can also decrease the required size of footings by acquiring a soil report, which may increase the allowable soil bearing pressure above the assumed value.

Footing design for a house is derived through a series of repetitions depending upon these few variable criteria. The foundation design can be as simple as a rectangle on the perimeter walls, with no interior footings if you use clear span trusses, or as complex as numerous interior point load footings to support interior bearing walls or concentrated loads. While design engineers use the method outlined above, the IRC establishes certain minimum thicknesses, widths and depth below natural grade. These minimum criteria are summarized in Table R403.1.

TABLE R403.1
MINIMUM WIDTH OF CONCRETE OR MASONRY FOOTINGS (inches)

	LOAD-BEARING VALUE OF SOIL (psf)					
	1,500	2,000	2,500	3,000	3,500	4,000
Conventional light-frame construction						
1-story	16	12	10	8	7	6
2-story	19	15	12	10	8	7
3-story	22	17	14	11	10	9
4-inch brick veneer over light frame or 8-inch hollow concrete masonry						
1-story	19	15	12	10	8	7
2-story	25	19	15	13	11	10
3-story	31	23	19	16	13	12
8-inch solid or fully grouted masonry						
1-story	22	17	13	11	10	9
2-story	31	23	19	16	13	12
3-story	40	30	24	20	17	15

For SI: 1 inch = 25.4 mm, 1 pound per square foot = 0.0479 kN/m^2.

Deformed steel rebars used as reinforcement in footings and foundation systems add tensile strength to this structural system. Steel withstands a great deal of tensile force. With deformed steel bars encased in concrete this structural system develops its strength. The concrete footing benefits by working in conjunction with the steel to resist a failure in soil stability. The code also establishes the minimum reinforcement required for footings within certain seismic zones. However, a good design would be to install at least two #4 reinforcement bars in the footing. Be sure to place them at least 3 inches above the bottom of the footing.

Foundation Wall Design

If you are not building a monolithic slab foundation, you will need to design a foundation wall. This wall may be as low as 8 inches high or it may extend to 8 feet for a full-height basement. The wall design may be either solid-grouted, steel reinforced concrete block, or steel reinforced concrete. Each requires a separate design.

Reinforced concrete masonry blocks (CMU) can be used to construct a very solid foundation wall. The block size that is the strongest for your purpose could vary in size between 8 inches to 12 inches in width. Foundation blocks have a hollow cavity, which allows reinforcement and a concrete grout mixture to harden and create a monolithic assembly that has great strength. These block are laid up together in one of many patterns using mortar to establish a bond between each block unit to create a complete wall. Vertical and horizontal reinforcement steel is installed during the block laying process. When the block wall is complete and the mortar has sufficiently cured, the hollow cavities are filled with concrete grout to complete the wall.

Poured in place, reinforced concrete wall is extremely solid and capable of serving as either a shorter foundation wall or a full-height basement wall. However, a basement wall that is subject to hydrostatic pressure from ground water requires an engineered design and cannot be built according to the conventional design outlined in these tables. A reinforced concrete wall design width may vary between 8 inches to 12 inches thick.

To achieve a wall that is smooth and uniformly thick, a substantial form must be used. There are several types of concrete forms that you can use. The most simple form is plywood acting as shoring. However, simply installing plywood is not enough. Concrete is very heavy, about 160 pounds per cubic foot. When fresh concrete is dropped down into a 4 foot high form, it can smash against the bottom of the form with tremendous force and may cause a *blow-out* or failure in the form. The form must be stabilized to prevent such a calamity.

A foundation wall composed of block must be grouted solid and may be required to be reinforced with deformed steel bars designed for that purpose. Foundations with stem walls must have one #4 bar at the top and the bottom of the wall. Slab on grade foundations must have one #4 bar at the top and bottom of the foundation. Each of these foundation reinforcement materials must be connected to the footing. The IRC establishes a minimum structural connection between the footing and the foundation wall. In certain seismic zones, a #4 bar must extend out of the footing into the foundation wall every four feet on center and have a 14 inch standard hook into the footing.

The design of these foundation walls is also a task generally performed by a structural engineer. However, the IRC provides a design for both of these wall systems based on some variables such as height, soil conditions, foundation thickness, and reinforcement. Basement walls may be considered as a foundation wall and designed as such under certain conditions.

Notice that depending upon the height of the wall, the difference in the height of earth on either side of the wall or unbalanced backfill height and the soil conditions, the required wall thickness can be derived for either poured concrete or laid-up concrete masonry block. Be sure to read the footnotes below the table if they apply to any column in the foundation wall thickness you have selected. These footnotes will either restrict your use of the material or require that certain other conditions be met. Your inspector will pick up on these distinctions.

Depending upon your seismic category, concrete and concrete block foundation walls may be required to be reinforced with deformed steel bars to increase their strength to resist lateral forces applied to the wall from the adjacent soil. Because of the weight of the soil and added moisture, the wall must resist these loads without failing. Tables R404.1.1(1), (2), (3), and (4) clarify these requirements. Each table is based on a different foundation wall thickness, either 8, 10, or 12 inches for either concrete or concrete block.

Another possible method of building a concrete wall is to use one of the many Insulated Concrete Forms, (ICF). These wall systems have an engineered design and are manufactured products having passed substantial testing. Even though there are numerous manufacturers, the products have some similarities. Some forms are molded, recycled plastic, which are lightweight. The form sections are

TABLE R404.1.1(1)
PLAIN CONCRETE AND PLAIN MASONRY FOUNDATION WALLS

MAXIMUM WALL HEIGHT (feet)	MAXIMUM UNBALANCED BACKFILL HEIGHT[c] (feet)	PLAIN CONCRETE MINIMUM NOMINAL WALL THICKNESS (inches)			PLAIN MASONRY[a] MINIMUM NOMINAL WALL THICKNESS (inches)		
		Soil classes[b]					
		GW, GP, SW and SP	GM, GC, SM, SM-SC and ML	SC, MH, ML-CL and inorganic CL	GW, GP, SW and SP	GM, GC, SM, SM-SC and ML	SC, MH, ML-CL and inorganic CL
5	4	6	6	6	6 solid[d] or 8	6 solid[d] or 8	6 solid[d] or 8
	5	6	6	6	6 solid[d] or 8	8	10
6	4	6	6	6	6 solid[d] or 8	6 solid[d] or 8	6 solid[d] or 8
	5	6	6	6	6 solid[d] or 8	8	10
	6	6	8[g]	8[g]	8	10	12
7	4	6	6	6	6 solid[d] or 8	8	8
	5	6	6	8[g]	6 solid[d] or 8	10	10
	6	6	8	8	10	12	10 solid[d]
	7	8	8	10	12	10 solid[d]	12 solid[d]
8	4	6	6	6	6 solid[d] or 8	6 solid[d] or 8	8
	5	6	6	8	6 solid[d] or 8	10	12
	6	8[h]	8	10	10	12	12 solid[d]
	7	8	10	10	12	12 solid[d]	Footnote e
	8	10	10	12	10 solid[d]	12 solid[d]	Footnote e
9	4	6	6	6	6 solid[d] or 8	6 solid[d] or 8	8
	5	6	8[g]	8	8	10	12
	6	8	8	10	10	12	12 solid[d]
	7	8	10	10	12	12 solid[d]	Footnote e
	8	10	10	12	12 solid[d]	Footnote e	Footnote e
	9	10	12	Footnote f	Footnote e	Footnote e	Footnote e

For SI: 1 inch = 25.4 mm, 1 foot = 304.8 mm, 1 pound per square inch = 6.895 Pa.

a. Mortar shall be Type M or S and masonry shall be laid in running bond. Ungrouted hollow masonry units are permitted except where otherwise indicated.

b. Soil classes are in accordance with the Unified Soil Classification System. Refer to Table R405.1.

c. Unbalanced backfill height is the difference in height of the exterior and interior finish ground levels. Where an interior concrete slab is provided, the unbalanced backfill height shall be measured from the exterior finish ground level to the top of the interior concrete slab.

d. Solid grouted hollow units or solid masonry units.

e. Wall construction shall be in accordance with Table R404.1.1(2) or a design shall be provided.

f. A design is required.

g. Thickness may be 6 inches, provided minimum specified compressive strength of concrete, f_c, is 4,000 psi.

usually very large compared to concrete block. Some are 2 feet high and 8 feet or 10 feet in length, but they are lightweight and easy to install. They connect to each other in multiple courses, much like giant building blocks.

As with block, reinforcement steel is added as the sections are added. When they are complete, and after inspection, concrete is poured through special pumps and hoses to fill the cavity. The result is a very strong wall with rigid insulation on either side, which usually complies with energy code requirements. Since this product is an engineered design, a *Special Inspector* may be required to watch this assembly and concrete pour. You may have to secure and pay for this service and submit engineering reports to the building department, so check with your product salesman in order to know what is required.

TABLE R404.1.1(2)
REINFORCED CONCRETE AND MASONRY[a] FOUNDATION WALLS

MAXIMUM WALL HEIGHT (feet)	MAXIMUM UNBALANCED BACKFILL HEIGHT[e] (feet)	MINIMUM VERTICAL REINFORCEMENT SIZE AND SPACING[b, c] FOR 8-INCH NOMINAL WALL THICKNESS		
		Soil classes[d]		
		GW, GP, SW and SP soils	GM, GC, SM, SM-SC and ML soils	SC, MH, ML-CL and inorganic CL soils
6	5	#4 at 48″ o.c.	#4 at 48″ o.c.	#4 at 48″ o.c.
	6	#4 at 48″ o.c.	#4 at 40″ o.c.	#5 at 48″ o.c.
7	4	#4 at 48″ o.c.	#4 at 48″ o.c.	#4 at 48″ o.c.
	5	#4 at 48″ o.c.	#4 at 48″ o.c.	#4 at 40″ o.c.
	6	#4 at 48″ o.c.	#5 at 48″ o.c.	#5 at 40″ o.c.
	7	#4 at 40″ o.c.	#5 at 40″ o.c.	#6 at 48″ o.c.
8	5	#4 at 48″ o.c.	#4 at 48″ o.c.	#4 at 40″ o.c.
	6	#4 at 48″ o.c.	#5 at 48″ o.c.	#5 at 40″ o.c.
	7	#5 at 48″ o.c.	#6 at 48″ o.c.	#6 at 40″ o.c.
	8	#5 at 40″ o.c.	#6 at 40″ o.c.	#6 at 24″ o.c.
9	5	#4 at 48″ o.c.	#4 at 48″ o.c.	#5 at 48″ o.c.
	6	#4 at 48″ o.c.	#5 at 48″ o.c.	#6 at 48″ o.c.
	7	#5 at 48″ o.c.	#6 at 48″ o.c.	#6 at 32″ o.c.
	8	#5 at 40″ o.c.	#6 at 32″ o.c.	#6 at 24″ o.c.
	9	#6 at 40″ o.c.	#6 at 24″ o.c.	#6 at 16″ o.c.

For SI: 1 inch = 25.4 mm, 1 foot = 304.8 mm.

a. Mortar shall be Type M or S and masonry shall be laid in running bond.

b. Alternative reinforcing bar sizes and spacings having an equivalent cross-sectional area of reinforcement per lineal foot of wall shall be permitted provided the spacing of the reinforcement does not exceed 72 inches.

c. Vertical reinforcement shall be Grade 60 minimum. The distance from the face of the soil side of the wall to the center of vertical reinforcement shall be at least 5 inches.

d. Soil classes are in accordance with the Unified Soil Classification System. Refer to Table R405.1.

e. Unbalanced backfill height is the difference in height of the exterior and interior finish ground levels. Where an interior concrete slab is provided, the unbalanced backfill height shall be measured from the exterior finish ground level to the top of the interior concrete slab.

TABLE R404.1.1(3)
REINFORCED CONCRETE AND MASONRY[a] FOUNDATION WALLS

MAXIMUM WALL HEIGHT (feet)	MAXIMUM UNBALANCED BACKFILL HEIGHT[e] (feet)	VERTICAL REINFORCEMENT SIZE AND SPACING[b,c] FOR 12-INCH NOMINAL WALL THICKNESS		
		Soil classes[d]		
		GW, GP, SW and SP soils	GM, GC, SM, SM-SC and ML soils	SC, MH, ML-CL and inorganic CL soils
7	4	#4 at 72″ o.c.	#4 at 72″ o.c.	#4 at 72″ o.c.
	5	#4 at 72″ o.c.	#4 at 72″ o.c.	#4 at 72″ o.c.
	6	#4 at 72″ o.c.	#4 at 64″ o.c.	#4 at 48″ o.c.
	7	#4 at 72″ o.c.	#4 at 48″ o.c.	#5 at 56″ o.c.
8	5	#4 at 72″ o.c.	#4 at 72″ o.c.	#4 at 72″ o.c.
	6	#4 at 72″ o.c.	#4 at 56″ o.c.	#5 at 72″ o.c.
	7	#4 at 64″ o.c.	#5 at 64″ o.c.	#4 at 32″ o.c.
	8	#4 at 48″ o.c.	#4 at 32″ o.c.	#5 at 40″ o.c.
9	5	#4 at 72″ o.c.	#4 at 72″ o.c.	#4 at 72″ o.c.
	6	#4 at 72″ o.c.	#4 at 56″ o.c.	#5 at 64″ o.c.
	7	#4 at 56″ o.c.	#4 at 40″ o.c.	#6 at 64″ o.c.
	8	#4 at 64″ o.c.	#6 at 64″ o.c.	#6 at 48″ o.c.
	9	#5 at 56″ o.c.	#7 at 72″ o.c.	#6 at 40″ o.c.

For SI: 1 inch = 25.4 mm, 1 foot = 304.8 mm.

a. Mortar shall be Type M or S and masonry shall be laid in running bond.

b. Alternative reinforcing bar sizes and spacing having an equivalent cross-sectional area of reinforcement per lineal foot of wall shall be permitted provided the spacing of the reinforcement does not exceed 72 inches.

c. Vertical reinforcement shall be Grade 60 minimum. The distance from the face of the soil side of the wall to the center of vertical reinforcement shall be at least 8.75 inches.

d. Soil classes are in accordance with the Unified Soil Classification System. Refer to Table R405.1.

e. Unbalanced backfill height is the difference in height of the exterior and interior finish ground levels. Where an interior concrete slab is provided, the unbalanced backfill height shall be measured from the exterior finish ground level to the top of the interior concrete slab.

TABLE R404.1.1(4)
REINFORCED CONCRETE AND MASONRY[a] FOUNDATION WALLS

MAXIMUM WALL HEIGHT (feet)	MAXIMUM UNBALANCED BACKFILL HEIGHT[e] (feet)	MINIMUM VERTICAL REINFORCEMENT SIZE AND SPACING[b,c] FOR 10-INCH NOMINAL WALL THICKNESS		
		Soil Classes[d]		
		GW, GP, SW and SP soils	GM, GC, SM, SM-SC and ML soils	SC, MH, ML-CL and inorganic CL soils
7	4	#4 at 56″ o.c.	#4 at 56″ o.c.	#4 at 56″ o.c
	5	#4 at 56″ o.c	#4 at 56″ o.c.	#4 at 56″ o.c.
	6	#4 at 56″ o.c.	#4 at 48″ o.c	#4 at 40″ o.c.
	7	#4 at 56″ o.c.	#5 at 56″ o.c.	#5 at 40″ o.c.
8	5	#4 at 56″ o.c.	#4 at 56″ o.c.	#4 at 48″ o.c.
	6	#4 at 56″ o.c	#4 at 48″ o.c	#5 at 56″ o.c
	7	#4 at 48″ o.c.	#4 at 32″ o.c.	#6 at 56″ o.c.
	8	#5 at 56″ o.c.	#5 at 40″ o.c.	#7 at 56″ o.c.
9	5	#4 at 56″ o.c.	#4 at 56″ o.c.	#4 at 48″ o.c.
	6	#4 at 56″ o.c	#4 at 40″ o.c	#4 at 32″ o.c
	7	#4 at 56″ o.c.	#5 at 48″ o.c.	#6 at 48″ o.c.
	8	#4 at 32″ o.c.	#6 at 48″ o.c.	#4 at 16″ o.c.
	9	#5 at 40″ o.c.	#6 at 40″ o.c.	#7 at 40″ o.c.

For SI: 1 inch = 25.4 mm, 1 foot = 304.8 mm.

a. Mortar shall be Type M or S and masonry shall be laid in running bond.

b. Alternative reinforcing bar sizes and spacings having an equivalent cross-sectional area of reinforcement per lineal foot of wall shall be permitted provided the spacing of the reinforcement does not exceed 72 inches.

c. Vertical reinforcement shall be Grade 60 minimum. The distance from the face of the soil side of the wall to the center of vertical reinforcement shall be at least 6.75 inches.

d. Soil classes are in accordance with the Unified Soil Classification System. Refer to Table R405.1.

e. Unbalanced backfill height is the difference in height of the exterior and interior finish ground levels. Where an interior concrete slab is provided, the unbalanced backfill height shall be measured from the exterior finish ground level to the top of the interior concrete slab.

Use the following checklist as a starting point in developing your foundation design. The sample plan illustrates the traditional appearance of a standard foundation plan.

Foundation Plan

___ Indicate that soil is protected against termite damage. *Section R324*

___ Show that foundations are capable of supporting all loads to the supporting soil. *Section R401.2*

___ Show that surface drainage is diverted, so as not to create a hazard. *Section 401.2*

___ In areas likely to have expansive, compressible, shifting, or similar soils, provide a soil report. *Section R401.4*

___ In lieu of a geotechnical report, show that soil design is according to *Table R401.4.1*

___ Show that concrete has a minimum strength as outlined in *Table R402.2*

___ Show that concrete exposed and subject to weathering is air-entrained. *Section 402.2*

___ Show that all exterior walls are supported on continuous solid concrete footings. *Section R403.1*

___ Show that footings are at least the minimum size and shape required in *Table 403.1*

___ Indicate braced walls required for lateral forces are supported by continuous reinforced footings. *Section R403.1.2*

___ Show that foundations with stem walls and slab-on-grade are reinforced with steel bars. *Section R403.1.3*

___ Show that exterior footings and foundation systems extend below the frost line or 12 inches below soil. *Section R403.1.4*

___ Show that shallow footings are protected from frost with insulation as specified by *Section R403.3 and Table R403.3.*

___ Show the top of footings are level and the bottom of footings do not have a slope exceeding 1:10. *Section R403.1.5*

___ Show that the wood frame wall is anchored to the foundation with 1/2 inch anchor bolts @ 6 feet o.c. *Section R403.1.6*

___ Show that these anchor bolts will be embedded 7 inches into concrete or masonry foundation. *Section R403.1.6*

___ Show any special anchorage required for seismic conditions. *Section R403.1.6.1*

___ Show that footings adjacent to slopes meet the setback requirements (3:1) of *Section R403.1.7.*

___ Show that foundation height is 12 inches higher than the street gutter of drainage discharge. *Section R4003.1.7.3*

___ Show that concrete or masonry foundation walls meet the size and shape requirements of *Table R404.1.1(1).*

___ Show that concrete or masonry foundation walls meet the reinforcement requirements of *Tables R404.1.1(2), (3).*

___ Show that concrete or masonry foundation walls in any seismic regions meet the requirements of *Section 404.1.4.*

___ Show that the finish elevation of foundation walls is at least 6 inches above adjacent grade. *Section R404.1.6*

___ Show that special foundation design meets the requirements of engineered design and manufacturer's specifications.

___ Show that concrete and masonry foundation walls are dampproofed to prevent moisture penetration. *Section R406*

___ Show crawl space is ventilated with openings in foundation wall with 1 square foot per 150 square feet of floor. *Section R408*

___ Show crawl space is provided with an access panel that is at least 18 inches by 24 inches. *Section R408.3*

___ Provide foundation details that are tied to foundation plan. *Figure R403.1(1)*

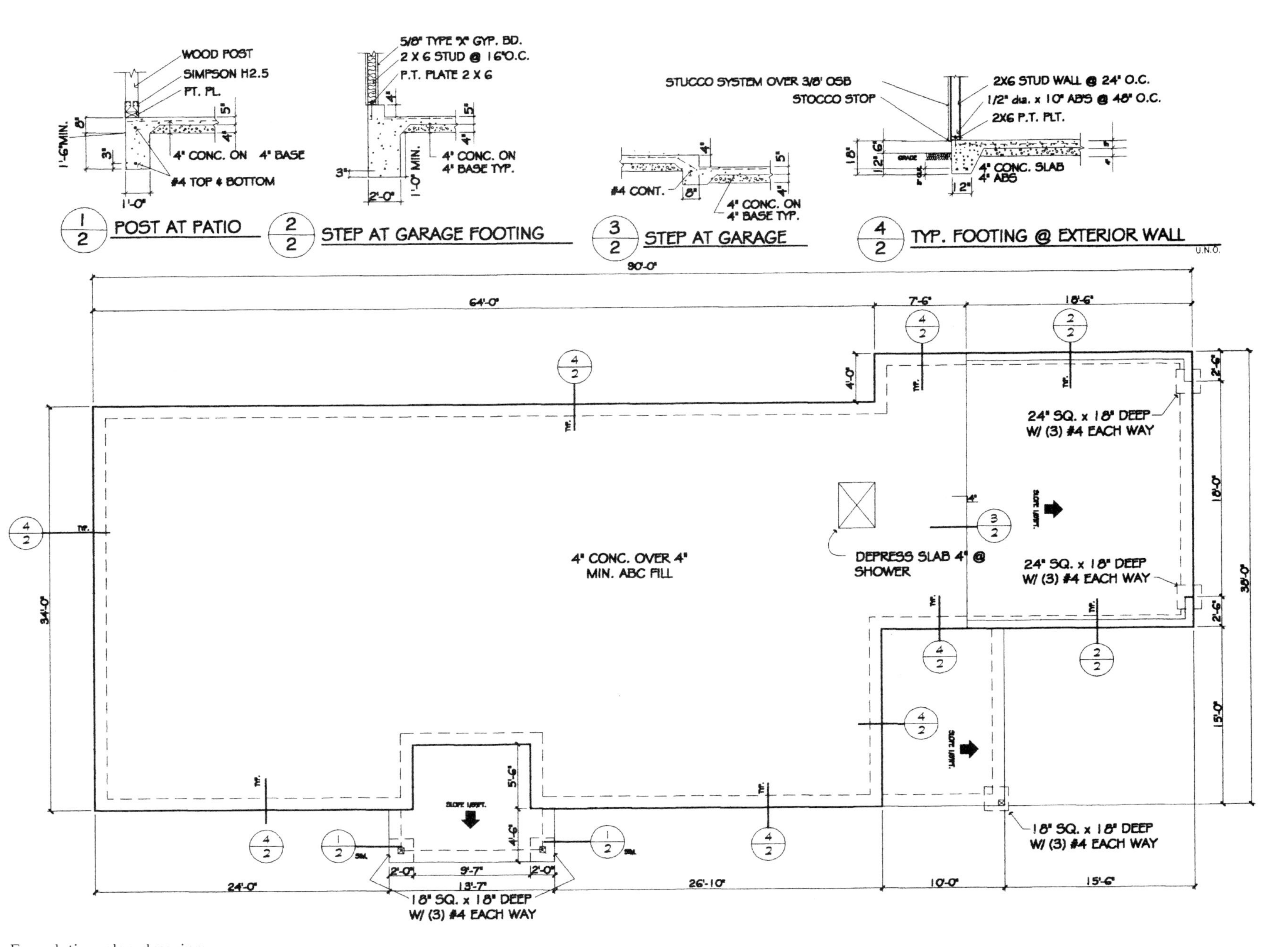

Foundation plan drawing.

Floor Plan

Once your family has settled on a basic floor plan, it is the first and the easiest to draw. Of course the floor plan will start with sketches on the kitchen table and will be the focus of the debate with your family about the look of your home. Use graph paper at first to verify if sizes and shapes work. It will evolve into a more skilled drawing as decisions are made. Use a ¼ inch per foot scale and make the drawing. Remember, because of drywall and siding, to allow 4½ inches width for interior 2 inch by 4 inch walls and 6½ inches for interior or exterior 2 inch by 6 inch walls. Include all the features discussed below.

The floor plan is the most important drawing of your set of plans. This plan will portray the floor layout including rooms, their uses, overall and segmented dimensions, doors—their swing and sizes—windows—their size and openings—cabinets, plumbing fixtures, mechanical equipment, interior and exterior wall types, stairs—their width and landings—and other architectural features. This plan is commonly used as the basis for other drawings such as the foundation, roof or floor framing, plumbing, mechanical, and electrical drawings. In fact, some computer aided drafting techniques use this plan's skeletal frame of walls as the framework for each of these other drawings. Since it sets the basis for your intent, its shape translates into these other drawings. If you do this, be sure to make copies of the drawing *before* you add details and notes unique to a floor plan.

Windows must be identified as to their overall opening size. For instance, if a window opens like a sliding glass door, it is a *horizontal slider*. If it opens with the lower frame upward as the movable element, the window is regarded as a *single hung* type. If the window completely opens outward like a door, swinging on hinges, it is regarded as a *casement* type. Window frames are made of wood, aluminum, or vinyl plastic. Windows also have different types and numbers of glazing panels. Some are single glazed, others are double or even triple glazed. Some have specialized surface coatings that help to resist heat flow. The multiple glazing tends to retard heat flow better. When considering where to locate a window, think about the room and where furniture will be located. Think about the desired view from each room. Be sure to consider security and privacy interests. Windows are identified as to size, shape, and operable or fixed. For example, a 3 x 5 XO implies a 3 foot wide, 5 foot high window with an operable panel. A 4 x 4 fixed is 4 feet wide and high, but with fixed glass and cannot be opened. Be sure to mark the window and door sizes on your plan.

If a window is located in a *hazardous* area, it must be safety-type glass. Being within 24 inches of a door's edge, or 18 inches of the floor, or within a shower or tub area in a bathroom below 60 inches of the floor are among the areas considered as hazardous. The approved types of glass permitted for use in hazardous areas include laminated glass, fully tempered glass, heat strengthened glass, wired glass and approved rigid plastics.

The minimum size of windows is regulated by the code. There are requirements for natural lighting and ventilation. Unless you provide artificial lighting or mechanical ventilation, each habitable room must be provided with windows that are at least 8 percent of the floor area, and half of that area (4 percent of floor area) must be operable. Stairways must be artificially lighted to allow safe use. Bathrooms must have either 3 square feet of glazing or artificial light. Certain interior rooms may enjoy the natural light from other adjoining rooms if the wall between them is 50 percent open. Skylights can be installed in these interior rooms that are away from exterior walls, such as a laundry room or bathroom. Skylights are available in metal or plastic frame assemblies. The lens of a skylight is available as transparent, clear, or opaque. Bedrooms must have windows or a door that opens onto the exterior. These are required for emergency egress and must meet four criteria:

1. The operable portion of the window must be at least 5 square feet
2. 20 inches wide

3. 24 inches high
4. The bottom sill of the window must not be higher than 44 inches above the floor

Exterior doors are usually either solid wood, metal-clad, or a composite material. At least one exterior door must be 36 inches wide and 80 inches tall. Exterior doors usually swing into the house, allowing a screen door to be added that will swing outward. Interior doors can be the pre-hung, hollow- core type. The door width to interior rooms is not regulated, but the wider, the better.

Habitable rooms must meet a minimum size requirement. According to the IRC, at least one room in a house must be 120 square feet. Other habitable rooms must be at least 70 square feet. A kitchen must be at least 50 square feet. Habitable rooms may not be less than 7 feet in any horizontal direction. Habitable rooms must have a minimum vertical ceiling height of 7 feet. Portions of the room with a sloping ceiling less than 5 feet or flat ceiling less than 7 feet are not considered as contributing to the minimum required floor area or horizontal dimension.

The dwelling must have bathrooms with toilet facilities including a water closet, bathtub or shower, and a lavatory. A house must also have a kitchen area and a sink. Every plumbing fixture must be connected to a public or private sewage disposal system. All plumbing fixtures must be connected to an approved water supply. Kitchen sinks, bathtubs, lavatories, showers, bidets, laundry tubs, and washing machine outlets must be connected to both hot and cold water.

There is a significant hazard for people inside a home from a garage where a car is parked. Gasoline vapor and fumes from oil can affect a person's health. Additionally, a fire can begin in a garage and spread into a home very rapidly and without much warning. Because of this, residential portions of a house must be separated from an attached garage. The required protection is a single layer of ½ inch drywall on the garage side. Any door opening from the garage into the dwelling must have a solid core. This door must be at least 1⅜ inches thick. In no case is an opening permitted between a sleeping room and a garage. Garage floors must be of noncombustible material such as concrete. The floor in the garage must slope so as to allow liquids to drain toward the main vehicle entry doorway. Carports are exempt from these requirements if two side of the carport are open. Smoke detectors are required to alert occupants in case of a fire. They are required to be placed inside each bedroom, outside each bedroom in the hallway that accesses the bedroom, and on each floor of the building.

Floor Plan

___ Show all rooms, and label their uses.

___ Show large equipment such as refrigerator, range/oven, dishwasher, clothes dryer, water heater, and furnace.

___ Show plumbing fixtures such as tubs, toilets, washing machines, and lavatory bases.

___ Show overall dimensions of structure and interior dimensions of each room.

___ Artificial or natural lighting (window sizes), for each room is at least 8 percent of floor area of that room. *Section R303.1*

___ Artificial or natural ventilation should be at least 4 percent of floor area for operable portions of windows. *Section R303.1*

___ At least one room must be 120 square feet Other rooms must be at least 70 square feet. *Section R304*

___ The minimum dimension should be at least 7 feet for any habitable room. *Section R304.3*

___ Ceiling height is at least 7 feet in habitable rooms. *Section R305*

___ Smoke detectors to be on each floor level, in each bedroom and each hallway leading to a bedroom. *Section R317*

___ Door and window sizes should be labeled for energy code compliance. *Section R308*

___ Each home must be equipped with facilities capable of heating room to 68 degrees F. *Section R308*

___ Attic access must be 22 inches by 30 inches where roof is at least 30 inches above ceiling or where equipment is in attic. *Section R807*

___ Attached garages must be separated from dwelling by ½ inch drywall and a rated or solid core door. *Section R309*

___ Stair location and shape. Show width, rise of step, length of run, handrail locations and landings. *Section R314*

___ Bathroom ventilation must be shown (natural or mechanical). *Section R303.3*

___ Provide egress from bedrooms with 5.0 square feet operable portion and show maximum sill height of 44 inches. *Section R310*

___ The walls and soffits of the enclosed space under stairs shall be protected with ½ inch gypsum board. *Section R314.8*

___ Note that shower enclosures shall be finished with a smooth, hard, nonabsorbent finish. *Section R307.2*

___ Note that shower enclosures have a minimum dimension of 30 inches and a floor area of 900 square inches. *Figure R307.2*

___ Note that all glass enclosures shall be tempered. This included windows within shower. *Section R308.4*

___ Show that each water closet has a clear space not less than 30 inches wide and not less than 24 inches in front. *Figure R307.2*

___ Fireplaces. *Section R1001*

  ___ indicate fireplace location, hearth size and materials

  ___ for factory built or zero clearance fireplaces: provide cut sheet, make, model number, and the approval number of the testing agency. Additionally, submit the required clearance to combustible material.

  ___ for masonry fireplaces: show cross section with vertical steel and solid grout specified.

___ Comply with requirements for landings at doors. *Section R312*

___ Appliances installed in garages shall be mounted on platforms at least 18 inches above the floor. *Section M1307.3*

___ Show locations of skylights. Provide manufacturer's listing on all skylights over 2 feet x 4 feet. *Section R106.1.2*

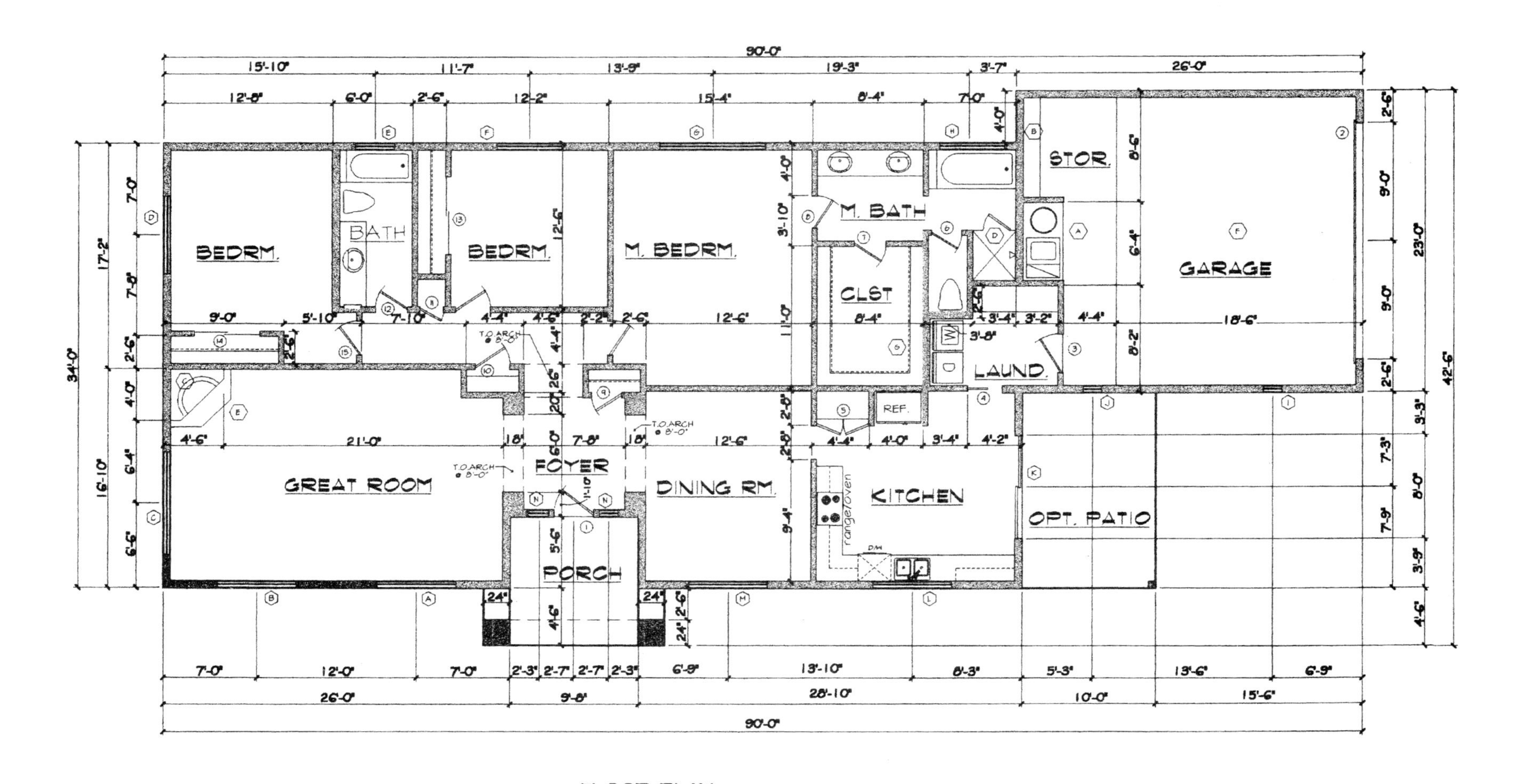

Floor plan drawing.

Stair Details

Stairs serve as a means of moving between floors. The slope, stepping area, headroom, handrail support and landing are all regulated by building codes. These codes hail from safety design standards, which are the result of accidents and injuries. The slope of the stair is defined by the maximum unit *rise* and minimum unit *run* at each stair section. You may need a separate detail to indicate how you intend to construct your stairs. If so, decide how you will build them and draw them. Straight-run stairs are the easiest, circular stairs are harder, and spiral stairs are generally manufactured. Spiral stairs are permitted only to serve smaller areas such as a loft. All have minimum limitations on their headroom, width of stair, width of tread, maximum rise, minimum run of tread, etc.

The IRC establishes a maximum rise and a minimum run. For residential houses the maximum riser height is 7¾ inches and the minimum tread run is 10 inches. Riser heights may be as short as 4 inches. Keep in mind that code minimum requirements for slope may be more steep than a comfortable riser height and tread width would provide. A very comfortable slope for a stair is 7 inches rise and 11 inches run. Because a normal person would predict the size of the next step, it is important to make all rises and runs the same. A variation of no more than ⅜ inch may occur between all tread widths and between all riser heights.

This *rise* and *run* is also a design tool to be used to determine the number of steps between floors. For instance, if the height between floor levels is exactly 8 feet 6½ inches, how will you determine the number of steps needed? Should you use the maximum rise permitted? In that case, their will be 102.5 divided by 7.75 = 13.22 or 14 steps, (it is obviously not possible to have a portion of a step). Should you make the slope of your stair more shallow? What if you have a limited length of overall run for the stair. Will that determine the slope?

Headroom in stairs is also regulated. To keep from banging your head and to accommodate moving bulky furniture between floors, there must be at least a clear headroom height of at least 6 feet 8 inches above the tread nosing. A stair's width must be at least 36 inches If you wish to move furniture between floors, it would be better to provide more width. Your stair must have handrails that allow for safety. Handrails have certain shape, size and location requirements as well. The cross sectional area must be between 1¼ inch and 2⅝ inches, or the shape must have an equivalent gripping feature.

A standard handrail, sold in lumber yards, should be acceptable. The handrail must be installed along one side of the wall between 34 inches to 38 inches above the nosing of the tread. At the top and bottom of stairs, there must be a *landing* that is at least as wide as the stair and as long as its length. For instance, if your stair is 42 inches wide, the landing must be 42 inches by 42 inches. This allows a person to embark onto and disembark from the flight of stairs in a safe manner. The best way to illustrate stair design is by an example. Follow the example below and watch how a stair design is determined:

Example:

Floor level difference = 8 feet 6½ inches
Maximum length available for stair = 13 feet 6 inches
Design a stair within code limitations, but with slightly less slope.

Solution:

Break the height into inches: (8 x 12) + 6½ = 102½ inches
Break the length available into inches: (13.5 x 12) = 162 inches
Determine code minimum: At 7¾ inches maximum rise and 10 inches minimum run, what would be the number of steps and length used? Height divided by unit minimum rise = steps needed:

102½ inches divided by 7¾ inches = 13.22; try using 14 steps
With 14 steps (slightly better than code), what is rise? Run?
102½ inches divided by 14 steps = 7.32 inches per rise. Using 10 inches and standard tread and knowing that there is one less tread than riser, the overall run at that rate of rise is (13 x 10) = 130 inches or 10 feet 10 inches

Straight run stairs are usually built from two larger, supporting members running along and supporting the stair treads. These support members are called *stringers*. Stringers are usually built from 2 inch by 12 inch framing lumber. Stairs can be built with an open stringer, where the stringer is cut in a sawtooth shape to allow the treads to rest on top of the sawtooth cuts, or closed stringer where the tread is cut into the sides of the stringers to accommodate the steps. If you are hesitant, you might consider having two metal stringers made at a welding shop. The assembly is easy, just secure treads on stringer supports.

Follow the checklist below for developing a design. Label as much of the drawing as possible. Follow the illustration for assistance.

Stair Details

___ Rise of step to be a maximum of 7¾ inches. *Section R314.2*

___ Run of tread to be a minimum of 10 inches. *Section R314.2*

___ Width of stair to be a minimum of 36 inches. *Section R314.1*

___ Headroom of stairwell to be 6 feet 8 inches. *Section R314.3*

___ Handrail to be between 1¼ inches and 2⅝ inches in diameter. *Section R315*

___ Handrail to be placed between 34 and 38 inches above nosing of the tread. *Section R315*

___ Stair landing to be as wide as stair and as long as its width. *Section R312.2*

___ Provide stairway connection details and stringer sizes.

___ Prefabricated metal stair shall have manufacturer's specifications attached to the plans.

___ Illumination must be provided for landings and treads. *Section R303.4*

___ Under-stair areas must be protected with gypsum wall board. *Section R314.8*

___ The maximum variation between all tread's width is ⅜ inches. *Section R314.2*

___ The maximum variation between all riser heights cannot exceed ⅜ inches. *Section R314.2*

Exits and Guardrail Requirements

Certain safety considerations may need to be demonstrated in more detail. These include those aspects of exit doors and guard rails. They are simple to present graphically and demonstrate your knowledge of safety. These details will show required exit doors, door swing, landings, guard rails, exit steps, and handrails as needed to provide for safety.

Exits and Guards Requirements

___ Required exit doorways must be 3 feet wide by 6 feet 8 inches tall. *Section R311.3*

___ Required guards must be 36 inches high. *Section R316*

___ Openings in guards must be such that a 4 inch orb (sphere) may not pass through the railing. *Section R316.2*

___ Guards may not be constructed with horizontal members that create a ladder effect. *Section R316*

___ Exterior stairway landings must be a minimum of 36 inches wide and 36 inches long. *Section R312.1*

___ Handrail for exterior steps to be between 34 inches to 38 inches above the nosing of the tread. *Section R315*

___ Ramps should be designed for maximum slope of 12.5 percent. *Section R313*

___ Ramps shall have a landing that is 36 inches by 36 inches at the top and bottom of ramp. *Section R313*

___ Handrails shall be provided for ramp if the slope exceeds 1:12 or 8.33 percent. *Section R313*

Electrical Plan

There are a few basic considerations that must be discussed prior to this plan drawing. Electrical design for a house is very clearly delineated by examples in the NEC. I encourage you to consult that code book for answers beyond the scope of this book.

First, you must determine the size of your primary service disconnecting equipment and meter base. There are basic methods of calculating the service equipment size for your house. For smaller services, you normally use the standard calculation. If the service is over 100 amps you may use the optional calculation. Since most services nowadays are over 100 amps, I'll review the optional calculation in the example below. Each outlet, switch or device consumes electricity from a branch circuit. Each of these branch circuits is limited in current capacity to a maximum, based on the wire size. Branch circuits are limited to a maximum number of outlets and devices based on demand from each and the branch circuit size. Table E3502.2 from the IRC will demonstrate the method for calculating the required service size. Standard electrical service panels include 100 amp, 125 amp, 150 amp, and 200 amp. So, when determining your minimum size electrical service size, you will select a service from among these standard sizes.

Example:

Assume a 1500 square foot house with the following loads. Two kitchen small appliance circuits, one laundry circuit, one 1200 watt dishwasher, one range/oven at 8000 watts, one water heater at 4500 watts, one clothes dryer at 5000 watts. Air conditioning (which in this case is larger than heating) at 10,000 watts.

1.	3 @ 1500	4500 watts	(Lighting and outlet load)
2.	1 @ 1500	1500 watts	(Bathroom outlet)

3.	2 @ 1500	3000 watts	(Small appliance circuits)
4.	1 @ 8000	8000 watts	(Range)
5.	1 @ 5000	5000 watts	(Clothes dryer)
6.	1 @ 1200	1200 watts	(Dishwasher)
7.	1 @ 4500	4500 watts	(Water heater)

Total: 27700 watts

1st 10,000 @ 100 percent 10000 watts
remainder @ 40 percent (17700 x .40) = 7080 watts

Total demand: 17080 watts

Divide by 240 volts (17080 divided by 240) = 71.16 amps
A/C (10,000 divided by 240) @ 100 percent = 42 amps
Total: 113 amps

Use the next available size: 125 amps service. Note: 200 amp service may actually be less expensive, because they are more common.

TABLE E3502.2
MINIMUM SERVICE LOAD CALCULATION LOADS AND PROCEDURE

LOADS AND PROCEDURE
3 volt-amperes per square foot of floor area for general lighting and convenience receptacle outlets.
Plus
1,500 volt-amperes × total number of 20-ampere-rated small appliance and laundry circuits.
Plus
The nameplate volt-ampere rating of all fastened-in-place, permanently connected or dedicated circuit-supplied appliances such as ranges, ovens, cooking units, clothes dryers and water heaters.
Apply the following demand factors to the above subtotal:
The minimum subtotal for the loads above shall be 100 percent of the first 10,000 volt-amperes of the sum of the above loads plus 40 percent of any portion of the sum that is in excess of 10,000 volt-amperes.
Plus the largest of the following:
Nameplate rating(s) of air-conditioning and cooling equipment including heat pump compressors.
Nameplate rating of the electric thermal storage and other heating systems where the usual load is expected to be continuous at the full nameplate value. Systems qualifying under this selection shall not be figured under any other category in this table.
Sixty-five percent of nameplate rating of central electric space-heating equipment, including integral supplemental heating for heat pump systems.
Sixty-five percent of nameplate rating(s) of electric space-heating units if less than four separately controlled units.
Forty percent of nameplate rating(s) of electric space-heating units of four or more separately controlled units.
The minimum total load in amperes shall be the volt-ampere sum calculated above divided by 240 volts.

Electric Service Size Calculation

Most jurisdictions will require you to provide an electrical load calculation as part of your submittal criteria. This sample will at least expose you to the concept of service size determination. For all practical purposes, the best bet will be for you to select a 200 amp main service, even if your demand load is far less. The relative cost is nil and you can plan for expansion.

Now that you have selected the service size, you must select the size of wire which will serve to supply electricity from the power company's transformer. Although the electric company will probably install the cable, you may have to design the wire size and type. Table E3503.1 from the IRC will help in this determination as well as the required grounding electrode conductor size.

TABLE E3503.1
SERVICE CONDUCTOR AND GROUNDING ELECTRODE CONDUCTOR SIZING CONDUCTOR TYPES AND SIZES—THHN, THHW, THW, THW-2, THWN, THWN-2, USE, XHHW, XHHW-2, SE, USE, USE-2

CONDUCTOR TYPES AND SIZES—THHW, THW, THWN, USE, XHHW (Parallel sets of 1/0 and larger conductors are permitted in either a single raceway or in separate raceways)		ALLOWABLE AMPACITY	MINIMUM GROUNDING ELECTRODE CONDUCTOR SIZE[a]	
Copper (AWG)	Aluminum and copper-clad aluminum (AWG)	Maximum load (amps)	Copper (AWG)	Aluminum (AWG)
4	2	100	8[b]	6[c]
3	1	110	8[b]	6[c]
2	1/0	125	8[b]	6[c]
1	2/0	150	6[c]	4
1/0	3/0	175	6[c]	4
2/0	4/0 or two sets of 1/0	200	4[d]	2[d]
3/0	250 kcmil or two sets of 2/0	225	4[d]	2[d]
4/0 or two sets of 1/0	300 kcmil or two sets of 3/0	250	2[d]	1/0[d]
250 kcmil or two sets of 2/0	350 kcmil or two sets of 4/0	300	2[d]	1/0[d]
350 kcmil or two sets of 3/0	500 kcmil or two sets of 250 kcmil	350	2[d]	1/0[d]
400 kcmil or two sets of 4/0	600 kcmil or two sets of 300 kcmil	400	1/0[d]	3/0[d]

For SI: 1 inch = 25.4 mm.

a. Where protected by a metal raceway, grounding electrode conductors shall be electrically bonded to the metal raceway at both ends.

b. No. 8 grounding electrode conductors shall be protected with metal conduit or nonmetallic conduit.

c. Where not protected, No. 6 grounding electrode conductors shall closely follow a structural surface for physical protection. The supports shall be spaced not more than 24 inches on center and shall be within 12 inches of any enclosure or termination.

d. Where the sole grounding electrode system is a ground rod or pipe as covered in Section E3508.2, the grounding electrode conductor shall not be required to be larger than No. 6 copper or No. 4 aluminum. Where the sole grounding electrode system is the footing steel as covered in Section E3508.1.2, the grounding electrode conductor shall not be required to be larger than No. 4 copper conductor.

Next, you must draw an electrical wiring diagram of your house. Here's how: First, start with a copy of the floor plan and begin locating electrical outlets, lights and their switches. There are a few rules to follow. First, every wall space must have an outlet such that an appliance with only a six foot cord could plug into the wall and sit anywhere along the wall. This means that every 12 feet along a wall space an outlet must be installed. Hallways must have at least one outlet if they are over 10 feet long.

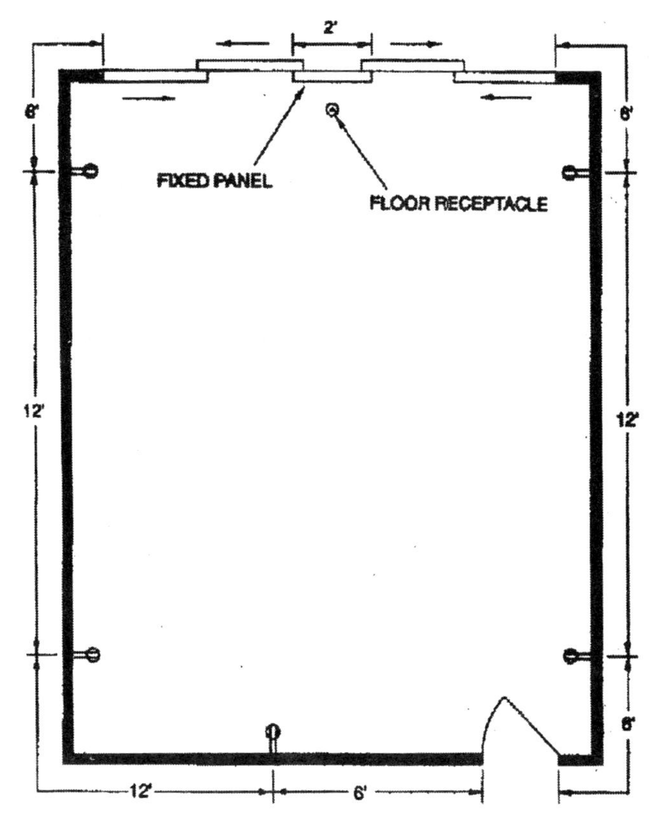

For SI: 1 foot = 304.8 mm.

FIGURE E3801.2
GENERAL USE RECEPTACLE DISTRIBUTION

Bathrooms must have outlets adjacent to each basin. They must be on a dedicated 20 amp circuit and be ground fault circuit interrupter (GFCI) type. In kitchens, along a counter space, an outlet must be placed so that no point along a wall space is more than 24 inches away from an outlet. Each section of countertop 12 inches or wider must have an outlet. A kitchen island or peninsula must have at least one outlet for appliances.

An outdoor outlet must be installed in front and back of the dwelling. These must also be GFCI type with a water-proof cover. At least one circuit must be installed for the laundry. At least one outlet must be installed in a basement and garage. Each of these must be a GFCI protected outlet.

In addition, you must provide service outlets for heating and cooling equipment within 25 feet of such appliances. At least one switched controlled light must be installed in each habitable room, in bathrooms, hallways, stairways, attached garages, detached garages, and at entrances or exits. Generally speaking, you may use a switched wall outlet in the bedrooms, living room, den, or family room in lieu of a lighting outlet. Three way switches are required for stairs. These allow you to turn a light on and off alternately from two remote locations at each level.

The next step will be to present a panel schedule. This is simply a list of what part of the house is controlled by each circuit breaker. It establishes which breakers will be protecting which circuits.

TABLE E3602.13
BRANCH-CIRCUIT REQUIREMENTS—SUMMARY[a,b]

	CIRCUIT RATING		
	15 amp	20 amp	30 amp
Conductors: Minimum size (AWG) circuit conductors	14	12	10
Maximum overcurrent-protection device rating Ampere rating	15	20	30
Outlet devices: Lampholders permitted	Any type	Any type	N/A
Receptacle rating (amperes)	15 maximum	15 or 20	30
Maximum load (amperes)	15	20	30

a. These gages are for copper conductors.
b. N/A means not allowed.

TABLE E3605.5.3
OVERCURRENT-PROTECTION RATING

COPPER		ALUMINUM OR COPPER-CLAD ALUMINUM	
Size (AWG)	Maximum overcurrent-protection-device rating[a] (amps)	Size (AWG)	Maximum overcurrent-protection-device rating[a] (amps)
14	15	12	15
12	20	10	25
10	30	8	30

a. The maximum overcurrent-protection-device rating shall not exceed the conductor allowable ampacity determined by the application of the correction and adjustment factors in accordance with Sections E3605.2 and E3605.3.

Again, the NEC establishes the exact criteria for overcurrent protection of wiring, however there are a few routine practices that will keep you out of trouble. The most common type of insulated cable is identified by the abbreviation: NM. Use the following table for wire size selection:

ELECTRICAL PANEL SCHEDULE • 200 AMP SERVICE PANEL

Circuit	Name	Breaker	Wire	Wire	Breaker	Name	Circuit
1	Heat Pump	2/50	6/2g	6/3g	2/50	Range/Oven	2
3	Heat Pump	2/50	6/2g	6/3g	2/50	Range/Oven	4
5	Water Heater	2/30	10/2g	10/3g	2/30	HVAC Blower	6
7	Water Heater	2/30	10/2g	10/3g	2/30	HVAC Blower	8
9	Dryer	2/30	10/3g	12/2g	20	Kitchen Appliance GFCI	10
11	Dryer	2/30	10/3g	12/2g	20	Kitchen Appliance GFCI	12
13	General Lighting	15	14/2g	12/2g	20	Bathroom Lavatory GFCI	14
15	General Lighting	15	14/2g	12/2g	20	Outdoor lighting/outlets GFCI	16
17	General Lighting	15	14/2g	12/2g	20	Outdoor lighting/outlets GFCI	18
19	General Lighting	15	14/2g	12/2g	20	General Outlets	20
21	General Outlets	20	12/2g	12/2g	20	General Outlets	22
23	General Outlets	20	12/2g	12/2g	20	General Outlets	24
25	General Outlets	20	12/2g	12/2g	20	General Outlets	26
27	Spare					Spare	28
29	Spare					Spare	30
31	Spare					Spare	32
33	Spare					Spare	34
35	Spare					Spare	36
37	Spare					Spare	38
39	Spare					Spare	40
41	Spare					Spare	42

Electrical Plans*

___ Location of main service disconnect. *Section E3501.6.2*

___ Location and sizes of other subpanels. Subpanels may not be in a bathroom or closet. *Section E3501.6.2*

___ Show all lights, outlets, and electrical devices with circuit designation numbers that correlate to panel schedule.

___ All GFCI outlets identified where required. *Section E3802*

___ Enlarged drawing of service panel and meter base, showing grounding, bonding. *Section E3507 and E3509*

___ Panel schedule showing circuit size, wire size, overcurrent protection and circuit number. *Table E3602.13*

___ Show 4-wire type cable serving range, oven, and clothes dryer outlets.

___ Load calculations, panel schedules and one-line diagrams shall be provided on plan. *Table E3502.2*

___ Note the service equipment manufacturer, type, and amp capacity of the panel. *Section E3501.6.1*

___ Note the grounding conductor is #4 bare copper wire connected to 20 feet footing steel. *Table E3503.1*

___ Note a bonding conductor: #4 copper wire connected between service panel and metal pipe. *Section E3509*

___ At least one switched light or outlet must be provided in bathrooms, hallways, stairways. *Section E3803*

___ At least one switch controlled light or outlet must be provided in all habitable rooms. *Section E3803*

___ At least one switch controlled light must be provided in attached garages, outdoor entrances. *Section E3803.3*

___ At least every twelve feet of wall space within a room is provided with an outlet. *Figure E3801.2*

___ Provide at least one receptacle outlet in hallways 10 or more feet in length. *Section E3801.10*

___ Provide an exterior weatherproof GFCI outlet at grade level at both the front and the back of house. *Section E3802.3*

___ GFCI receptacle outlets shall be provided at each kitchen counter space wider than 12 inches. *Section E3802.6*

___ GFCI receptacle outlet shall be provided adjacent to the basin in each bathroom. *Section E3802.1*

___ Bathroom outlets shall be on a separate 20 amp circuit with no other outlets. *Section E3603.4*

___ Receptacles in bathrooms, kitchens, unfinished basements are GFCI or Arc fault type. *Section E3802*

___ Receptacles in garages or carports are GFCI or Arc fault type. *Section E3802*

___ Note that two appliance circuits are provided to serve the kitchen, breakfast and dining room. *Section E3603.2*

___ Such circuits shall have no other outlets. *Section E3603.2*

___ Note that at least one 20 amp branch circuit shall be installed to serve the laundry room. *Section E3603.3*

___ Note that this laundry room circuit shall have no other outlets. *Section E3603.3*

___ Indicate the location of all air conditioning, heating units, air handlers, compressors and electrical disconnects.

___ Indicate a convenience outlet and light switch for attic heating and A/C equipment. *Section M1305.1.4.3*

___ Provide service outlet for HVAC equipment within 25 feet of equipment. *Section M1305.1.4.3*

___ Where ceiling fans are specified, indicate that approved outlet boxes are supported. *Section E3904.4*

___ Note that light fixtures in clothes closets meet clearance requirements (18 inches from shelf). *Figure E3903.11*

___ Note that smoke detectors are wired to building power by noting circuit designation number. *Section R317*

* *Assumption is the installation of a maximum 200 amp service panel for a standard size house with limited electrical demand.*

Plumbing Plan

A plumbing expert began a lecture on plumbing with a noteworthy quip. He said, "Plumbing is not a science, but an art." He elaborated a little further by saying, "There are as many ways to properly plumb a house as there are qualified, competent plumbers willing to perform the job." With the variety of materials and fittings on the market and the possible combinations of those fittings it is reasonable to predict a myriad possible permutations of plumbing installations. You may get lots of good, yet contradictory advicc on this aspect of construction.

Common banter among plumbers is that there are only three rules to drainage plumbing: sewage waste flows downhill, hot water is on the left and payday's Friday. This bit of lucid wit illustrates a point. The object to drainage plumbing is simple: remove waste! You can get caught up in rules to a point where you sometimes lose sight of the objective. The simplified strategy is that if water is brought into a building, it must be taken out of a building.

The plumbing plan includes water supply piping and drain-waste and vent piping. These can all be incorporated in a single plan. I will discuss them one at a time. This and all other utility plans such as the mechanical and electrical plan can be drawn onto a copy of the floor plan. The plan should show where each fixture is located and the routing and size of pipe to those fixtures.

First, drain-waste and vent piping (DWV) drains away wastes and vents sewer gases away from inside the home. DWV piping consists of vent piping, waste drain piping, and traps. The way drainage works is that water or soil flows out from a *fixture* such as a sink or shower through a *trap* which prevents waste gases from entering the home. After passing through this trap, the effluent

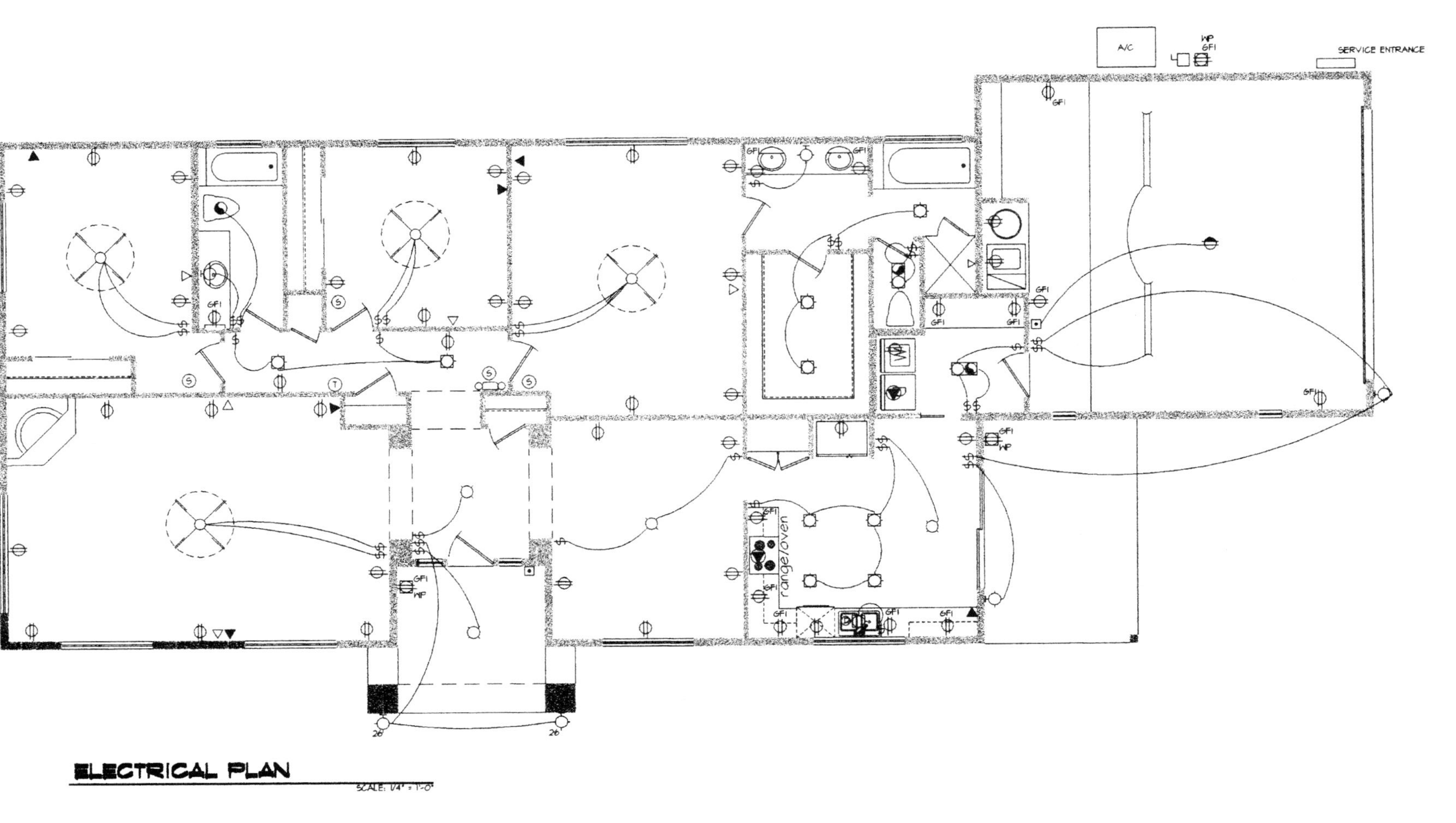

Electrical plan drawing.

travels through a trap arm toward a drainage pipe which has a vent. At this point a vent pipe will jut upward to the outside to allow for ventilation of the waste drainage pipe. This vent pipe prevents siphonage and facilitates the gravity drainage of waste. From here, the waste within the pipe will flow downward toward drainage pipes, then to the exterior of the house where the sewage will drain toward either a private septic tank or public sewer connection.

The materials most commonly used for drainage piping are ABS (Acrylonitrile Butadiene Styrene) and PVC (Polyvinyl Chloride). Various fittings and pipe lengths are available to make *any* variation of plumbing installation permitted. Each pipe material is glued together with special cement. Cast iron used to be the most commonly used material for DWV piping. It is precise, demanding work, limited to professional installers. Since the advent of plastic piping, plumbing is less demanding.

The size of drainage and vent piping is based on the amount of fluids and waste being disposed and is available in standard diameter sizes, 1¼, 1½ inches, 2 inches, 3 inches and 4 inches. There are other sizes, but these are the most commonly used. The basis for pipe sizing is the *drainage fixture unit* measure. Each drainage fixture is assigned a value based upon its capacity to purge waste water. There are both *waste fixture units* and *supply fixture units*. The IRC establishes the value for these fixture units. They are enumerated in two tables. Drainage fixture units are listed in this table: Table P3004.1. Water supply fixture units are listed in this table: Table P2903.6.

Consider that a drain, waste, and vent system looks much like a tree. It has a main trunk with numerous branches, and each of these can have yet smaller branches. The main drain correlates with the tree's trunk. Just as the tree's trunk must be large enough to support all of the branches, the plumbing main drain must be large enough to collect all of the waste from its branches. Both the main branch and each of the individual branch lines are designed for the total drainage of fixture units that they serve.

Every time the plumbing drain line turns, the change in direction must be made with the appropriate use of fittings, including sanitary tees, wyes, sweeps, bends or any combination thereof. The particular type of fitting and its commensurate change in direction is represented in Table P3005.1, derived from the IRC.

TABLE P2903.6
WATER-SUPPLY FIXTURE-UNIT VALUES FOR VARIOUS PLUMBING FIXTURES AND FIXTURE GROUPS

TYPE OF FIXTURES OR GROUP OF FIXTURES	WATER-SUPPLY FIXTURE-UNIT VALUE (w.s.f.u.)		
	Hot	Cold	Combined
Bathtub (with/without overhead shower head)	1.0	1.0	1.4
Clothes washer	1.0	1.0	1.4
Dishwasher	1.4	—	1.4
Hose bibb (sillcock)[a]	—	2.5	2.5
Kitchen sink	1.0	1.0	1.4
Lavatory	0.5	0.5	0.7
Laundry tub	1.0	1.0	1.4
Shower stall	1.0	1.0	1.4
Water closet (tank type)	—	2.2	2.2
Full-bath group with bathtub (with/without shower head) or shower stall	1.5	2.7	3.6
Half-bath group (water closet and lavatory)	0.5	2.5	2.6
Kitchen group (dishwasher and sink with/without garbage grinder)	1.9	1.0	2.5
Laundry group (clothes washer standpipe and laundry tub)	1.8	1.8	2.5

For SI: 1 gallon per minute = 3.785 L/m.

a. The fixture unit value 2.5 assumes a flow demand of 2.5 gpm, such as for an individual lawn sprinkler device. If a hose bibb/sill cock will be required to furnish a greater flow rate, the equivalent fixture-unit value may be obtained from Table P2903.6 or Table P2903.7.

TABLE P3004.1
DRAINAGE FIXTURE UNIT (d.f.u.) VALUES FOR VARIOUS PLUMBING FIXTURES

TYPE OF FIXTURE OR GROUP OF FIXTURES	DRAINAGE FIXTURE UNIT VALUE (d.f.u.)[a]
Bar sink	1
Bathtub (with or without shower head and/or whirlpool attachments)	2
Bidet	1
Clothes washer standpipe	2
Dishwasher	2
Floor drain[b]	0
Kitchen sink	2
Lavatory	1
Laundry tub	2
Shower stall	2
Water closet (1.6 gallons per flush)	3
Water closet (greater than 1.6 gallons per flush)	4
Full-bath group with bathtub (with 1.6 gallon per flush water closet, and with or without shower head and/or whirlpool attachment on the bathtub or shower stall)	5
Full-bath group with bathtub (water closet greater than 1.6 gallon per flush, and with or without shower head and/or whirlpool attachment on the bathtub or shower stall)	6
Half-bath group (1.6 gallon per flush water closet plus lavatory)	4
Half-bath group (water closet greater than 1.6 gallon per flush plus lavatory)	5
Kitchen group (dishwasher and sink with or without garbage grinder)	3
Laundry group (clothes washer standpipe and laundry tub)	3
Multiple-bath groups[c]:	
1.5 baths	7
2 baths	8
2.5 baths	9
3 baths	10
3.5 baths	11

For SI: 1 gallon = 3.785 L.

a. For a continuous or semicontinuous flow into a drainage system, such as from a pump or similar device, 1.5 fixture units shall be allowed per gpm of flow. For a fixture not listed, use the highest d.f.u. value for a similar listed fixture.

b. A floor drain itself adds no hydraulic load. However, where used as a receptor, the fixture unit value of the fixture discharging into the receptor shall be applicable.

c. Add 2 d.f.u. for each additional full bath.

Design of the drainage system begins with the smaller branch lines at the furthest source from the sewer connection or the septic tank. These represent clusters of plumbing fixtures within your home. Let's say that a bathroom is at one end of your home. A branch line that collects waste from this bathroom must be large enough for all of the fixtures, including a water closet or toilet, a shower, and two lavatories. Then consider that this branch line intersects with a branch line from another bathroom that has a water closet, a bathtub and a single lavatory. At the intersection of the two, the new larger branch line must be large enough to carry sewage from two water closets, a bathtub, a shower and three lavatories. This process of designing for more effluent continues until you reach the main plumbing drain line, which must be large enough for *all* of the fixtures within your home.

The next step to designing the plumbing system is applying the fixture unit load for your branches with the maximum unit loading and maximum length of a specific size of drainage and vent pipe. Review Table P3005.4.1 and Table P3005.4.2 from the IRC. Use these tables to select the proper pipe size for each of these drainage branches and main drain. Notice an important point: A water closet must

TABLE P3005.1
FITTINGS FOR CHANGE IN DIRECTION

TYPE OF FITTING PATTERN	CHANGE IN DIRECTION		
	Horizontal to vertical[c]	Vertical to horizontal	Horizontal to horizontal
Sixteenth bend	X	X	X
Eighth bend	X	X	X
Sixth bend	X	X	X
Quarter bend	X	X[a]	X[a]
Short sweep	X	X[a,b]	X[a]
Long sweep	X	X	X
Sanitary tee	X[c]		
Wye	X	X	X
Combination wye and eighth bend	X	X	X

For SI: 1 inch = 25.4 mm.

a The fittings shall only be permitted for a 2-inch or smaller fixture drain.

b Three inches and larger.

c For a limitation on multiple connection fittings, see Section P3005.1.1.

TABLE P3005.4.2
MAXIMUM NUMBER OF FIXTURE UNITS ALLOWED TO BE CONNECTED TO THE BUILDING DRAIN, BUILDING DRAIN BRANCHES OR THE BUILDING SEWER

DIAMETER OF PIPE (inches)	SLOPE PER FOOT		
	1/8 inch	1/4 inch	1/2 inch
1 1/2 [a,b]	—	—[a]	—[a]
2[b]	—	21	27
2 1/2[b]	—	24	31
3	36	42	50
4	180	216	250

For SI: 1 inch = 25.4 mm, 1 foot = 304.8 mm.

a. 1 1/2-inch pipe size limited to a building drain branch serving not more than two waste fixtures, or not more than one waste fixture if serving a pumped discharge fixture or garbage grinder discharge.

b. No water closets.

be on at least a 3 inch drainage line. So even though a single water closet has only 4 fixture units, it must be served by a 3 inch waste pipe. While sizing the minimum pipe size, be sure to remember that each fixture has a minimum trap size. A bathroom group includes the most common three fixtures: a water closet, bathtub and a lavatory, and at a reduced fixture unit demand.

Each plumbing fixture has a trap to prevent gases from entering the house. The minimum size of these traps are listed in Table P3201.7. Each plumbing trap must be served with a vent to allow sewage to drain and prevent syphonage. Each of these vents must be sized as well. The table shows the maximum fixture unit loading permitted for a particular size vent pipe. Several fixtures may share a vent with the appropriate fitting. Since a bathroom group has a reduced fixture unit loading, vent pipes can be smaller. For instance, a bathroom that has a water closet, a bathtub and a lavatory normally represents eight drainage fixture units. However, when collected together as a bathroom group, the value is reduced to five. This would permit the use of a 1½ inch vent pipe. The IRC establishes the minimum size for traps. Review Table P3201.7 for the required trap size for each fixture in your home.

An air admittance valve is a one-way valve designed to allow air into the plumbing drainage system when a negative pressure develops in the piping. In other words, if circumstances prevent it, you can eliminate the need to extend vents through the roof. These air admittance valves come in handy if you have a bathroom under a stair where venting becomes a problem.

The last aspect for the DWV plan is the requirement for drainage cleanouts. Plumbing pipes can become clogged from time to time with everything from rubber balls to massive amounts of toilet paper. That eventuality has led to the requirement for installing a plumbing fitting that can be easily opened to allow a *plumbing snake* to push the clog out toward the sewer. The location for these cleanout fittings is based on the degree of difficulty for a snake to effectively be routed through the drainage system. A cleanout is needed for every change in direction of the drain pipe that exceeds 135 degrees. One is also required at the end of each branch line. A kitchen sink must always be provided with a cleanout due to the excessive use and disposal of food products. With some exceptions,

TABLE P3005.4.1
MAXIMUM FIXTURE UNITS ALLOWED TO BE CONNECTED TO BRANCHES AND STACKS

NOMINAL PIPE SIZE (inches)	ANY HORIZONTAL FIXTURE BRANCH	ANY ONE VERTICAL STACK OR DRAIN
$1^1/_4$[a]	—	—
$1^1/_2$[b]	3	4
2[b]	6	10
$2^1/_2$[b]	12	20
3	20	48
4	160	240

For SI: 1 inch = 25.4 mm.

a. $1^1/_4$-inch pipe size limited to a single-fixture drain or trap arm. See Table P3201.7.

b. No water closets.

TABLE P3201.7
SIZE OF TRAPS AND TRAP ARMS FOR PLUMBING FIXTURES

PLUMBING FIXTURE	TRAP SIZE MINIMUM (inches)
Bathtub (with or without shower head and/or whirlpool attachments)	$1^1/_2$
Bidet	$1^1/_4$
Clothes washer standpipe	2
Dishwasher (on separate trap)	$1^1/_2$
Floor drain	2
Kitchen sink (one or two traps, with or without dishwasher and garbage grinder)	$1^1/_2$
Laundry tub (one or more compartments)	$1^1/_2$
Lavatory	$1^1/_4$
Shower	2
Water closet	a

For SI: 1 inch = 25.4 mm.

a. Consult fixture standards for trap dimensions of specific bowls.

these cleanouts must be the same size as the pipe. Table 3005.2.9 from the IRC lists the size cleanout for the particular pipe size.

After designing the drainage system, you must design the water supply system. Plumbing fixtures consume water at various rates. These rates are reflected by the water supply fixture unit. You must calculate the total supply fixture unit demand for your residence by adding each of the individual fixture units for each fixture. These are outlined in Table P2903.6 from the IRC. Calculate your total fixture units by multiplying the quantities of each fixture by its fixture unit value.

The total fixture unit calculation will permit you to establish the minimum size for the main water supply pipe. The size of the pipe is based on the available water pressure, the meter size and the total developed length of the pipe between the house and the meter. Table 2903.7 from the IRC will help you determine the required water supply pipe size based on the available water pressure and total developed length of the pipe. Check your available water pressure by using a water pressure gauge, which simply connects to a hose bib to verify pressure. If there is no way to check existing water pressure, the water company should be able to tell you the pressure at your site.

TABLE P3005.2.9
CLEANOUTS

PIPE SIZE (inches)	CLEANOUT SIZE (inches)
$1^1/_2$	$1^1/_2$
2	$1^1/_2$
3	$2^1/_2$
4 and larger	$3^1/_2$

For SI: 1 inch = 25.4 mm.

TABLE P2903.7
MINIMUM SIZE OF WATER METERS, MAINS AND DISTRIBUTION PIPING BASED ON WATER SUPPLY FIXTURE UNIT VALVES (w.s.f.u.)

Pressure Range—30 to 39 psi

METER AND SERVICE PIPE (inches)	DISTRIBUTION PIPE (inches)	MAXIMUM DEVELOPMENT LENGTH (feet)									
		40	60	80	100	150	200	250	300	400	500
3/4	1/2[a]	2.5	2	1.5	1.5	1	1	.5	.5	0	0
3/4	3/4	9.5	7.5	6	5.5	4	3.5	3	2.5	2	1.5
3/4	1	32	25	20	16.5	11	9	7.5	6.5	5.5	4.5
1	1	32	32	27	21	13.5	10	8	7	5.5	5
3/4	1 1/4	32	32	32	32	30	24	20	17	13	10.5
1	1 1/4	80	80	70	61	45	34	27	22	16	12
1 1/2	1 1/4	80	80	80	75	54	40	31	25	17.5	13
1	1 1/2	87	87	87	87	84	73	74	56	45	36
1 1/2	1 1/2	151	151	151	151	117	92	79	69	54	43

a. Minimum size for building supply is 3/4-inch pipe.

Pressure Range—40 to 49 psi

METER AND SERVICE PIPE (inches)	DISTRIBUTION PIPE (inches)	MAXIMUM DEVELOPMENT LENGTH (feet)									
		40	60	80	100	150	200	250	300	400	500
3/4	1/2[a]	3	2.5	2	1.5	1.5	1	1	.5	.5	.5
3/4	3/4	9.5	9.5	8.5	7	5.5	4.5	3.5	3	2.5	2
3/4	1	32	32	32	26	18	13.5	10.5	9	7.5	6
1	1	32	32	32	32	21	15	11.5	9.5	7.5	6.5
3/4	1 1/4	32	32	32	32	32	32	32	27	21	16.5
1	1 1/4	80	80	80	80	65	52	42	35	26	20
1 1/2	1 1/4	80	80	80	80	75	59	48	39	28	21
1	1 1/2	87	87	87	87	87	87	87	78	65	55
1 1/2	1 1/2	151	151	151	151	151	130	109	93	75	63

a. Minimum size for building supply is 3/4-inch pipe.

Pressure Range—50 to 60 psi

METER AND SERVICE PIPE (inches)	DISTRIBUTION PIPE (inches)	MAXIMUM DEVELOPMENT LENGTH (feet)									
		40	60	80	100	150	200	250	300	400	500
3/4	1/2[a]	3	3	2.5	2	1.5	1	1	1	.5	.5
3/4	3/4	9.5	9.5	9.5	8.5	6.5	5	4.5	4	3	2.5
3/4	1	32	32	32	32	25	18.5	14.5	12	9.5	8
1	1	32	32	32	32	30	22	16.5	13	10	8
3/4	1 1/4	32	32	32	32	32	32	32	32	29	24
1	1 1/4	80	80	80	80	80	68	57	48	35	28
1 1/2	1 1/4	80	80	80	80	80	75	63	53	39	29
1	1 1/2	87	87	87	87	87	87	87	87	82	70
1 1/2	1 1/2	151	151	151	151	151	151	139	120	94	79

a. Minimum size for building supply is 3/4-inch pipe.

TABLE P2903.7—continued
MINIMUM SIZE OF WATER METERS, MAINS AND DISTRIBUTION PIPING BASED ON WATER SUPPLY FIXTURE UNIT VALVES (w.s.f.u.)

Pressure Range—over 60 psi

METER AND SERVICE PIPE (inches)	DISTRIBUTION PIPE (inches)	MAXIMUM DEVELOPMENT LENGTH (feet)									
		40	60	80	100	150	200	250	300	400	500
3/4	1/2[a]	3	3	3	2.5	2	1.5	1.5	1	1	.5
3/4	3/4	9.5	9.5	9.5	9.5	7.5	6	5	4.5	3.5	3
3/4	1	32	32	32	32	32	24	19.5	15.5	11.5	9.5
1	1	32	32	32	32	32	28	22	17	12	9.5
3/4	1 1/4	32	32	32	32	32	32	32	32	32	30
1	1 1/4	80	80	80	80	80	80	69	60	46	36
1 1/2	1 1/4	80	80	80	80	80	80	76	65	50	38
1	1 1/2	87	87	87	87	87	87	87	87	87	84
1 1/2	1 1/2	151	151	151	151	151	151	151	144	114	94

For SI: 1 inch = 25.4 mm, 1 foot = 304.8 mm.

a. Minimum size for building supply is 3/4-inch pipe.

In addition, the IRC requires that each fixture must have an adequate supply of water flow rate. Table P2903.1 from the IRC lists these rates. However, to prevent from wasting the precious resource, the IRC also limits the maximum flow rate from certain fixtures. Table P2903.2 lists these amounts. So, when you are shopping for fixtures, keep these limitations in mind.

TABLE P2903.1
REQUIRED CAPACITIES AT POINT OF OUTLET DISCHARGE

FIXTURE AT POINT OF OUTLET	FLOW RATE (gpm)	FLOW PRESSURE (psi)
Bathtub	4	8
Bidet	2	4
Dishwasher	2.75	8
Laundry tub	4	8
Lavatory	2	8
Shower	3	8
Shower, temperature controlled	3	20
Sillcock, hose bibb	5	8
Sink	2.5	8
Water closet, flushometer tank	1.6	15
Water closet, tank, close coupled	3	8
Water closet, tank, one-piece	6	20

For SI: 1 gallon per minute = 3.785 L/m, 1 pound per square inch = 6.895 kPa.

TABLE P2903.2
MAXIMUM FLOW RATES AND CONSUMPTION FOR PLUMBING FIXTURES AND FIXTURE FITTINGS[b]

PLUMBING FIXTURE OR FIXTURE FITTING	PLUMBING FIXTURE OR FIXTURE FITTING
Lavatory faucet	2.2 gpm at 60 psi
Shower head[a]	2.5 gpm at 80 psi
Sink faucet	2.2 gpm at 60 psi
Water closet	1.6 gallons per flushing cycle

For SI: 1 gallon per minute = 3.785 L/m, 1 pound per square inch = 6.895 kPa.

a. A handheld shower spray is also a shower head.

b. Consumption tolerances shall be determined from referenced standards.

Plumbing Plans

___ Water service pipe material, size and distance to water meter. *Section P2904*

___ Location of all plumbing fixtures, piping material, sizes and layout. *Chapter 27*

___ Show that water supply pressure is at least 40 pounds per square inch (psi). *Section P2903.3*

___ Show that water supply pressure is less than a maximum of 80 pounds per square inch (psi). *Section P2903.1*

___ Water heater installation. *Chapter 28*

___ Show that a pan is provided under water heater and has a discharge pipe. *Section P2801.5*

___ Provide pressure and temperature relief valve and discharge line at water heater. *Section P2803*

___ Water softener, floor drains, clothes, and dishwasher and all similar fixtures. *Section P2904*

___ Provide an isometric of drainage piping. This is a 3D view of drain, waste and vent piping installation.

___ Show that pipe is protected from damage. *Section P2603*

___ Show that pipes will be supported as required. *Section P2605*

___ ABS or PVC used in DWV system must be schedule 40. *Section P3002*

___ Show that drainage pipe fittings meet the standards for change of direction. *Section P3005 and Table 3005.1*

___ Show that drainage pipe size meet the limitations for maximum fixture units. *Section P3005.4*

___ Plumbing relief vents shall be a minimum of 10 feet away from all air intakes. *Section P3103.5*

___ Show that vent pipes meet the minimum size requirements based on drainage pipe size. *Section P3113*

___ Copper tubing used in water piping must be specified type M. *Section P2904.5.1*

___ Water pipe installed under a slab shall have only approved joints. *Section P2904.5.1*

___ Water lines provide total fixture demand and maximum developed length. *Section P2903.7*

___ Demonstrate that water supply pipe size is according to *Table 2903.7.*

___ Showers and tub-shower combinations shall be provided a thermostatic mixing valve. *Section P2708.3*

___ Hose bibs must be located on the plan and noted to have back-flow prevention. *Section P2902.2.4*

___ Specify the waste line is as high as possible under the countertop for dishwashers. *Section P2717*

___ Specify that shower be at least 30 inches in any dimension and provide shut-off valves for all fixtures. *Figure 307.2*

___ Show the size of trap for each fixture and demonstrate compliance with *Table P3201.7.*

___ Specify that the distance between traps and vents is within the limitations of *Table P3105.1.*

TYPICAL FIXTURE UNIT CALCULATION WORKSHEET FOR RESIDENTIAL PLUMBING PIPE DESIGN

Fixture	Quantity	Drainage FU	Total	Water FU	Total
Water closet (1.6 gallon flush)	3	3	9	2.2	6.6
Bathtub	2	2	4	1[1]	2
Shower	1	2	2	1[1]	1
Lavatory	4	1	4	.5[1]	2
Kitchen sink	1	2	2	1[1]	1
Clothes washer	1	2	2	1[1]	1
Full bathroom group[2]		5		1.5 hot 2.7 cold[1]	
Building total			23		13.6

[1] *Assumes a demand from either hot or cold water supply*

[2] *See special limitations on use in Chapter 29 and 30 of the International Residential Code*

Mechanical Plan

Besides keeping the rain and snow off your head, the whole point of your home is to provide a safe, comfortable, *conditioned* place to live, sleep and keep your belongings. The most efficient, cost-effective method of providing this environmental condition is the responsibility of the mechanical system. Aspects such as heating and cooling equipment type, efficiency rating, fuel type, thermal envelope integrity, volume of conditioned space, location of heating and cooling equipment, and duct materials and routing all affect the quality of the indoor environment and its cost. The IRC will require that you provide a permanent means of providing heat in all habitable rooms such as living room, dining room, kitchen and bedrooms. Most other rooms not normally occupied such as garages, storage rooms, closets, hallways or utility spaces are not considered habitable for the purposes of required heating. It is important to satisfy the code requirement that every habitable room or space must be provided with means of elevating temperature to at least 68 degrees at a point three feet from the floor and two feet from exterior walls.

The mechanical plan will describe the method for heating, cooling and ventilating your home. Additionally, any pipe or duct supplying or exhausting air of any kind will be depicted in the plan. For example, not only will you draw duct work for supply air and return air, but you will also need to show an exhaust duct for a clothes dryer vent, bathroom and laundry room exhaust fans, and a range hood vent.

Choices for HVAC

The decision of how to heat (and cool) your home should be based on practical considerations. Which type of heating is most cost effective? Which type of heating (or cooling) costs the most initially? Which lasts the longest? Which maintains the highest efficiency for the longest time? Which type of equipment is most common where you live? How prepared are you to install gas heating equipment? If you need help, are there qualified tradesmen available to install your chosen system? What is the safest, most dependable, least hazardous HVAC system? Which has the best cost-recovery based on energy savings?

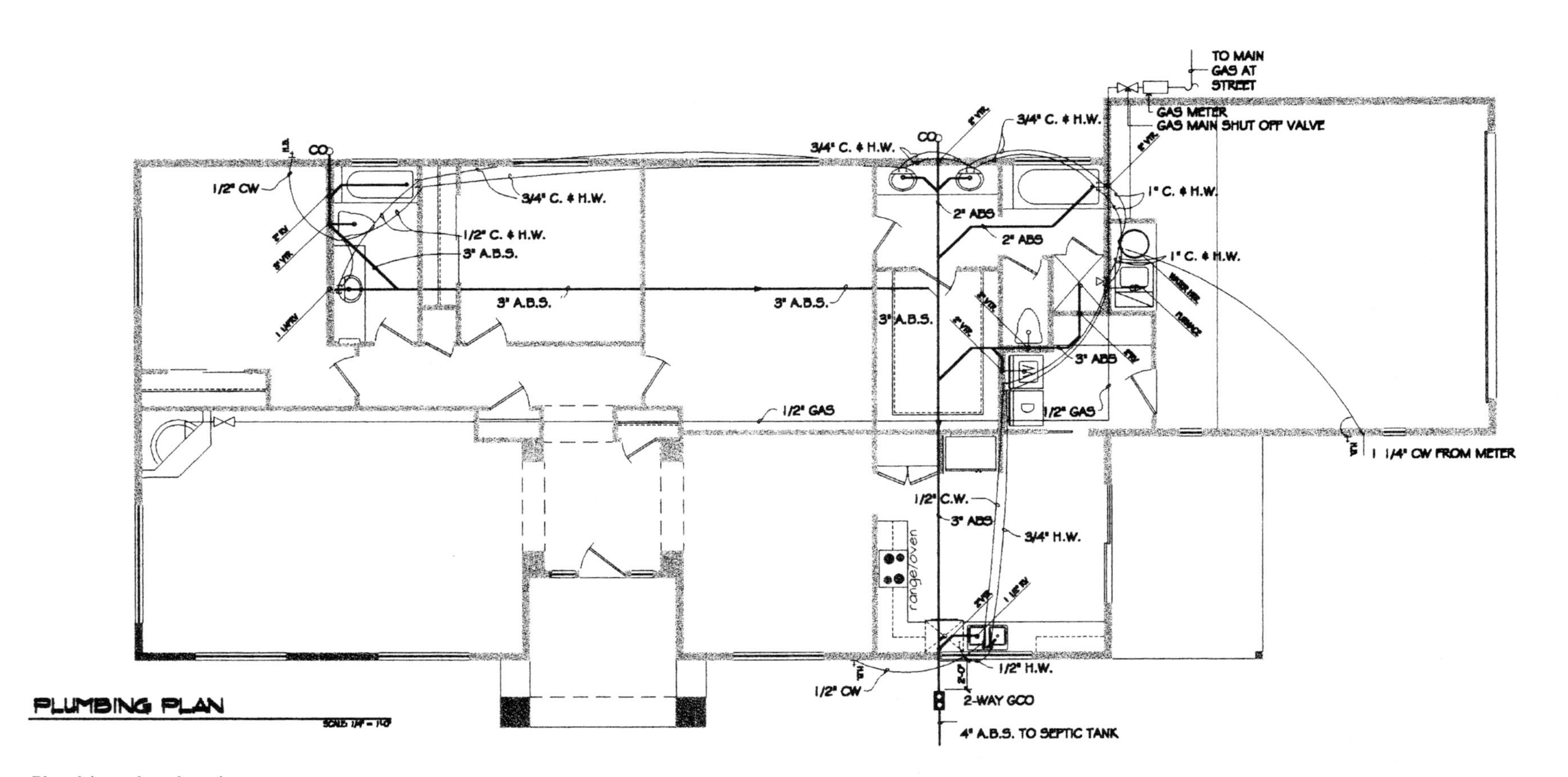

Plumbing plan drawing.

The heat and warmth of the gas flame is nearest to that of a fireplace. However, if you are not in an area served by a natural gas purveyor, liquid petroleum may be problematic for your building design. If you have a heated basement, LP gas heating systems may not be permitted, due to the weight of LP in relation to air and the location of the furnace.

You may have similar concerns. I suggest that you thoroughly explore all options and consider carefully your comfort level with installing a particular heating system. The design criteria presented below is general and will work for most systems. I recommend that you consider one of the more efficient type *heat pumps*, that are easy to install and highly efficient. There is the added advantage of providing both heating or cooling at the flip of a thermostat.

In areas where providing cooling is not important, baseboard electric resistant heating strips are a simple method of providing ample heat. In addition to these baseboard heaters, electric resistant wall heaters are another excellent option. Radiant heating systems that use heated water in pipes in the floor may be a method to consider. Wood stoves and fireplaces will provide necessary heat, but may not be approved by a building inspector, since the convective heat tends to stay in one or two rooms.

Equipment Choices

Equipment size, unit efficiency, ease of installation, availability and cost will affect your decision for equipment selection. A heat pump's size is rated in tons of cooling potential. A ton of cooled air is roughly equivalent to 12,000 Btu's or approximately 400 cubic feet per minute. Size selection is based on a number of factors including the volume of conditioned space, climate, heating degree days and indoor design temperature. The volume of conditioned space is simply the total volume in the interior of your home. Most weather service agencies will have data regarding climate and the prevailing outdoor temperature during the heating season. For those who are interested in the design methods that engineers use to calculate energy loss, Appendix B illustrates some common methods.

In its heating mode, the heat pump has an efficiency rating that exceeds 100 percent. It can supply more energy than it consumes. It extracts even modest amounts of heat from outdoor air to transfer the heating needed for the conditioned space. In essence, more energy is provided in heat form than is required by electricity to operate the refrigeration cycle. The measure of this efficiency is called the *coefficient of performance,* or COP. This efficiency rating is simply the ratio of the heat supplied, divided by the electrical current. The Seasonal Energy Efficiency Rating, or SEER, is an efficiency rating method for cooling. It is established by a similar method. It is the total cooling of the heat pump divided by the total electrical energy input in watt-hours during a similar period. The efficiency will vary depending on geographic locations.

To select a proper size heat pump as a heating system, divide the total btu's lost in your home by 12,000 (Btu's/ton). Then divide this number by the COP of the particular unit you plan to buy. For example, a house is being proposed that has a heat loss calculated at 120,000 Btu per hour. Here is how to find the correct unit size.

A heat pump with a COP of 2.0
120,000 Btuh / [12,000 Btu/ton] = 10 tons
10 tons divided by 2.0 = 5 tons

The point behind the calculation methods in Appendix B is that in order to maintain a design temperature of 68 degrees, every Btu lost through walls, roof or floors must be replaced. Design professionals can crank out an exact number of Btu's for your particular house. But this calculated heat loss value is based on certain assumptions. Most heating and cooling equipment dealers will use a rule of thumb method to select the appropriate size for you. Efficiency rating on mechanical equipment can diminish with time and use. Mechanical equipment dealers may recommend that you purchase a unit slightly larger than you need because of these variables. Another useful means of selecting the right size equipment is to determine what others have done. Find a house near the same size and configuration as yours in a housing development, maybe even in construction stage. Look at that HVAC system. Compare it with your design. Ask the owners about their comfort level and their utility bills.

Duct Work

In order to deliver the conditioned air from your fan coil/blower assembly to the rooms you want to heat and cool, ductwork must be installed. Ducts are simply a conduit for air to move to a space. Some ducts are rigid metal and look like miniature hallways. Other ducts are flexible and look like a large garden hose. Some rigid ducts are rectangular or square and others are round or even oval in shape. Rigid metal ductwork is usually manufactured by a sheet metal specialist in pieces and assembled, in place, by a mechanical subcontractor. Flexible ductwork is a manufactured product, which is listed and tested for specific uses. Some products are permitted to be used as supply and return air ducts for residential heat pumps.

The flexible duct is round in shape and is made in a variety of sizes ranging from 3 inches to 90 inches in diameter. The most common sizes you will need range between 6 inches to 12 inches for supply air ducts and 20 inches to 30 inches for the return air duct. Size selection for ducts are based on required air volume needed for the room and the design air velocity of the conditioned air through the duct. Flexible duct material may need to be cleaned periodically to maintain a dust-free air supply.

The volume of conditioned air that is required to be supplied to the space for the necessary heating (or cooling) is based on the Btu's needed to achieve the design temperature. Each room or space will have a distinct heat loss (or gain) based on its aggregate insulation values in the exterior wall, attic and floor as well as exterior glazing of that room. Review the calculation for heat loss for the whole house, which is outlined in Appendix B. Use that same formula for each room and solve for Btu's needed to maintain the desired temperature within each room. Since a ton of cooling is equivalent to 12,000 Btu's, and since there are about 400 cfm per ton, for every cubic foot per minute (CFM) of heat delivered to a space, approximately 30 Btu's are added. Thus, you can calculate the actual cfm required for each room.

A good rule of thumb for the required volume of heated air to a space is about 2 CFM per square foot of floor area. For instance, a bedroom has 144 square feet of floor area and would need around 288 CFM of conditioned air. The rate of Btu's provided would be 288 x 30 = 8,640 Btuh. The size of the duct for a given space is regulated by the code. If heating is being provided by a heat pump, for each 1000 Btuh (Btu's per hour), your duct must be at least 6 square inches. So for instance, if the small bedroom we considered needed 8,640 Btuh to maintain the desired temperature, the cross sectional area of the duct would need to be at least:

8,640 Btuh/1000 x 6 = 51.8 square inches

Area = π^2

51.8 in^2 = 3.14159 r^2

$r = [\sqrt{51.8 / 3.14159}] = 4.06$ inches

diameter = 2r = [4.06 x 2] = 8.1 inches use next larger size:

Use a 9 inch round duct

A 9 inch diameter duct is the minimum trade size available which exceeds the minimum required in this case. There are other factors based on noise, climate, and environment which affect duct size. I recommend that you consult your local mechanical equipment dealer to assist you in duct size selection for each room.

Other Choices

If you are in an area where cooling is not important, you may be able to install electric resistant heating. Although it is not as efficient as a heat pump, it is extremely easy to install. These electric resistant heaters are designed as either baseboard heaters or wall heaters. Wall heaters are built into the wall frame before drywall is installed. They are available in different power outputs. Baseboard heaters are installed at the lowest level along a wall next to the baseboard, hence their name. The baseboard heaters come in different lengths as well as different power outputs. Electric resistant heating will supply 3413 Btu's for every kilowatt hour of electric current. Usually each separate room or space will need at least one heater. To determine the size needed, each room's heat loss should be estimated with the calculations outlined earlier. For example, if a bedroom has a heat loss of 8640 btuh, the estimate of the energy needed is:

$$8640\,\text{Btuh} \times \left[\frac{1000\,\text{watts}}{3413\,\text{Btu}}\right] = 2531\,\text{watts}$$

A 3000 watt or two 1500 watt heaters should be adequate for this room.

Other Mechanical Systems

Certain rooms must be provided with a means of exhausting vapor or noxious smells. For instance, a laundry room will accumulate moisture vapor in the ambient air unless an exhaust vent with a mechanical fan removes it to the outside. In order to remove the same moisture vapor and smells in bathrooms or toilet rooms a similar fan must be installed on the ceiling. Each of these exhaust fans must be connected to an appropriate size duct that extends to the outdoors. Usually a 3 inch duct is adequate to carry moisture and smells away from these rooms.

A clothes dryer must be equipped with an exhaust vent that is at least a 4 inch round metal duct. The maximum this duct can extend is related to the number of bends and the material type. If there are no more than four 90 degree bends, the exhaust duct can stretch a maximum of 14 feet. The dryer duct must extend to the outdoors and must not have screens or filters on the terminal. These will collect debris and could damage the equipment or even cause a fire.

A domestic kitchen range is not required to have a means of exhausting grease and smoke or other cooking odors. However, if you install one, it must meet similar exhaust duct requirements. The exhaust duct must be made of metal and have a smooth interior to prevent obstruction.

Energy Code

More than likely, you will need to demonstrate compliance with energy efficiency standards in your new home. The *International Energy Conservation Code* published by the ICC is the most commonly used document to set minimum energy conservation standards in residential construction. The code uses many of the methods outlined above to establish heat flow. The minimum energy conservation standards can be met in one of two methods; through component performance methods or through prescriptive criteria.

The component performance method allows for various trade-offs to compensate for an overall effectiveness to the resistance of heat flow. For instance if you have more windows, you might compensate by adding more ceiling insulation. The prescriptive criteria sets some basic minimum standards, which will allow you to select within the energy efficiency components of your structure. This method hinges on the climate in your particular region. This historical climatic record is based on the amount of heating needed to maintain a comfortable temperature within your home.

Begin with the premise that a dwelling must be able to maintain a temperature of 68 degrees. This is based on Section R303.6 in the IRC. Then consider that in the coldest winter evening in your area, the overnight temperature reaches a low of 15 degrees. The difference in those two temperatures is 53 degrees; (68 minus 15). However, during the more mild winter evenings, the temperature only reaches 48 degrees. The difference between those two temperatures is 20 degrees; (68 minus 48). During an entire heating season the accumulated differences in temperatures is defined as heating degree days. This number is available for numerous common locations throughout the United States in Table N1101.2 in the IRC. If your home town is not within this table, you may be able to acquire the number from your local branch of the National Weather Service.

Using the heating degree days for your area, enter Table N1102.1 from the IRC, and determine the minimum standards for meeting the prescriptive requirements for the energy code.

TABLE N1102.1
SIMPLIFIED PRESCRIPTIVE BUILDING ENVELOPE THERMAL COMPONENT CRITERIA
MINIMUM REQUIRED THERMAL PERFORMANCE (*U*-FACTOR AND *R*-VALUE)

HDD	MAXIMUM GLAZING U-FACTOR [Btu / (hr · ft² · °F)]	MINIMUM INSULATION *R*-VALUE [(hr·ft²·°F) / Btu]					
		Ceilings	Walls	Floors	Basement walls	Slab perimeter *R*-value and depth	Crawl space walls
0-499	Any	R-13	R-11	R-11	R-0	R-0	R-0
500-999	0.90	R-19	R-11	R-11	R-0	R-0	R-4
1,000-1,499	0.75	R-19	R-11	R-11	R-0	R-0	R-5
1,500-1,999	0.75	R-26	R-13	R-11	R-5	R-0	R-5
2,000-2,499	0.65	R-30	R-13	R-11	R-5	R-0	R-6
2,500-2,999	0.60	R-30	R-13	R-19	R-6	R-4, 2 ft.	R-7
3,000-3,499	0.55	R-30	R-13	R-19	R-7	R-4, 2 ft.	R-8
3,500-3,999	0.50	R-30	R-13	R-19	R-8	R-5, 2 ft.	R-10
4,000-4,499	0.45	R-38	R-13	R-19	R-8	R-5, 2 ft.	R-11
4,500-4,999	0.45	R-38	R-16	R-19	R-9	R-6, 2 ft.	R-17
5,000-5,499	0.45	R-38	R-18	R-19	R-9	R-6, 2 ft.	R-17
5,500-5,999	0.40	R-38	R-18	R-21	R-10	R-9, 4 ft.	R-19
6,000-6,499	0.35	R-38	R-18	R-21	R-10	R-9, 4 ft.	R-20
6,500-6,999	0.35	R-49	R-21	R-21	R-11	R-11, 4 ft.	R-20
7,000-8,499	0.35	R-49	R-21	R-21	R-11	R-13, 4 ft.	R-20
8,500-8,999	0.35	R-49	R-21	R-21	R-18	R-14, 4 ft.	R-20
9,000-12,999	0.35	R-49	R-21	R-21	R-19	R-18, 4 ft.	R-20

For SI: 1 Btu/(hr · ft² · °F) = 5.68W/m² · K, 1 (hr · ft² · °F)/Btu = 0.176m² · K/W.

All that's left now is to put down on paper what you have designed. Use the checklist below to make sure that the needed information is provided. Use the sample mechanical plan as a guide.

Mechanical Plans

___ Provide a mechanical plan that is separate from other plans such as plumbing or electrical plans.

___ Designate the locations and size of the heating and air conditioning equipment. *Section M1301.3*

___ Show that heating or cooling equipment is properly sized according to recognized standards. *Section M1401.3*

___ Show that heat pump and other mechanical equipment is labeled. *Section M1303*

___ Show that appliance installation methods comply with manufacturer's listing and *Section M1307.*

___ Show that appliances in garage or carport are protected from damage due to impact. *Section M1307.3.1*

___ Show that appliances are installed with a clearance from combustible materials. *Section M1306*

___ Any baseboard heaters must be installed per manufacturer's specifications. *Section M1405*

___ Fireplace or wood stoves must be installed according to *Section M1414* and manufacturer's listings.

___ In certain seismic zones, appliances must be anchored to prevent collapse. *Section M1307.2*

___ Show that heat pump equipment installed outside is supported at least 3 inches above grade. *Section M1403.2*

___ Designate the locations of each supply register, return air grill and all duct work. *Section M1403*

___ Show material, size and routing of all ductwork. *Section M1601*

___ Show that ducts meet the minimum size requirements of *Sections M1602* for return air and *M1603* for supply air.

___ Show that any solar energy systems are installed per Chapter 23.

___ Show evaporative cooler location and installation methods. *Section M1413*

___ Indicate access, clearances, and working space around HVAC equipment. *Section M1305*

___ Show condensation drains for HVAC units. *Section M1411.3*

___ Show provisions for and route of secondary condensate drain if needed. *Section M1411.3.1*

___ Mechanical exhaust vents shall terminate not less than the distance permitted by *Section M1804.2.*

___ Show exhaust fan size and locations for bathrooms, water closet compartments, and laundry rooms. *Section R303.3*

___ Provide a minimum 4 inch diameter moisture exhaust vent for dryer. *Section R303.3 and M1501*

___ Show that any range hood is exhausted to atmosphere according to the standards of *Section M1502.*

___ Show that ducts for exhaust vents for range hoods comply with the material limitations. *Section M1502.2*

___ Show that all exhaust vents terminate to atmosphere. *Chapter 15*

___ If heating or air conditioning equipment is in attic, show access per *Section M1305.1.3*

___ If heating or air conditioning equipment is in attic, show cat walk and working platform. *Section M1305.1.3*

___ If heating or air conditioning equipment is in attic, show switched light and service outlet. *Section M1305.1.3.1*

Structural Floor or Roof

The presentation of the structural design is a very important part of the overall plan. This plan should completely describe how the building stands and is able to resist natural forces such as wind and earthquakes, as well as working loads imposed by living such as pianos, pool tables, and refrigerators. The plan must include detail cuts or sectional views. Connectors, braces, straps, and even nailing schedule should be on such a plan.

How Wood Behaves Structurally

Trees grow upward, and it is in the length of the tree that wood has its strength. Long, interconnected fibers join together to maintain the tree's structural integrity. The tree ages from the inside out. Anyone who has seen a giant tree sliced in a museum can count rings back to an important date in history. These rings roughly portray multiple growing seasons. The tree siphons water to its branches and leaves through the living portion of the tree. That living portion is in the outer ring, just under the bark. Within that living area, the tree may be dead but is still very strong.

Wood is cut from trees lengthwise, in different geometric shapes, including squares and rectangles. The shape can be cut in various locations within the cross sectional area of the tree. This location will have a certain effect on the grade of lumber and strength of the particular piece. Branches grow from the outer perimeter of the tree. Wood cut from this area of the tree will have more knots due to the branches. Conversely, wood from the heart of the tree will be clear and free of knots and receive a better grade. The tree's species is another determinate of wood's strength. Oak is a stronger wood than pine for instance, because it has a greater ability to resist loads due to its strength and density.

Wood can withstand a variety of different kinds of loading conditions. It can take exceptional compressive loads. An example of these would be vertical loads pushing down on a post. For instance, a 4 inch x 4 inch Douglas fir #1 post, which is 8 feet tall and standing upright, can support almost 6000 pounds before failing. Wood can support a substantial bending load as well. For instance, a 6 inch x 8 inch Douglas fir #1 beam, spanning 10 feet, can support almost 4000 pounds before exceeding its design limits. A tree develops elastic strength as well. Wind can blow a tree back and forth throughout its lifetime. A tree overcomes this swaying force and develops elastic strength. Each species of tree will have a different *modulus of elasticity*, or elastic value. The direction a board is placed to resist an imposed load has an effect on its ability to resist a flexural load. For instance, if a 2 inch x 12 inch board is placed upright, its strength is far greater than if it were turned flat.

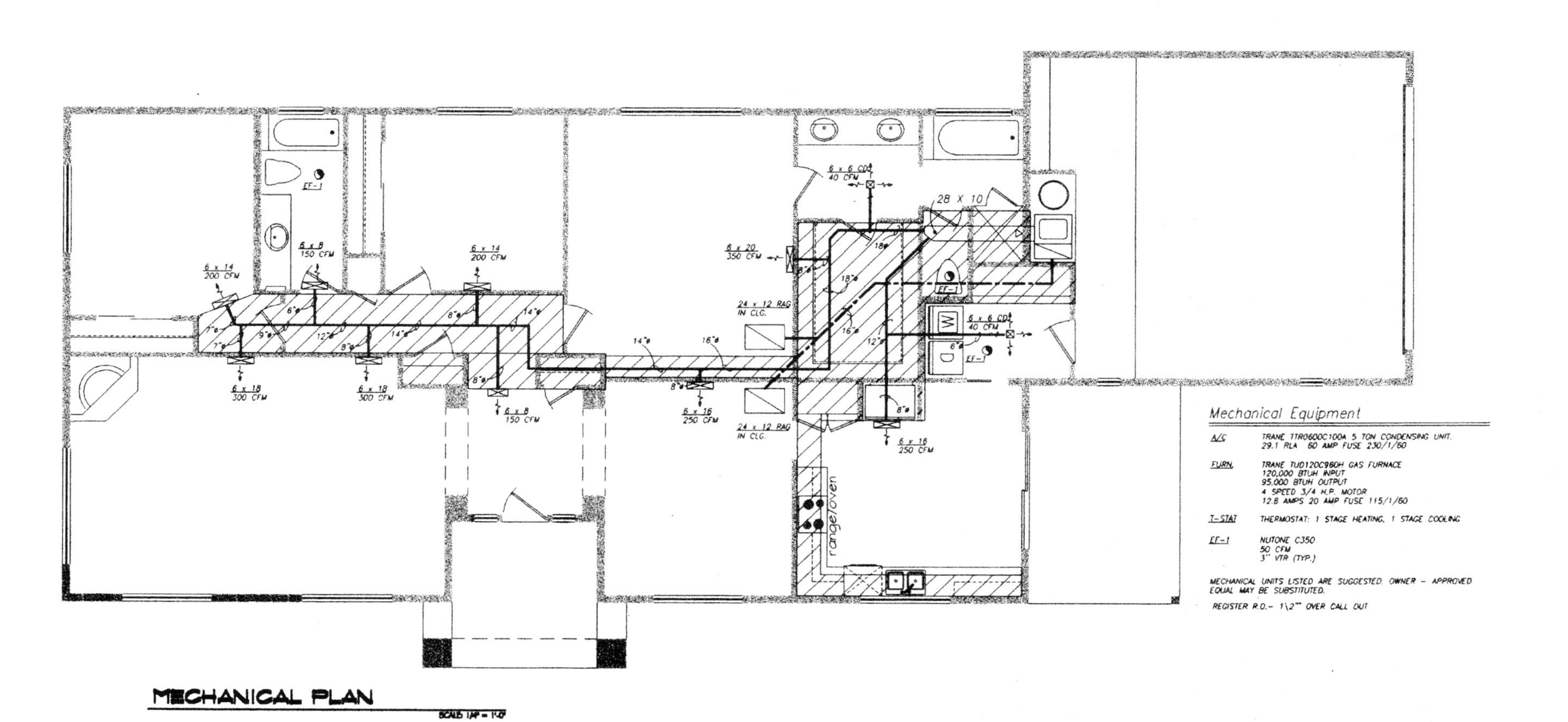

Mechanical plan drawing.

Trees are filled with water when they are harvested. Nearly half of a living tree's weight is water. When harvested, and over time, the tree will eventually dry out enough to use as a structural member. In some cases, structural lumber is dried in kilns at the mill. This is a more controlled method of removing moisture. The code establishes that the moisture content of lumber shall be less than 19 percent by weight before it may be used. Lumber shrinks when it dries. If it is installed wet with nails and bolts, then later dries out, the resulting shrinkage can result in nail withdrawal and connection failure.

At the lumber mill, where trees are cut into boards, they are graded by an approved grading agency. A *grader* will watch lumber pass by on a conveyor, then determine its grade and mark it with a grade stamp. You can see these grade stamps on each piece of lumber you buy. This grade stamp will include the species of the wood as well. Wood beams and boards receive a grade of either Select, 1, 2, or 3. Select is better than 1, 1 is better than 2, and so on. Many things affect individual board grade, including the number and location of knots, checks and splits.

Wood is cut into boards with shapes that refer to nominal size. However, the descriptive size is not a true representation of the actual size. The *nominal size* refers to the size a board would have been before it was trimmed or *dressed* to achieve a uniform shape. The following list portrays the difference between nominal and actual size for some commonly used types of wood framing members.

BOARDS		BEAMS	
Nominal Size	**Actual Size**	**Nominal Size**	**Actual Size**
2" x 4"	1½" x 3½"	4" x 4"	3½" x 3½"
2" x 6"	1½" x 5½"	4" x 6"	3½" x 5½"
2" x 8"	1½" x 7½"	4" x 8"	3½" x 7½"
2" x 10"	1½" x 9½"	6" x 8"	5½" x 7½"
2" x 12"	1½" x 11½"	8" x 10"	7½" x 9½"

The nominal board and beam sizes shown in the table will be refered to throughout the text in simplified forms using common construction terms. For example; a two inch by four inch board size will be shown as a 2 x 4, and described as a "two by four." A two inch by six inch board will be shown as a 2 x 6, and called a "two by six," and so on. Various lengths of boards and beams will be described in feet.

Wood Beams, Floor Joists and Roof Rafters

Wood beams support trusses, rafters and floor joists when openings are desired within bearing walls. Shorter openings or *spans* can be accommodated with headers. These headers are site-built, composite beams. They are built with multiple layers of dimensional lumber laminated together with nails to form a beam. For longer spans, solid, homogeneous wood beams may be more appropriate. The required size of these beams is derived through calculations based on engineering principles. These principles and sample calculations are offered for your review at the latter portion of this book (Appendix C*)*. However, the IRC has a table that simplifies header selection.

Table R502.5(1) illustrates that header and girder selection is based on several criteria, including local snow loading conditions, overall building width, the use of the header such as roof or floor support and the span of the header. This table is extremely useful and helps avoid costly engineering fees.

The way to use the table is to first determine your local snow load, which may be either 30 pounds or as much as 50 pounds per square foot. Ask your local building department for this information. Next, using the overall width of your house that is bearing on the header, follow the column down to the type of support intended. The type load could be any of the following:

1. A roof load only: wall of single story roof or wall of upper story of two story
2. A center-bearing floor and roof load: lower wall with intermediate support of floor and clear span of roof for two story
3. A clear span floor and roof load: lower wall of clear span floor and roof of two story
4. Two center-bearing floors and a roof load: interior, lower wall supporting two floors and roof of three story

Within the appropriate row, select a header size within the span limitations for your wall opening. Tables R502.3.1(1) and (2) allow you to select floor joists in a similar manner. To select a floor joist, you must know the design live load for the use intended. The IRC prescribes two different live load categories to use within a residence. In bedrooms or sleeping areas the live load permitted is 30 pounds per square foot (psf). In other areas of the house the live load must be designed at 40 pounds per square foot (psf). Table R502.3.1(1) is used for the 30 pounds per square foot (psf) sleeping areas, and Table R502.3.1(2) is used to design other areas of the house with a live load designed at 40 pounds per square foot (psf).

Dead load is the weight of structural framework, decking, insulations, drywall and other portions of the structure. The option for dead load in these tables is either 10 or 20 pounds per square foot (psf). Your building department may make this determination, but I would suggest using the larger load to provide a conservative and safe design. With these two loading conditions established, enter the table from the first column, which is the center spacing desired for the floor joists. Then select the species and grade of the wood you intend to use. Follow that row over to the joist span, which is the distance between supports such as bearing walls or even headers for your floor. That column has a heading that shows the minimum size floor joist permitted for those variables.

Engineered Trusses

A truss is an assembly of individual components, which, when properly connected, will provide structural support. The advantage of trusses over comparable size, solid sawn lumber is that of increased strength with a longer possible span. There are several types of trusses that use numerous types of materials such as wood and steel. Wood trusses are regarded as an engineered design due to the complexity of connection locations, forces and member sizes. Hence, trusses must be designed and sealed by a registered professional engineer, using the forces intended for your unique loading conditions.

These drawings, along with the design, must be submitted to the building department for their review. The drawings must include the information listed below.

- Slope or depth, span and spacing of trusses
- Location of all joints
- Required bearing widths
- Design loads such as top and bottom chord live and dead load, concentrated loads, lateral loads
- Adjustments to lumber and joint connector design values for conditions of use

- Each reaction force with appropriate direction
- Joint connector type and description
- Lumber size, species and grade for each member
- Connection requirements for truss to girder truss, truss ply to ply, field splices
- Calculated deflection ratio and/or maximum description for live and dead load
- Maximum axial compression forces in the truss members
- Required permanent truss member bracing location

When you purchase trusses from a manufacturer, be sure to acquire all of the submittal requirements for these trusses. Remember that each of these truss drawings must be sealed by a registered engineer of record with a seal from your state. Ask the salesman for documents that include specific provisions for the truss design, as outlined above. If you do not get this information at the time of sale, it may be harder to get it later.

Homemade Trusses

This is a reminder that trusses are an engineered design and a manufactured product. Although a truss design can be duplicated and still work, it is not insured by any design and is not accepted by most building departments. This is an aspect of construction that you should leave to professionals. You could be penny-wise and pound-foolish if you try to build your own trusses. There are just too many variables that could cause general structural failure, if not immediately, then certainly within time. The expense in setting up a jig for each truss could cost more than the purchase of all the trusses you need.

Alterations to Trusses

Since trusses are an engineered design and a manufactured product, you must not alter them during installation. This includes cutting, notching, boring holes, splicing or otherwise diminishing their strength. To avoid the need to alter trusses, take your entire building plan to the truss manufacturer and discuss the project, including mechanical, plumbing and electrical installations. Find areas that may cause conflicts and design the trusses to accommodate these aspects. If you have no choice but to alter a truss, you must consult both the manufacturer and the registered engineer who performed the design. They may be able to offer an approved *fix* for the truss, one that will ensure safety and satisfy the building department.

Anchoring Roof System To Bearing Wall

Trusses must be secured to walls to prevent them from being carried away by wind. If you are in a windy region, this is especially pertinent. The type and strength of the connector is subject to the force of wind against the roof. The calculation of this force is outlined in figure R301.2.(4) and Table R301.2(2) from the IRC 2000 edition (not included). When you determine the amount of force (in pounds per square foot), you can use Table R802.11 to determine the strength capacity of the anchor or hold-down. There are numerous manufacturers of metal anchors and hold-down devices from which you can choose. Be sure and read the manufacturer's specification and verify the required strength.

TABLE R502.5(1)
GIRDER SPANS[a] AND HEADER SPANS[a] FOR EXTERIOR BEARING WALLS
(Maximum header spans for douglas fir-larch, hem-fir, southern pine and spruce-pine-fir[b] and required number of jack studs)

HEADERS SUPPORTING	SIZE	GROUND SNOW LOAD (psf)[e]											
		30						50					
		Building width[c] (feet)											
		20		28		36		20		28		36	
		Span	NJ[d]	Span	NJ[d]	Span	NJ[d]	Span	NJ[d]	Span	NJ[d]	Span	NJ[d]
Roof and ceiling	2-2×4	3-6	1	3-2	1	2-10	1	3-2	1	2-9	1	2-6	1
	2-2×6	5-5	1	4-8	1	4-2	1	4-8	1	4-1	1	3-8	2
	2-2×8	6-10	1	5-11	2	5-4	2	5-11	2	5-2	2	4-7	2
	2-2×10	8-5	2	7-3	2	6-6	2	7-3	2	6-3	2	5-7	2
	2-2×12	9-9	2	8-5	2	7-6	2	8-5	2	7-3	2	6-6	2
	3-2×8	8-4	1	7-5	1	6-8	1	7-5	1	6-5	2	5-9	2
	3-2×10	10-6	1	9-1	2	8-2	2	9-1	2	7-10	2	7-0	2
	3-2×12	12-2	2	10-7	2	9-5	2	10-7	2	9-2	2	8-2	2
	4-2×8	7-0	1	6-1	2	5-5	2	6-1	2	5-3	2	4-8	2
	4-2×10	11-8	1	10-6	1	9-5	2	10-6	1	9-1	2	8-2	2
	4-2×12	14-1	1	12-2	2	10-11	2	12-2	2	10-7	2	9-5	2
Roof, ceiling and one center-bearing floor	2-2×4	3-1	1	2-9	1	2-5	1	2-9	1	2-5	1	2-2	1
	2-2×6	4-6	1	4-0	1	3-7	2	4-1	1	3-7	2	3-3	2
	2-2×8	5-9	2	5-0	2	4-6	2	5-2	2	4-6	2	4-1	2
	2-2×10	7-0	2	6-2	2	5-6	2	6-4	2	5-6	2	5-0	2
	2-2×12	8-1	2	7-1	2	6-5	2	7-4	2	6-5	2	5-9	3
	3-2×8	7-2	1	6-3	2	5-8	2	6-5	2	5-8	2	5-1	2
	3-2×10	8-9	2	7-8	2	6-11	2	7-11	2	6-11	2	6-3	2
	3-2×12	10-2	2	8-11	2	8-0	2	9-2	2	8-0	2	7-3	2
	4-2×8	5-10	2	5-2	2	4-8	2	5-3	2	4-7	2	4-2	2
	4-2×10	10-1	1	8-10	2	8-0	2	9-1	2	8-0	2	7-2	2
	4-2×12	11-9	2	10-3	2	9-3	2	10-7	2	9-3	2	8-4	2
Roof, ceiling and one clear span floor	2-2×4	2-8	1	2-4	1	2-1	1	2-7	1	2-3	1	2-0	1
	2-2×6	3-11	1	3-5	2	3-0	2	3-10	2	3-4	2	3-0	2
	2-2×8	5-0	2	4-4	2	3-10	2	4-10	2	4-2	2	3-9	2
	2-2×10	6-1	2	5-3	2	4-8	2	5-11	2	5-1	2	4-7	3
	2-2×12	7-1	2	6-1	3	5-5	3	6-10	2	5-11	3	5-4	3
	3-2×8	6-3	2	5-5	2	4-10	2	6-1	2	5-3	2	4-8	2
	3-2×10	7-7	2	6-7	2	5-11	2	7-5	2	6-5	2	5-9	2
	3-2×12	8-10	2	7-8	2	6-10	2	8-7	2	7-5	2	6-8	2
	4-2×8	5-1	2	4-5	2	3-11	2	4-11	2	4-3	2	3-10	2
	4-2×10	8-9	2	7-7	2	6-10	2	8-7	2	7-5	2	6-7	2
	4-2×12	10-2	2	8-10	2	7-11	2	9-11	2	8-7	2	7-8	2
Roof, ceiling and two center-bearing floors	2-2×4	2-7	1	2-3	1	2-0	1	2-6	1	2-2	1	1-11	1
	2-2×6	3-9	2	3-3	2	2-11	2	3-8	2	3-2	2	2-10	2
	2-2×8	4-9	2	4-2	2	3-9	2	4-7	2	4-0	2	3-8	2
	2-2×10	5-9	2	5-1	2	4-7	3	5-8	2	4-11	2	4-5	3
	2-2×12	6-8	2	5-10	3	5-3	3	6-6	2	5-9	3	5-2	3
	3-2×8	5-11	2	5-2	2	4-8	2	5-9	2	5-1	2	4-7	2
	3-2×10	7-3	2	6-4	2	5-8	2	7-1	2	6-2	2	5-7	2
	3-2×12	8-5	2	7-4	2	6-7	2	8-2	2	7-2	2	6-5	3
	4-2×8	4-10	2	4-3	2	3-10	2	4-9	2	4-2	2	3-9	2
	4-2×10	8-4	2	7-4	2	6-7	2	8-2	2	7-2	2	6-5	2
	4-2×12	9-8	2	8-6	2	7-8	2	9-5	2	8-3	2	7-5	2

For SI: 1 inch = 25.4 mm, 1 pound per square foot = 0.0479 kN/m^2.

a. Spans are given in feet and inches.

b. Tabulated values assume #2 grade lumber.

c. Building width is measured perpendicular to the ridge. For widths between those shown, spans are permitted to be interpolated.

d. NJ - Number of jack studs required to support each end. Where the number of required jack studs equals one, the header are permitted to be supported by an approved framing anchor attached to the full-height wall stud and to the header.

e. Use 30 psf ground snow load for cases in which ground snow load is less than 30 psf and the roof live load is equal to or less than 20 psf.

TABLE R502.5(2)
GIRDER SPANS[a] AND HEADER SPANS[a] FOR INTERIOR BEARING WALLS
(Maximum header spans for douglas fir-larch, hem-fir, southern pine and spruce-pine-fir[b] and required number of jack studs)

HEADERS AND GIRDERS SUPPORTING	SIZE	BUILDING WIDTH[c] (feet)					
		20		28		36	
		Span	NJ[d]	Span	NJ[d]	Span	NJ[d]
One floor only	2-2×4	3-1	1	2-8	1	2-5	1
	2-2×6	4-6	1	3-11	1	3-6	1
	2-2×8	5-9	1	5-0	2	4-5	2
	2-2×10	7-0	2	6-1	2	5-5	2
	2-2×12	8-1	2	7-0	2	6-3	2
	3-2×8	7-2	1	6-3	1	5-7	2
	3-2×10	8-9	1	7-7	2	6-9	2
	3-2×12	10-2	2	8-10	2	7-10	2
	4-2×8	5-10	1	5-1	2	4-6	2
	4-2×10	10-1	1	8-9	1	7-10	2
	4-2×12	11-9	1	10-2	2	9-1	2
Two floors	2-2×4	2-2	1	1-10	1	1-7	1
	2-2×6	3-2	2	2-9	2	2-5	2
	2-2×8	4-1	2	3-6	2	3-2	2
	2-2×10	4-11	2	4-3	2	3-10	3
	2-2×12	5-9	2	5-0	3	4-5	3
	3-2×8	5-1	2	4-5	2	3-11	2
	3-2×10	6-2	2	5-4	2	4-10	2
	3-2×12	7-2	2	6-3	2	5-7	3
	4-2×8	4-2	2	3-7	2	3-2	2
	4-2×10	7-2	2	6-2	2	5-6	2
	4-2×12	8-4	2	7-2	2	6-5	2

For SI: 1 inch = 25.4 mm, 1 foot = 304.8 mm.

a. Spans are given in feet and inches.

b. Tabulated values assume #2 grade lumber.

c. Building width is measured perpendicular to the ridge. For widths between those shown, spans are permitted to be interpolated.

d. NJ - Number of jack studs required to support each end. Where the number of required jack studs equals one, the headers are permitted to be supported by an approved framing anchor attached to the full-height wall stud and to the header.

TABLE R502.3.1(1)
FLOOR JOIST SPANS FOR COMMON LUMBER SPECIES
(Residential sleeping areas, live load=30 psf, L/Δ=360)

		DEAD LOAD = 10 psf				DEAD LOAD = 20 psf			
		2x6	2x8	2x10	2x12	2x6	2x8	2x10	2x12
		Maximum floor joist spans							
JOIST SPACING (inches)	SPECIE AND GRADE	(ft.- in.)	(ft.- in.)	(ft.- in.)	(ft.- in.)	(ft.- in.)	(ft.- in.)	(ft.- in.)	(ft.- in.)
12	Douglas fir-larch SS	12-6	16-6	21-0	25-7	12-6	16-6	21-0	25-7
	Douglas fir-larch #1	12-0	15-10	20-3	24-8	12-0	15-7	19-0	22-0
	Douglas fir-larch #2	11-10	15-7	19-10	23-0	11-6	14-7	17-9	20-7
	Douglas fir-larch #3	9-8	12-4	15-0	17-5	8-8	11-0	13-5	15-7
	Hem-fir SS	11-10	15-7	19-10	24-2	11-10	15-7	19-10	24-2
	Hem-fir #1	11-7	15-3	19-5	23-7	11-7	15-2	18-6	21-6
	Hem-fir #2	11-0	14-6	18-6	22-6	11-0	14-4	17-6	20-4
	Hem-fir #3	9-8	12-4	15-0	17-5	8-8	11-0	13-5	15-7
	Southern pine SS	12-3	16-2	20-8	25-1	12-3	16-2	20-8	25-1
	Southern pine #1	12-0	15-10	20-3	24-8	12-0	15-10	20-3	24-8
	Southern pine #2	11-10	15-7	19-10	18-8	11-10	15-7	18-7	21-9
	Southern pine #3	10-5	13-3	15-8	18-8	9-4	11-11	14-0	16-8
	Spruce-pine-fir SS	11-7	15-3	19-5	23-7	11-7	15-3	19-5	23-7
	Spruce-pine-fir #1	11-3	14-11	19-0	23-0	11-3	14-7	17-9	20-7
	Spruce-pine-fir #2	11-3	14-11	19-0	23-0	11-3	14-7	17-9	20-7
	Spruce-pine-fir #3	9-8	12-4	15-0	17-5	8-8	11-0	13-5	15-7
16	Douglas fir-larch SS	11-4	15-0	19-1	23-3	11-4	15-0	19-1	23-0
	Douglas fir-larch #1	10-11	14-5	18-5	21-4	10-8	13-6	16-5	19-1
	Douglas fir-larch #2	10-9	14-1	17-2	19-11	9-11	12-7	15-5	17-10
	Douglas fir-larch #3	8-5	10-8	13-0	15-1	7-6	9-6	11-8	13-6
	Hem-fir SS	10-9	14-2	18-0	21-11	10-9	14-2	18-0	21-11
	Hem-fir #1	10-6	13-10	17-8	20-9	10-4	13-1	16-0	18-7
	Hem-fir #2	10-0	13-2	16-10	19-8	9-10	12-5	15-2	17-7
	Hem-fir #3	8-5	10-8	13-0	15-1	7-6	9-6	11-8	13-6
	Southern pine SS	11-2	14-8	18-9	22-10	11-2	14-8	18-9	22-10
	Southern pine #1	10-11	14-5	18-5	22-5	10-11	14-5	17-11	21-4
	Southern pine #2	10-9	14-2	18-0	21-1	10-5	13-6	16-1	18-10
	Southern pine #3	9-0	11-6	13-7	16-2	8-1	10-3	12-2	14-6
	Spruce-pine-fir SS	10-6	13-10	17-8	21-6	10-6	13-10	17-8	21-4
	Spruce-pine-fir #1	10-3	13-6	17-2	19-11	9-11	12-7	15-5	17-10
	Spruce-pine-fir #2	10-3	13-6	17-2	19-11	9-11	12-7	15-5	17-10
	Spruce-pine-fir #3	8-5	10-8	13-0	15-1	7-6	9-6	11-8	13-6
19.2	Douglas fir-larch SS	10-8	14-1	18-0	21-10	10-8	14-1	18-0	21-0
	Douglas fir-larch #1	10-4	13-7	16-9	19-6	9-8	12-4	15-0	17-5
	Douglas fir-larch #2	10-1	12-10	15-8	18-3	9-1	11-6	14-1	16-3
	Douglas fir-larch #3	7-8	9-9	11-10	13-9	6-10	8-8	10-7	12-4
	Hem-fir SS	10-1	13-4	17-0	20-8	10-1	13-4	17-0	20-7
	Hem-fir #1	9-10	13-0	16-4	19-0	9-6	12-0	14-8	17-0
	Hem-fir #2	9-5	12-5	15-6	17-1	8-11	11-4	13-10	16-1
	Hem-fir #3	7-8	9-9	11-10	13-9	6-10	8-8	10-7	12-4
	Southern pine SS	10-6	13-10	17-8	21-6	10-6	13-10	17-8	21-6
	Southern pine #1	10-4	13-7	17-4	21-1	10-4	13-7	16-4	19-6
	Southern pine #2	10-1	13-4	16-5	19-3	9-6	12-4	14-8	17-2
	Southern pine #3	8-3	10-6	12-5	14-9	7-4	9-5	11-1	13-2
	Spruce-pine-fir SS	9-10	13-0	16-7	20-2	9-10	13-0	16-7	19-6
	Spruce-pine-fir #1	9-8	12-9	15-8	18-3	9-1	11-6	14-1	16-3
	Spruce-pine-fir #2	9-8	12-9	15-8	18-3	9-1	11-6	14-1	16-3
	Spruce-pine-fir #3	7-8	9-9	11-10	13-9	6-10	8-8	10-7	12-4
24	Douglas fir-larch SS	9-11	13-1	16-8	20-3	9-11	13-1	16-2	18-9
	Douglas fir-larch #1	9-7	12-4	15-0	17-5	8-8	11-0	13-5	15-7
	Douglas fir-larch #2	9-1	11-6	14-1	16-3	8-1	10-3	12-7	14-7
	Douglas fir-larch #3	6-10	8-8	10-7	12-4	6-2	7-9	9-6	11-0
	Hem-fir SS	9-4	12-4	15-9	19-2	9-4	12-4	15-9	18-5
	Hem-fir #1	9-2	12-0	14-8	17-0	8-6	10-9	13-1	15-2
	Hem-fir #2	8-9	11-4	13-10	16-1	8-0	10-2	12-5	14-4
	Hem-fir #3	6-10	8-8	10-7	12-4	6-2	7-9	9-6	11-0
	Southern pine SS	9-9	12-10	16-5	19-11	9-9	12-10	16-5	19-11
	Southern pine #1	9-7	12-7	16-1	19-6	9-7	12-4	14-7	17-5
	Southern pine #2	9-4	12-4	14-8	17-2	8-6	11-0	13-1	15-5
	Southern pine #3	7-4	9-5	11-1	13-2	6-7	8-5	9-11	11-10
	Spruce-pine-fir SS	9-2	12-1	15-5	18-9	9-2	12-1	15-0	17-5
	Spruce-pine-fir #1	8-11	11-6	14-1	16-3	8-1	10-3	12-7	14-7
	Spruce-pine-fir #2	8-11	11-6	14-1	16-3	8-1	10-3	12-7	14-7
	Spruce-pine-fir #3	6-10	8-8	10-7	12-4	6-2	7-9	9-6	11-0

For SI: 1 inch = 25.4 mm, 1 foot = 304.8 mm, 1 pound per square foot = 0.0479 kN/m^2.

NOTE: Check sources for availability of lumber in lengths greater than 20 feet.

TABLE R502.3.1(2)
FLOOR JOIST SPANS FOR COMMON LUMBER SPECIES (Residential living areas, live load=40 psf, L/Δ=360)

			DEAD LOAD = 10 psf				DEAD LOAD = 20 psf			
			2x6	2x8	2x10	2x12	2x6	2x8	2x10	2x12
			Maximum floor joist spans							
JOIST SPACING (inches)	SPECIE AND GRADE		(ft.- in.)	(ft.- in.)	(ft.- in.)	(ft.- in.)	(ft.- in.)	(ft.- in.)	(ft.- in.)	(ft.- in.)
12	Douglas fir-larch	SS	11-4	15-0	19-1	23-3	11-4	15-0	19-1	23-3
	Douglas fir-larch	#1	10-11	14-5	18-5	22-0	10-11	14-2	17-4	20-1
	Douglas fir-larch	#2	10-9	14-2	17-9	20-7	10-6	13-3	16-3	18-10
	Douglas fir-larch	#3	8-8	11-0	13-5	15-7	7-11	10-0	12-3	14-3
	Hem-fir	SS	10-9	14-2	18-0	21-11	10-9	14-2	18-0	21-11
	Hem-fir	#1	10-6	13-10	17-8	21-6	10-6	13-10	16-11	19-7
	Hem-fir	#2	10-0	13-2	16-10	20-4	10-0	13-1	16-0	18-6
	Hem-fir	#3	8-8	11-0	13-5	15-7	7-11	10-0	12-3	14-3
	Southern pine	SS	11-2	14-8	18-9	22-10	11-2	14-8	18-9	22-10
	Southern pine	#1	10-11	14-5	18-5	22-5	10-11	14-5	18-5	22-5
	Southern pine	#2	10-9	14-2	18-0	21-9	10-9	14-2	16-11	19-10
	Southern pine	#3	9-4	11-11	14-0	16-8	8-6	10-10	12-10	15-3
	Spruce-pine-fir	SS	10-6	13-10	17-8	21-6	10-6	13-10	17-8	21-6
	Spruce-pine-fir	#1	10-3	13-6	17-3	20-7	10-3	13-3	16-3	18-10
	Spruce-pine-fir	#2	10-3	13-6	17-3	20-7	10-3	13-3	16-3	18-10
	Spruce-pine-fir	#3	8-8	11-0	13-5	15-7	7-11	10-0	12-3	14-3
16	Douglas fir-larch	SS	10-4	13-7	17-4	21-1	10-4	13-7	17-4	21-0
	Douglas fir-larch	#1	9-11	13-1	16-5	19-1	9-8	12-4	15-0	17-5
	Douglas fir-larch	#2	9-9	12-7	15-5	17-10	9-1	11-6	14-1	16-3
	Douglas fir-larch	#3	7-6	9-6	11-8	13-6	6-10	8-8	10-7	12-4
	Hem-fir	SS	9-9	12-10	16-5	19-11	9-9	12-10	16-5	19-11
	Hem-fir	#1	9-6	12-7	16-0	18-7	9-6	12-0	14-8	17-0
	Hem-fir	#2	9-1	12-0	15-2	17-7	8-11	11-4	13-10	16-1
	Hem-fir	#3	7-6	9-6	11-8	13-6	6-10	8-8	10-7	12-4
	Southern pine	SS	10-2	13-4	17-0	20-9	10-2	13-4	17-0	20-9
	Southern pine	#1	9-11	13-1	16-9	20-4	9-11	13-1	16-4	19-6
	Southern pine	#2	9-9	12-10	16-1	18-10	9-6	12-4	14-8	17-2
	Southern pine	#3	8-1	10-3	12-2	14-6	7-4	9-5	11-1	13-2
	Spruce-pine-fir	SS	9-6	12-7	16-0	19-6	9-6	12-7	16-0	19-6
	Spruce-pine-fir	#1	9-4	12-3	15-5	17-10	9-1	11-6	14-1	16-3
	Spruce-pine-fir	#2	9-4	12-3	15-5	17-10	9-1	11-6	14-1	16-3
	Spruce-pine-fir	#3	7-6	9-6	11-8	13-6	6-10	8-8	10-7	12-4
19.2	Douglas fir-larch	SS	9-8	12-10	16-4	19-10	9-8	12-10	16-4	19-2
	Douglas fir-larch	#1	9-4	12-4	15-0	17-5	8-10	11-3	13-8	15-11
	Douglas fir-larch	#2	9-1	11-6	14-1	16-3	8-3	10-6	12-10	14-10
	Douglas fir-larch	#3	6-10	8-8	10-7	12-4	6-3	7-11	9-8	11-3
	Hem-fir	SS	9-2	12-1	15-5	18-9	9-2	12-1	15-5	18-9
	Hem-fir	#1	9-0	11-10	14-8	17-0	8-8	10-11	13-4	15-6
	Hem-fir	#2	8-7	11-3	13-10	16-1	8-2	10-4	12-8	14-8
	Hem-fir	#3	6-10	8-8	10-7	12-4	6-3	7-11	9-8	11-3
	Southern pine	SS	9-6	12-7	16-0	19-6	9-6	12-7	16-0	19-6
	Southern pine	#1	9-4	12-4	15-9	19-2	9-4	12-4	14-11	17-9
	Southern pine	#2	9-2	12-1	14-8	17-2	8-8	11-3	13-5	15-8
	Southern pine	#3	7-4	9-5	11-1	13-2	6-9	8-7	10-1	12-1
	Spruce-pine-fir	SS	9-0	11-10	15-1	18-4	9-0	11-10	15-1	17-9
	Spruce-pine-fir	#1	8-9	11-6	14-1	16-3	8-3	10-6	12-10	14-10
	Spruce-pine-fir	#2	8-9	11-6	14-1	16-3	8-3	10-6	12-10	14-10
	Spruce-pine-fir	#3	6-10	8-8	10-7	12-4	6-3	7-11	9-8	11-3
24	Douglas fir-larch	SS	9-0	11-11	15-2	18-5	9-0	11-11	14-9	17-1
	Douglas fir-larch	#1	8-8	11-0	13-5	15-7	7-11	10-0	12-3	14-3
	Douglas fir-larch	#2	8-1	10-3	12-7	14-7	7-5	9-5	11-6	13-4
	Douglas fir-larch	#3	6-2	7-9	9-6	11-0	5-7	7-1	8-8	10-1
	Hem-fir	SS	8-6	11-3	14-4	17-5	8-6	11-3	14-4	16-10[a]
	Hem-fir	#1	8-4	10-9	13-1	15-2	7-9	9-9	11-11	13-10
	Hem-fir	#2	7-11	10-2	12-5	14-4	7-4	9-3	11-4	13-1
	Hem-fir	#3	6-2	7-9	9-6	11-0	5-7	7-1	8-8	10-1
	Southern pine	SS	8-10	11-8	14-11	18-1	8-10	11-8	14-11	18-1
	Southern pine	#1	8-8	11-5	14-7	17-5	8-8	11-3	13-4	15-11
	Southern pine	#2	8-6	11-0	13-1	15-5	7-9	10-0	12-0	14-0
	Southern pine	#3	6-7	8-5	9-11	11-10	6-0	7-8	9-1	10-9
	Spruce-pine-fir	SS	8-4	11-0	14-0	17-0	8-4	11-0	13-8	15-11
	Spruce-pine-fir	#1	8-1	10-3	12-7	14-7	7-5	9-5	11-6	13-4
	Spruce-pine-fir	#2	8-1	10-3	12-7	14-7	7-5	9-5	11-6	13-4
	Spruce-pine-fir	#3	6-2	7-9	9-6	11-0	5-7	7-1	8-8	10-1

Check sources for availability of lumber in lengths greater than 20 feet.
For SI: 1 inch = 25.4 mm, 1 foot = 308.4 mm, 1 pound per square foot = 0.0479 kN/m.
a. End bearing length shall be increased to 2 inches.

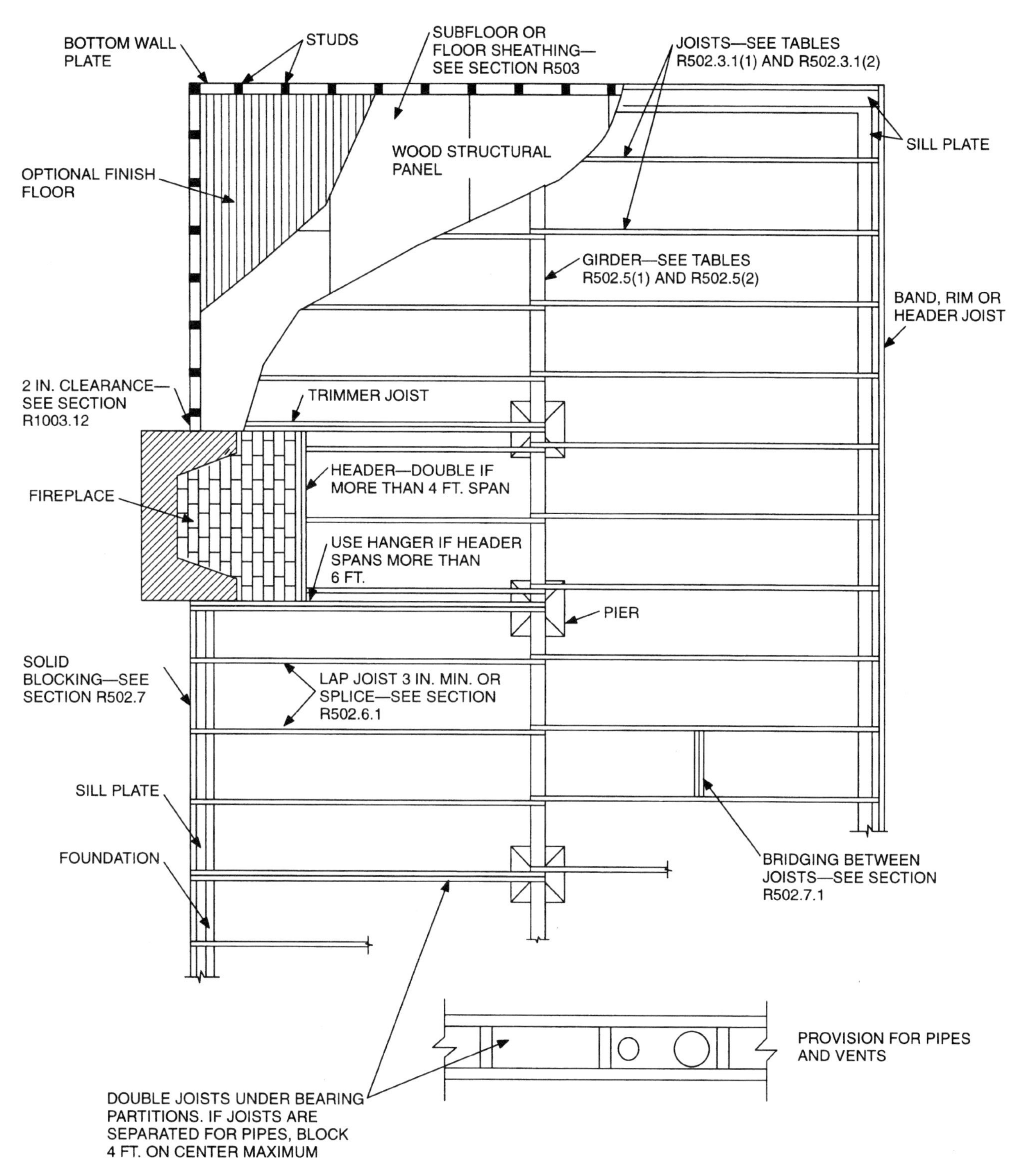

For SI: 1 inch = 25.4 mm, 1 foot = 304.8 mm.

FIGURE R502.2
FLOOR CONSTRUCTION

Roof Rafters

If you are experienced in building frame roof structures from solid dimensional lumber, and the design you have chosen is less appropriate for trusses, rafter framing is the best choice. Only experienced carpenters should attempt to build this type of roof. Since this book is targeted toward the apprentice carpenter, it will not thoroughly explore the particulars of this subject. Keep in mind that the IRC has specific requirements and limitations on this type of framing. If you desire to build such a roof, I strongly recommend that you purchase a copy of this code book to analyze all of the requirements before you begin your design. This code book is filled with sketches, drawings, tables and charts, which thoroughly illustrate the requirements.

However, to illustrate the basics of rafter design, let's touch on some aspects of the design. First, a pitched roof is designed with one of two basic methods: *ridge beam support* or *ridge board support*. The ridge beam is a large wood beam that spans between bearing walls and supports rafters that frame into the side or rest on the top. This design allows for a vaulted ceiling design. A ridge board is a much smaller dimensional wood joist that also spans between bearing walls with rafters framing into each side. The dead and live loads above the rafters attempt to force the two opposing rafters together at the ridge board. The compressive forces holds the rafters upright. However, the forces resolve themselves at the bearing walls supporting the rafters opposite each other. The forces here attempt to push the bearing walls further apart from each other. These forces would succeed in pushing the walls down unless something held them together. The structural member that does this is called a ceiling joist. The ceiling joist performs two tasks: it supports the drywall on the ceiling, and it ties the lower ends of the rafters together at the bearing wall. This forms a triangle frame assembly, which is very stable.

There are several rules that regulate this type of framing, including location of ceiling joists, maximum slope of rafters, required blocking, connections between rafters and ceiling joists, and allowable span for rafters and ceiling joists. The maximum slope permissible using this type of framing is 3:12. The rafters must be equal lengths with the ridge board in the exact center. The ceiling joists must be as low as possible along the rafter legs and rest on the bearing walls. The connections between the rafters and the ceiling joists must be sufficient to resist the tension loads imposed upon them. Rafter ties must be provided every 4 feet o.c. in addition to the ceiling joists. Ceiling joists must be continuous or securely joined together over interior partitions. The maximum span for rafters and ceiling joists are based on the loads they carry and the strength of the particular species of wood. Tables R802.5.1(1) and (2) show some of the typical spans based on different loading conditions. Others are elsewhere within the IRC.

Shear Walls

Wind (and earthquakes) affect the lateral stability of a structure by applying natural energy from the side. The structure remains stable by resisting that energy through walls adjacent to the wall facing the wind or seismic event. That's why the wind does not blow a house over on its side.

Set a box on your kitchen table. Now, gently push on one side of the box. Notice that the box moves away from where it was. Now, imagine that you have applied glue to the perimeter of the bottom of the box and set it on the table. When the glue has dried, try to push the box again. Your pushing represents the wind. What holds your home in place from a wind load? The answer to that question is complicated. Wind applies its energy to both the walls and roof of your home, usually from one side. How is that energy resisted? Let's follow the lines of force of the wind's energy. The roof and wall facing the wind absorbs the energy first. With the plywood sheathing applied to both the wall and roof

TABLE R802.5.1(1)
RAFTER SPANS FOR COMMON LUMBER SPECIES
(Roof live load=20 psf, ceiling not attached to rafters, L/Δ=180)

RAFTER SPACING (inches)	SPECIE AND GRADE	DEAD LOAD = 10 psf					DEAD LOAD = 20 psf				
		2x4	2x6	2x8	2x10	2x12	2x4	2x6	2x8	2x10	2x12
		Maximum rafter spans[a]									
		(feet - inches)	(feet - inches)	(feet - inches)	(feet - inches)	(feet - inches)	(feet - inches)	(feet - inches)	(feet - inches)	(feet - inches)	(feet - inches)
12	Douglas fir-larch SS	11-6	18-0	23-9	Note[b]	Note[b]	11-6	18-0	23-5	Note[b]	Note[b]
	Douglas fir-larch #1	11-1	17-4	22-5	Note[b]	Note[b]	10-6	15-4	19-5	23-9	Note[b]
	Douglas fir-larch #2	10-10	16-7	21-0	25-8	Note[b]	9-10	14-4	18-2	22-3	25-9
	Douglas fir-larch #3	8-7	12-6	15-10	19-5	22-6	7-5	10-10	13-9	16-9	19-6
	Hem-fir SS	10-10	17-0	22-5	Note[b]	Note[b]	10-10	17-0	22-5	Note[b]	Note[b]
	Hem-fir #1	10-7	16-8	21-10	Note[b]	Note[b]	10-3	14-11	18-11	23-2	Note[b]
	Hem-fir #2	10-1	15-11	20-8	25-3	Note[b]	9-8	14-2	17-11	21-11	25-5
	Hem-fir #3	8-7	12-6	15-10	19-5	22-6	7-5	10-10	13-9	16-9	19-6
	Southern pine SS	11-3	17-8	23-4	Note[b]	Note[b]	11-3	17-8	23-4	Note[b]	Note[b]
	Southern pine #1	11-1	17-4	22-11	Note[b]	Note[b]	11-1	17-3	21-9	25-10	Note[b]
	Southern pine #2	10-10	17-0	22-5	Note[b]	Note[b]	10-6	15-1	19-5	23-2	Note[b]
	Southern pine #3	9-1	13-6	17-2	20-3	24-1	7-11	11-8	14-10	17-6	20-11
	Spruce-pine-fir SS	10-7	16-8	21-11	Note[b]	Note[b]	10-7	16-8	21-9	Note[b]	Note[b]
	Spruce-pine-fir #1	10-4	16-3	21-0	25-8	Note[b]	9-10	14-4	18-2	22-3	25-9
	Spruce-pine-fir #2	10-4	16-3	21-0	25-8	Note[b]	9-10	14-4	18-2	22-3	25-9
	Spruce-pine-fir #3	8-7	12-6	15-10	19-5	22-6	7-5	10-10	13-9	16-9	19-6
16	Douglas fir-larch SS	10-5	16-4	21-7	Note[b]	Note[b]	10-5	16-0	20-3	24-9	Note[b]
	Douglas fir-larch #1	10-0	15-4	19-5	23-9	Note[b]	9-1	13-3	16-10	20-7	23-10
	Douglas fir-larch #2	9-10	14-4	18-2	22-3	25-9	8-6	12-5	15-9	19-3	22-4
	Douglas fir-larch #3	7-5	10-10	13-9	16-9	19-6	6-5	9-5	11-11	14-6	16-10
	Hem-fir SS	9-10	15-6	20-5	Note[b]	Note[b]	9-10	15-6	19-11	24-4	Note[b]
	Hem-fir #1	9-8	14-11	18-11	23-2	Note[b]	8-10	12-11	16-5	20-0	23-3
	Hem-fir #2	9-2	14-2	17-11	21-11	25-5	8-5	12-3	15-6	18-11	22-0
	Hem-fir #3	7-5	10-10	13-9	16-9	19-6	6-5	9-5	11-11	14-6	16-10
	Southern pine SS	10-3	16-1	21-2	Note[b]	Note[b]	10-3	16-1	21-2	Note[b]	Note[b]
	Southern pine #1	10-0	15-9	20-10	25-10	Note[b]	10-0	15-0	18-10	22-4	Note[b]
	Southern pine #2	9-10	15-1	19-5	23-2	Note[b]	9-1	13-0	16-10	20-1	23-7
	Southern pine #3	7-11	11-8	14-10	17-6	20-11	6-10	10-1	12-10	15-2	18-1
	Spruce-pine-fir SS	9-8	15-2	19-11	25-5	Note[b]	9-8	14-10	18-10	23-0	Note[b]
	Spruce-pine-fir #1	9-5	14-4	18-2	22-3	25-9	8-6	12-5	15-9	19-3	22-4
	Spruce-pine-fir #2	9-5	14-4	18-2	22-3	25-9	8-6	12-5	15-9	19-3	22-4
	Spruce-pine-fir #3	7-5	10-10	13-9	16-9	19-6	6-5	9-5	11-11	14-6	16-10
19.2	Douglas fir-larch SS	9-10	15-5	20-4	25-11	Note[b]	9-10	14-7	18-6	22-7	Note[b]
	Douglas fir-larch #1	9-5	14-0	17-9	21-8	25-2	8-4	12-2	15-4	18-9	21-9
	Douglas fir-larch #2	8-11	13-1	16-7	20-3	23-6	7-9	11-4	14-4	17-7	20-4
	Douglas fir-larch #3	6-9	9-11	12-7	15-4	17-9	5-10	8-7	10-10	13-3	15-5
	Hem-fir SS	9-3	14-7	19-2	24-6	Note[b]	9-3	14-4	18-2	22-3	25-9
	Hem-fir #1	9-1	13-8	17-4	21-1	24-6	8-1	11-10	15-0	18-4	21-3
	Hem-fir #2	8-8	12-11	16-4	20-0	23-2	7-8	11-2	14-2	17-4	20-1
	Hem-fir #3	6-9	9-11	12-7	15-4	17-9	5-10	8-7	10-10	13-3	15-5
	Southern pine SS	9-8	15-2	19-11	25-5	Note[b]	9-8	15-2	19-11	25-5	Note[b]
	Southern pine #1	9-5	14-10	19-7	23-7	Note[b]	9-3	13-8	17-2	20-5	24-4
	Southern pine #2	9-3	13-9	17-9	21-2	24-10	8-4	11-11	15-4	18-4	21-6
	Southern pine #3	7-3	10-8	13-7	16-0	19-1	6-3	9-3	11-9	13-10	16-6
	Spruce-pine-fir SS	9-1	14-3	18-9	23-11	Note[b]	9-1	13-7	17-2	21-0	24-4
	Spruce-pine-fir #1	8-10	13-1	16-7	20-3	23-6	7-9	11-4	14-4	17-7	20-4
	Spruce-pine-fir #2	8-10	13-1	16-7	20-3	23-6	7-9	11-4	14-4	17-7	20-4
	Spruce-pine-fir #3	6-9	9-11	12-7	15-4	17-9	5-10	8-7	10-10	13-3	15-5

(continued)

TABLE R802.5.1(1)—continued
RAFTER SPANS FOR COMMON LUMBER SPECIES
(Roof live load=20 psf, ceiling not attached to rafters, L/Δ=180)

RAFTER SPACING (inches)	SPECIE AND GRADE		DEAD LOAD = 10 psf					DEAD LOAD = 20 psf				
			2x4	2x6	2x8	2x10	2x12	2x4	2x6	2x8	2x10	2x12
			Maximum rafter spans[a]									
			(feet - inches)	(feet - inches)	(feet - inches)	(feet - inches)	(feet - inches)	(feet - inches)	(feet - inches)	(feet - inches)	(feet - inches)	(feet - inches)
24	Douglas fir-larch	SS	9-1	14-4	18-10	23-4	23-4	8-11	13-1	16-7	20-3	23-5
	Douglas fir-larch	#1	8-7	12-6	15-10	19-5	19-5	7-5	10-10	13-9	16-9	19-6
	Douglas fir-larch	#2	8-0	11-9	14-10	18-2	18-2	6-11	10-2	12-10	15-8	18-3
	Douglas fir-larch	#3	6-1	8-10	11-3	13-8	13-8	5-3	7-8	9-9	11-10	13-9
	Hem-fir	SS	8-7	13-6	17-10	22-9	22-9	8-7	12-10	16-3	19-10	23-0
	Hem-fir	#1	8-4	12-3	15-6	18-11	18-11	7-3	10-7	13-5	16-4	19-0
	Hem-fir	#2	7-11	11-7	14-8	17-10	17-10	6-10	10-0	12-8	15-6	17-11
	Hem-fir	#3	6-1	8-10	11-3	13-8	13-8	5-3	7-8	9-9	11-10	13-9
	Southern pine	SS	8-11	14-1	18-6	23-8	23-8	8-11	14-1	18-6	22-11	Note[b]
	Southern pine	#1	8-9	13-9	17-9	21-1	21-1	8-3	12-3	15-4	18-3	21-9
	Southern pine	#2	8-7	12-3	15-10	18-11	18-11	7-5	10-8	13-9	16-5	19-3
	Southern pine	#3	6-5	9-6	12-1	14-4	14-4	5-7	8-3	10-6	12-5	14-9
	Spruce-pine-fir	SS	8-5	13-3	17-5	21-8	21-8	8-4	12-2	15-4	18-9	21-9
	Spruce-pine-fir	#1	8-0	11-9	14-10	18-2	18-2	6-11	10-2	12-10	15-8	18-3
	Spruce-pine-fir	#2	8-0	11-9	14-10	18-2	18-2	6-11	10-2	12-10	15-8	18-3
	Spruce-pine-fir	#3	6-1	8-10	11-3	13-8	13-8	5-3	7-8	9-9	11-10	13-9

For SI: 1 inch = 25.4 mm, 1 foot = 304.8 mm, 1 pound per square foot = 0.0479 kN/m^2.

a. The tabulated rafter spans assume that ceiling joists are located at the bottom of the attic space or that some other method of resisting the outward push of the rafters on the bearing walls, such as rafter ties, is provided at that location. When ceiling joists or rafter ties are located higher in the attic space, the rafter spans shall be multiplied by the factors given below:

surface, the energy is transferred to the adjacent walls, which run the same direction as the wind is blowing. The top plates of each of these walls transfer the energy into the exterior plywood sheathing via studs, which are part of the wall frame. The wall frame and the plywood sheathing act in concert to transfer the forces. The wind energy is then absorbed into the bottom plate and ultimately into the foundation. The dead load of the foundation restrains the wall from overturning. The corners are this wall's weakest link. The wind's energy is trying to shear the roof off the wall corner, and the wall off the foundation. The wall resists these lateral forces and is acting as a shear wall.

Notice that the rectangular box has two sides, which vary in length. Now, imagine the direction from which wind is presumably pushing. The resisting wall that is shorter will absorb the same amount of wind energy as the long wall. However, this shearing force is applied over a smaller section. Short wall sections that absorb a shearing force such as wind represent the most critical element in a lateral force resisting element.

Seismic forces work in reverse of wind forces, transferring energy from the foundation, through the shear walls, into the roof sheathing. In lower hazard seismic areas, light-frame buildings absorb more shear forces from wind than from earthquakes. Therefore, if you meet the conventional design requirements for lateral design, your building should be stable. However, check with your local jurisdiction regarding the conditions at your location.

Braced wall lines are required by most codes for lateral stability, which includes both wind and seismic forces. Wood framed walls sheathed with traditional hardboard siding or plywood are considered to meet the conventional provisions of required bracing as long as certain conditions are met. First, a required brace panel must be provided at *every* corner and every 25 foot length along a wall. This means that a single, full size (uncut) 4 foot wide sheet of plywood or hard board siding must be installed at every corner in each direction and every 25 feet of wall length.

TABLE R802.5.1(2)
RAFTER SPANS FOR COMMON LUMBER SPECIES
(Roof live load=20 psf, ceiling attached to rafters, L/Δ=240)

RAFTER SPACING (inches)	SPECIE AND GRADE	DEAD LOAD = 10 psf					DEAD LOAD = 20 psf				
		2x4	2x6	2x8	2x10	2x12	2x4	2x6	2x8	2x10	2x12
		Maximum rafter spans[a]									
		(feet - inches)	(feet - inches)	(feet - inches)	(feet - inches)	(feet - inches)	(feet - inches)	(feet - inches)	(feet - inches)	(feet - inches)	(feet - inches)
12	Douglas fir-larch SS	10-5	16-4	21-7	Note[b]	Note[b]	10-5	16-4	21-7	Note[b]	Note[b]
	Douglas fir-larch #1	10-0	15-9	20-10	Note[b]	Note[b]	10-0	15-4	19-5	23-9	Note[b]
	Douglas fir-larch #2	9-10	15-6	20-5	25-8	Note[b]	9-10	14-4	18-2	22-3	25-9
	Douglas fir-larch #3	8-7	12-6	15-10	19-5	22-6	7-5	10-10	13-9	16-9	19-6
	Hem-fir SS	9-10	15-6	20-5	Note[b]	Note[b]	9-10	15-6	20-5	Note[b]	Note[b]
	Hem-fir #1	9-8	15-2	19-11	25-5	Note[b]	9-8	14-11	18-11	23-2	Note[b]
	Hem-fir #2	9-2	14-5	19-0	24-3	Note[b]	9-2	14-2	17-11	21-11	25-5
	Hem-fir #3	8-7	12-6	15-10	19-5	22-6	7-5	10-10	13-9	16-9	19-6
	Southern pine SS	10-3	16-1	21-2	Note[b]	Note[b]	10-3	16-1	21-2	Note[b]	Note[b]
	Southern pine #1	10-0	15-9	20-10	Note[b]	Note[b]	10-0	15-9	20-10	25-10	Note[b]
	Southern pine #2	9-10	15-6	20-5	Note[b]	Note[b]	9-10	15-1	19-5	23-2	Note[b]
	Southern pine #3	9-1	13-6	17-2	20-3	24-1	7-11	11-8	14-10	17-6	20-11
	Spruce-pine-fir SS	9-8	15-2	19-11	25-5	Note[b]	9-8	15-2	19-11	25-5	Note[b]
	Spruce-pine-fir #1	9-5	14-9	19-6	24-10	Note[b]	9-5	14-4	18-2	22-3	25-9
	Spruce-pine-fir #2	9-5	14-9	19-6	24-10	Note[b]	9-5	14-4	18-2	22-3	25-9
	Spruce-pine-fir #3	8-7	12-6	15-10	19-5	22-6	7-5	10-10	13-9	16-9	19-6
16	Douglas fir-larch SS	9-6	14-11	19-7	25-0	Note[b]	9-6	14-11	19-7	24-9	Note[b]
	Douglas fir-larch #1	9-1	14-4	18-11	23-9	Note[b]	9-1	13-3	16-10	20-7	23-10
	Douglas fir-larch #2	8-11	14-1	18-2	22-3	25-9	8-6	12-5	15-9	19-3	22-4
	Douglas fir-larch #3	7-5	10-10	13-9	16-9	19-6	6-5	9-5	11-11	14-6	16-10
	Hem-fir SS	8-11	14-1	18-6	23-8	Note[b]	8-11	14-1	18-6	23-8	Note[b]
	Hem-fir #1	8-9	13-9	18-1	23-1	Note[b]	8-9	12-11	16-5	20-0	23-3
	Hem-fir #2	8-4	13-1	17-3	21-11	25-5	8-4	12-3	15-6	18-11	22-0
	Hem-fir #3	7-5	10-10	13-9	16-9	19-6	6-5	9-5	11-11	14-6	16-10
	Southern pine SS	9-4	14-7	19-3	24-7	Note[b]	9-4	14-7	19-3	24-7	Note[b]
	Southern pine #1	9-1	14-4	18-11	24-1	Note[b]	9-1	14-4	18-10	22-4	Note[b]
	Southern pine #2	8-11	14-1	18-6	23-2	Note[b]	8-11	13-0	16-10	20-1	23-7
	Southern pine #3	7-11	11-8	14-10	17-6	20-11	6-10	10-1	12-10	15-2	18-1
	Spruce-pine-fir SS	8-9	13-9	18-1	23-1	Note[b]	8-9	13-9	18-1	23-0	Note[b]
	Spruce-pine-fir #1	8-7	13-5	17-9	22-3	25-9	8-6	12-5	15-9	19-3	22-4
	Spruce-pine-fir #2	8-7	13-5	17-9	22-3	25-9	8-6	12-5	15-9	19-3	22-4
	Spruce-pine-fir #3	7-5	10-10	13-9	16-9	19-6	6-5	9-5	11-11	14-6	16-10
19.2	Douglas fir-larch SS	8-11	14-0	18-5	23-7	Note[b]	8-11	14-0	18-5	22-7	Note[b]
	Douglas fir-larch #1	8-7	13-6	17-9	21-8	25-2	8-4	12-2	15-4	18-9	21-9
	Douglas fir-larch #2	8-5	13-1	16-7	20-3	23-6	7-9	11-4	14-4	17-7	20-4
	Douglas fir-larch #3	6-9	9-11	12-7	15-4	17-9	5-10	8-7	10-10	13-3	15-5
	Hem-fir SS	8-5	13-3	17-5	22-3	Note[b]	8-5	13-3	17-5	22-3	25-9
	Hem-fir #1	8-3	12-11	17-1	21-1	24-6	8-1	11-10	15-0	18-4	21-3
	Hem-fir #2	7-10	12-4	16-3	20-0	23-2	7-8	11-2	14-2	17-4	20-1
	Hem-fir #3	6-9	9-11	12-7	15-4	17-9	5-10	8-7	10-10	13-3	15-5
	Southern pine SS	8-9	13-9	18-1	23-1	Note[b]	8-9	13-9	18-1	23-1	Note[b]
	Southern pine #1	8-7	13-6	17-9	22-8	Note[b]	8-7	13-6	17-2	20-5	24-4
	Southern pine #2	8-5	13-3	17-5	21-2	24-10	8-4	11-11	15-4	18-4	21-6
	Southern pine #3	7-3	10-8	13-7	16-0	19-1	6-3	9-3	11-9	13-10	16-6
	Spruce-pine-fir SS	8-3	12-11	17-1	21-9	Note[b]	8-3	12-11	17-1	21-0	24-4
	Spruce-pine-fir #1	8-1	12-8	16-7	20-3	23-6	7-9	11-4	14-4	17-7	20-4
	Spruce-pine-fir #2	8-1	12-8	16-7	20-3	23-6	7-9	11-4	14-4	17-7	20-4
	Spruce-pine-fir #3	6-9	9-11	12-7	15-4	17-9	5-10	8-7	10-10	13-3	15-5

TABLE R802.5.1(2)—continued
RAFTER SPANS FOR COMMON LUMBER SPECIES
(Roof live load=20 psf, ceiling attached to rafters, L/Δ=240)

RAFTER SPACING (inches)	SPECIE AND GRADE		DEAD LOAD = 10 psf					DEAD LOAD = 20 psf				
			2x4	2x6	2x8	2x10	2x12	2x4	2x6	2x8	2x10	2x12
			Maximum rafter spans[a]									
			(feet - inches)	(feet - inches)	(feet - inches)	(feet - inches)	(feet - inches)	(feet - inches)	(feet - inches)	(feet - inches)	(feet - inches)	(feet - inches)
24	Douglas fir-larch	SS	8-3	13-0	17-2	21-10	Note[b]	8-3	13-0	16-7	20-3	23-5
	Douglas fir-larch	#1	8-0	12-6	15-10	19-5	22-6	7-5	10-10	13-9	16-9	19-6
	Douglas fir-larch	#2	7-10	11-9	14-10	18-2	21-0	6-11	10-2	12-10	15-8	18-3
	Douglas fir-larch	#3	6-1	8-10	11-3	13-8	15-11	5-3	7-8	9-9	11-10	13-9
	Hem-fir	SS	7-10	12-3	16-2	20-8	25-1	7-10	12-3	16-2	19-10	23-0
	Hem-fir	#1	7-8	12-0	15-6	18-11	21-11	7-3	10-7	13-5	16-4	19-0
	Hem-fir	#2	7-3	11-5	14-8	17-10	20-9	6-10	10-0	12-8	15-6	17-11
	Hem-fir	#3	6-1	8-10	11-3	13-8	15-11	5-3	7-8	9-9	11-10	13-9
	Southern pine	SS	8-1	12-9	16-10	21-6	Note[b]	8-1	12-9	16-10	21-6	Note[b]
	Southern pine	#1	8-0	12-6	16-6	21-1	25-2	8-0	12-3	15-4	18-3	21-9
	Southern pine	#2	7-10	12-3	15-10	18-11	22-2	7-5	10-8	13-9	16-5	19-3
	Southern pine	#3	6-5	9-6	12-1	14-4	17-1	5-7	8-3	10-6	12-5	14-9
	Spruce-pine-fir	SS	7-8	12-0	15-10	20-2	24-7	7-8	12-0	15-4	18-9	21-9
	Spruce-pine-fir	#1	7-6	11-9	14-10	18-2	21-0	6-11	10-2	12-10	15-8	18-3
	Spruce-pine-fir	#2	7-6	11-9	14-10	18-2	21-0	6-11	10-2	12-10	15-8	18-3
	Spruce-pine-fir	#3	6-1	8-10	11-3	13-8	15-11	5-3	7-8	9-9	11-10	13-9

For SI: 1 inch = 25.4 mm, 1 foot = 304.8 mm, 1 pound per square foot = 0.0479 kN/m^2.

a. The tabulated rafter spans assume that ceiling joists are located at the bottom of the attic space or that some other method of resisting the outward push of the rafters on the bearing walls, such as rafter ties, is provided at that location. When ceiling joists or rafter ties are located higher in the attic space, the rafter spans shall be multiplied by the factors given below:

H_C/H_R	Rafter Span Adjustment Factor
2/3 or greater	0.50
1/2	0.58
1/3	0.67
1/4	0.76
1/5	0.83
1/6	0.90
1/7.5 and less	1.00

where: H_C = Height of ceiling joists or rafter ties measured vertically above the top of the rafter support walls.

H_R = Height of roof ridge measured vertically above the top of the rafter support walls.

b. Check sources for availability of lumber in lengths greater than 20 feet.

There are other methods provided by the IRC that meet the requirement, but structural wood panels such as plywood or hardboard siding work best and serve other purposes as well, such as a finish exterior or a substrate on which to apply stucco or masonry veneer. Installing these structural panels does not require too much accommodation, unless you want a corner window or door. Therefore, your design should consider this requirement. Try to keep windows and doors at least 4 feet away from corners. If you cannot, be sure at least to begin the brace wall panel within 12½ feet of the corner. These plywood panels must have specific nails and certain patterns to achieve a rigid shear wall. Usually, a 8d nail @ 6 inches o.c. on the outside boundary studs and an 8d nail @ 12 inches o.c.

in the interior studs of the plywood will suffice. The structural brace panel must be supported by at least 3 studs within the wall frame.

These braced wall panels become a part of the overall shear wall, which works in harmony with other walls to hold the building to the foundation. The connection that holds the brace wall panels to the foundation are the standard anchor bolts that are explained in the foundation design. These anchor bolts are regarded as sufficient to withstand nominal wind and seismic forces. They are at least ½ inch in diameter and are at least 10 inches long. They must be embedded in the foundation, and serve to anchor the wall frame.

Structural Floor Framing Plan

___ Layout and identification of structural support system including girders, joists, posts, beams, and decking.

___ Indicate the species of wood being used, its strength properties and grade. *Section R502.1*

___ Identify bearing wall locations. Verify that these walls match foundation plans for footing requirements.

___ Show center spacing of repetitive members such as floor joists. *Table R502.3(1)*

___ Show headers in wall frames. *Table R502.5(1)*

___ Show minimum bearing for floor joists. *Section R502.6*

___ Show that joists are prevented from rotating with blocking at points of support. *Section R502.1.2*

___ Show that joists under bearing partitions are doubled. *Section R502.4*

___ Show that joists are within span limitations of *Tables R502.3.1(1) and (2).*

___ Show that girders and beams are within span limitations of *Tables R502.5.1(1) and (2).*

___ Specify framing anchors, clips and other structural connectors for joists framing into girders. *Section R502.6.2*

Structural Roof Framing

___ Layout and identification of structural support system, including trusses, rafters, posts, beams, headers and decking.

___ Indicate the type of wood being used, its strength properties and grade. *Section R802.1*

___ Identify bearing wall locations. *Sections R601.2 and R602.2*

___ Verify that these walls match foundation plans for footing requirements. *Sections R601.2 and R602.2*

___ Show center spacing of repetitive members such as rafters. *Sections R802.5.1(1) through (7)*

___ Show that rafters are prevented from rotating with blocking at points of support. *Section R802.1.1*

___ Show that rafters are within span limitations of *Tables R802.5.1(1) through (8).*

___ Show that girders and beams are within span limitations of *Tables R502.5.1(1) and (2).*

Structural Wall Framing

___ Indicate the type of wood being used, its strength properties and grade. *Section R802.1 and 802.8*

___ Wall bracing must be indicated for exterior bearing and nonbearing walls. *Section R602.10*

___ Indicate type and location of wall bracing is according to *Section R602.10.*

___ Show that shear wall panels are provided at every corner and every 25 foot interval. *Section R602.10.1*

___ Specify on the plans that any trusses will be engineered and factory assembled. *Section R802.10.2*

___ Show stud size and spacing and wall height. *Table R602.3(5)*

___ Show shear walls and nailing schedule for walls designed to resist wind. *Table R602.3(1)*

___ Show required blocking that prevents joists or trusses from rotating at points of support. *Section R802.8*

___ Show required fireblocking. *Section R602.8*

Elevation Views

After your house is completed, if you stood in front of all four sides of your home and took photographs and then made line-drawings of the pictures, those would be elevation views. You must decide exactly how the exterior walls, roof structure, windows, doors and other features will appear. Sometimes, this drawing takes on the appearance of an artistic view. These views will show critical detail, which will help clarify your goal.

Elevation Views (All Sides)

___ Show view from grade to sky of all sides of proposed structure.

___ Show windows, doors, roof lines, chimney and all other features visible from a side view.

___ Show dimensions for height, width and length of the overall structure.

___ Show the material and the type of exterior finish proposed. *Section R703.3*

___ Show the required lateral bracing (shear walls). *Section R602.10*

___ Show all structures such as porches, balconies, roof overhangs, and parapets.

___ Show roof covering material such as shingles and roll roofing. *Chapter 9*

___ Note and specify all roof slopes that are dependent upon roofing material type. *Chapter 9*

___ Note and size all roof drains or scuppers on flat roofed areas. *Section R903.4*

___ Note and dimension that the fireplace chimney must terminate at least two feet above roof. *Section R1001.6*

___ Roof vent area must be 1/150 of roof area. Note location of all vents. *Section 807*

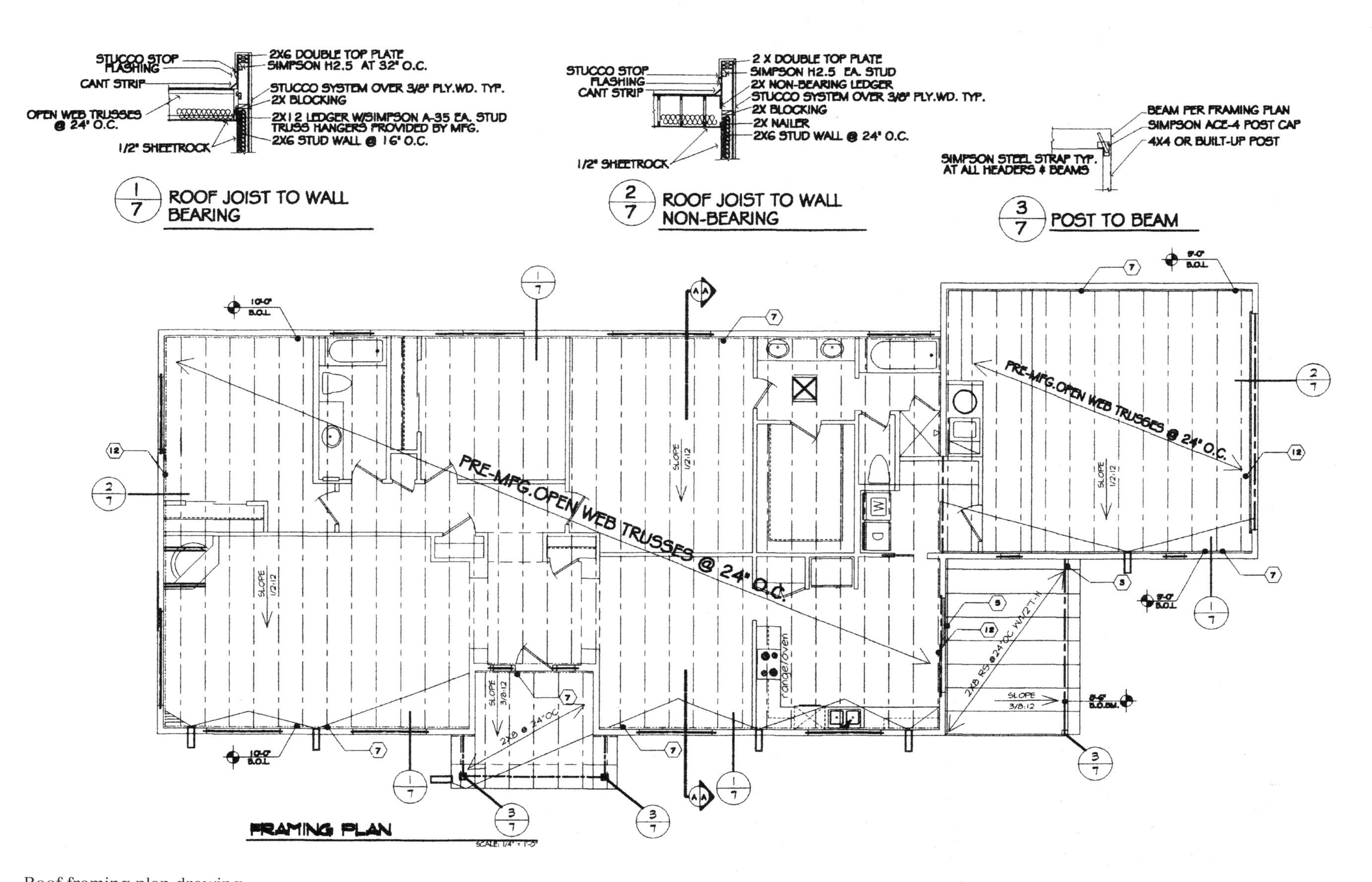

Roof framing plan drawing.

___ Note that glass in hazardous areas, within 24 inches of door or 18 inches of the floor, is safety glass. *Section R308*

___ For veneer, note and specify type. *Section R703.6*

___ Provide detail and specify anchoring method for any wall veneer, backing and vapor barrier. *Section R703.7.2*

___ For thin-coat stucco systems over foam board, note and specify the system. *Section R703.6*

___ Note and specify the vapor barrier under stucco and over siding. *Section R703.2*

Cross Sectional Views and Details

Imagine that you could slice your prospective home with a giant knife. Now position yourself in front of the slice and take a picture much like you did for the elevation view. The specific detail that you could see would tell how the building is built. That is the purpose of the cross sectional view. Cross sectional views include details of connections as well. Imagine that you could expand your view to a portion of a critical joint. You could see exactly how that joint was built. This is exactly the purpose of the cross sectional detail. There are some specific concepts you can demonstrate in a drawing such as this. These are enumerated below.

Note that the cross sectional views include section cuts. These section cuts reveal more detail of a particular connection between materials. These details are portrayed in an exploded view, which reveal methods of connections and joints.

Cross Sectional View and Details

___ Show wall section with details of construction such as treated bottom and double top plate. *Chapter 6*

___ Show wall sheathing, interior drywall, insulation, connection of wall to rafter or truss. Section *R702.3, 802.11*

___ Draft stops and fire blocking in walls to prevent smoke and fire in walls from entering attic. *Section R602.8*

___ Attic ventilation should be an opening that is at least 1/150 of roof area. *Section R807*

___ Foundation ventilation should be an opening that is at least 1/150 of floor area. *Section R408.2*

___ Attic access should be 22 inches by 30 inches for roofs with at least 30 inches of clear height. *Section R806*

___ Foundation crawl space access of 18 inches by 24 inches should be provided for areas under floor. *Section R408.3*

___ Structural wood located closer than 18 inches to the ground should be treated to resist decay. *Section R323.1*

___ Show roof pitch and roof covering material. *Chapter 9*

___ Show roof drains and overflows. *Section R903.4*

___ Show grades adjacent to exterior walls. *Section R404.1.6*

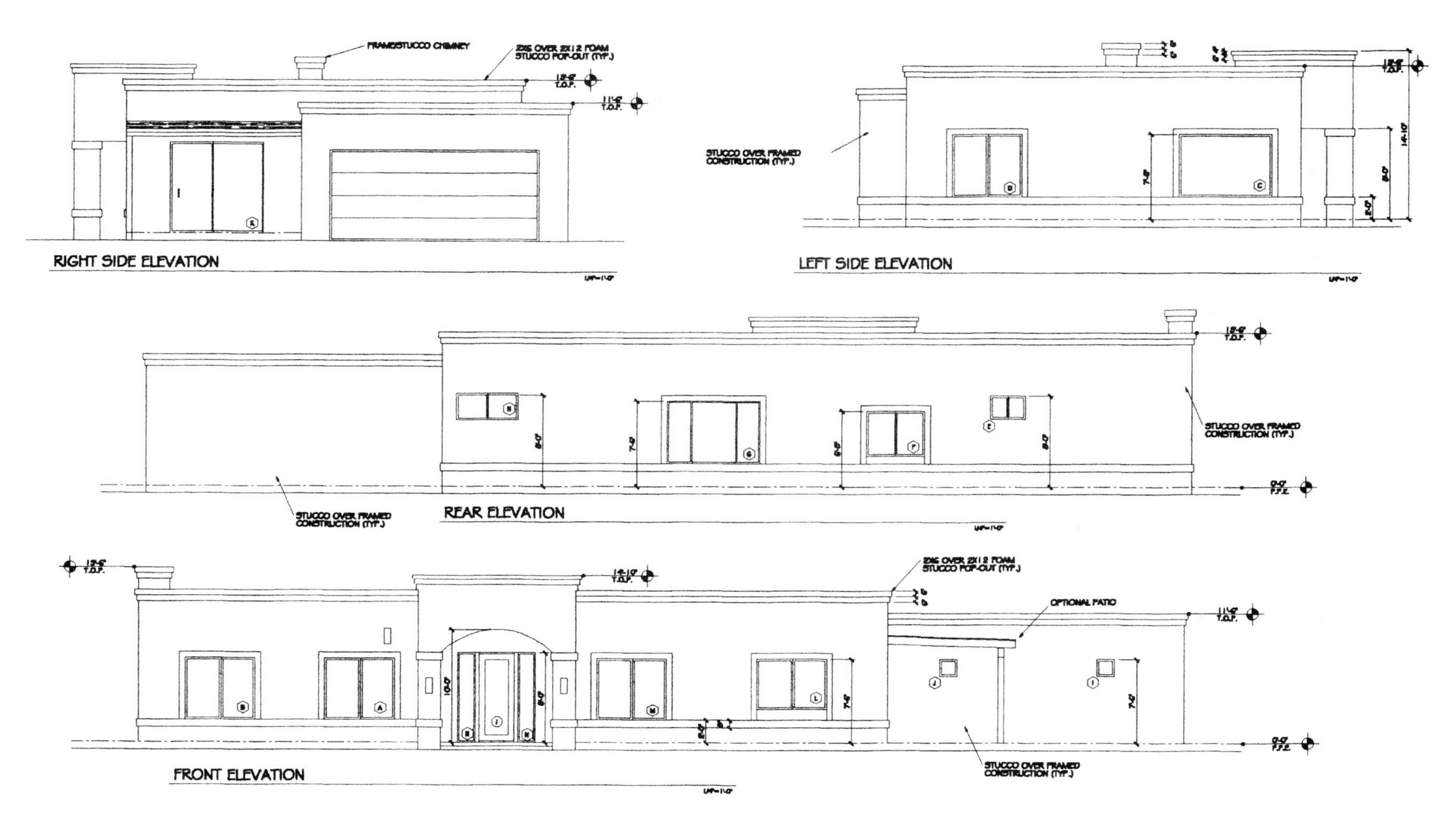

Typical elevation views drawing.

___ Show type of anchorage of veneers to wall frame. *Section R703.7*

___ Show details of moisture barrier and weather separation and flashing around planter boxes. *Section R703.2*

___ Cross sections shall be cross-referenced to the floor plan and framing plans.

___ Insulation required for energy code compliance. *Table N1101.2*

___ Show termite protection. *Section R324*

___ Show flood resistant construction. *Section R327*

___ Show interior drywall attachment method. *Section R702.3*

___ Show treated plate in wall framing. *Section R602.3.4*

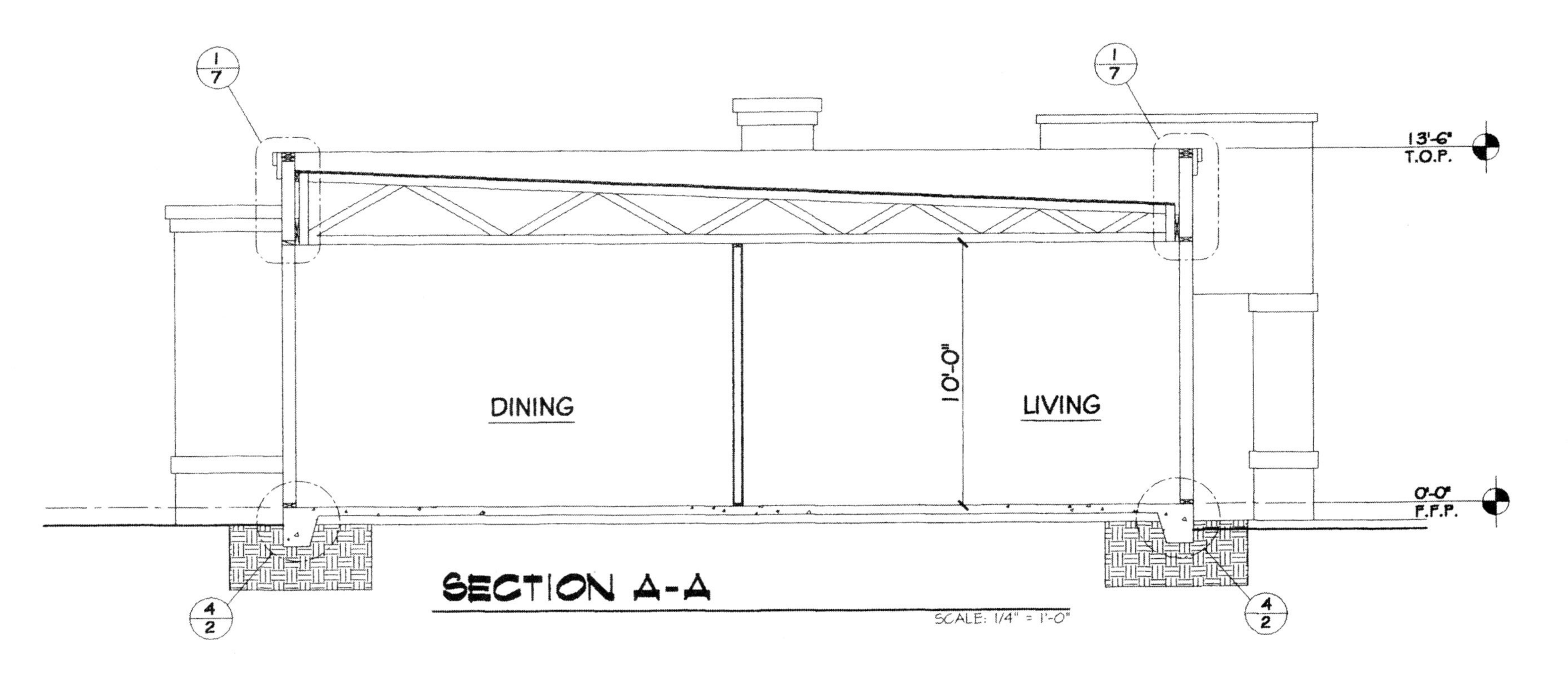

Typical cross sectional view drawing.

Part Two

BUILDING AND CONSTRUCTION

The Construction Process

Okay, you've done lots of planning. After months of brainstorming and compromise, your family has settled on a design. You have drawn the plans and gone through the building permit process, survived and acquired a permit. You have done cost estimates and know what your dream home will cost. You have determined an estimated time schedule with which you can live. Guess what? Now you get to go to work!

You are about to enter what could be the most rewarding experience of your life. It is also likely to be the most demanding, exhaustive effort you will ever face. Even if you have worked in construction before, this will be a unique event. This challenging opportunity you are about to embark upon will be personal. Some aspects of the construction process will fly by, and others will seem to be interminable. You will face daily decisions, which will both influence and be influenced by your character. "Should I tear that wall section down and start over again?" or "How can I fix this mistake?" will cross your thought processes, if not your lips. You will be faced with many choices. "Should I just live with the room being 12 inches shorter, since I measured wrong, or start over?" You will consider inviting friends to help and then question their advice. It's all part of being close to the job both physically and emotionally.

Learn to listen to experienced, qualified advice. Use all of the safety tips outlined earlier. Be prepared for accidents and injury. Keep telephone numbers of an emergency nature handy. Keep an eye on your kids and those of the neighborhood. You would never enjoy a dream home where a child was injured due to an accident. Be prepared for the unexpected event. Think about the vagaries of nature and human interactions which can wreck havoc on your work. Protect your investment, even during the construction stage.

Develop relationships with those around you. Get to know your neighbors and become one of them. Help an elderly family in need. Baby-sit for a young family with a special need for help. Assist in community affairs. Volunteer for the local fire department or public school system. Get to know your mailman and paperboy on a first name basis. Bake cookies for them on Christmas, Hanukkah, or New Years.

The construction of your home will be a positive event if you set a positive tone for your family to follow. Be optimistic! Look forward to the future. Always work to remind each other how good it will be to have a completed home, paid for. Living on-site will be an endurance test for your family. It is not an easy thing to do. You will cook three meals every day for your family in a camping oven in a travel trailer. You will wash clothes and iron on a porch that you would not normally consider adequate for pets. You will have to store many of your personal belongings away from your living site during the construction process. Since a travel trailer is a bit flimsy, nature will seem too close during every season. Try to focus on solutions to problems, instead of each other. Collectively make fun of the trailer, instead of blaming the decision to buy it on one or the other. When the kids get sick of living in a can, take them out for a pizza and ice cream.

Winners in every field have one thing in common: they never give up. Keep your eye on the prize. Remember the ones for whom you are doing this. Your family will remember your manner and attitude long after they have forgotten your achievements. Be an inspirational leader. Change attitudes for the better. You selected this path a few months ago because of a craving for independence for you and your family. Your family is counting on you. Believe me, they want you to succeed. Some of the younger members will whine and cry from time to time. You and your mate may even feel the same from time to time. They are just verbally expressing their short-term feelings and frustrations. Your job is keep them focused on the long-term objective.

YOU CAN DO IT!

Chapter 4

SOIL EXCAVATION AND SITE DEVELOPMENT

Building Site Layout

When you have settled on a site and are satisfied that the soil conditions are adequate, it is time to begin site development. Decide where on the property the house should be located. High ground is desirable in wet locations, and good cover is desirable in high wind regions. Even aspects of street access or view may be important considerations. After you know approximately where the house will set, verify that it meets all of the legal conditions through careful measurement.

If your property has not had a recent survey, or you are unsure where the property corners are located, you should hire a licensed land surveyor and pay for a survey to locate property corners. This will cost a few hundred dollars but will be worth it to know that you are building on the right location. The surveyor will provide you with a legal document indicating the shape of your lot, lineal dimensions and angles of corners. In addition, he (or she) will install stakes to mark property corners. For a few dollars more, they will mark the proposed building site. This will save you from doing that task.

Before you begin any excavation, you must be sure that you're in the proper location. The layout for the building is very important because mistakes in measurements will repeat themselves throughout the construction process. If you are too close to the property line, your entire project may be subject to abatement by the local jurisdiction. This will cost you in fines and delays, and in the extreme circumstance, will require you to relocate your building.

If you chose not to acquire a survey, rent a building level with a tripod and establish the exact property line. The builder's level will not turn angles, but when set up properly, will establish an accurate line of site between property corners. You begin by setting up the builder's level on top of one of the property corners. Using a plumb bob suspended from the center of the level, wiggle the tripod directly above the property corner using the plumb bob as a guide. Then, using the four leveling thumb screws, adjust the instrument so that it is level with the earth. Check the bubble located within the instrument to verify an accurate level. Next, using line of site with someone on the other end holding a pole over the other property corners, establish the true property line between the two corners. Mark strategic points with wood stakes to identify the property line. Continue until all property lines are determined and identified.

After carefully locating property corners, tie a string line to establish the front line of property. Then, carefully measure from corners along front lot line and then at right angles to a distance at least equal to the building setback. You may want your house to set back further than required. This is your first house corner. Do the same thing from the other corners and lot lines and verify layout. Mark the building layout with spray paint or other marking technique.

Tree or Stump Removal

If trees or large bushes need to be removed, do so prior to excavation. You can rent or buy chain saws and do this yourself. Some timber is marketable even while standing. It might save you the labor to have them removed. If the tree is large, you may want to consult professionals before you try removal. Removing the debris away from the building site will be less complicated if you have a destination in mind before you begin removal. Otherwise, you will move it twice. Try to find something useful to do with the material. If you have a fireplace or wood stove, you may even cut it into firewood. If left in a pile, rodents, and other creatures will assume that you are building them a home too, so store firewood well away from your future home.

Fill Material

Depending upon the desired or required height of the building pad, your site may need to be raised above the prevailing terrain. In this case, added earth must be brought in to the site to elevate the building pad. This earth is normally designed for the site by a registered engineer. The fill must be of

Fill material delivered by dump truck.

good quality without deleterious matter such as roots or large boulders. The fill material is brought in by dump trucks and deposited in mounds. The material is then compacted with tractors. A field compaction test is required for fill material. This test is performed by a registered engineer. The test will verify that the soil has been compacted sufficiently to support your home.

Excavation

Until now, everything you have done has been based on conception and planning. If your building site is high enough to satisfy the conditions of water runoff, you will not need fill material. However, you may need to level the site to provide a flat building pad. A building pad or level site must be created prior to foundation preparation. Clearing and grubbing the building site will transform your raw land into your vision of a home. It will begin to look like a construction site. Remember that a separate permit for excavation may be required in addition to a building permit. Get the permit and post it at the site before work begins. Also, remember that any heavy equipment that you rent or borrow, may need a moving permit.

Arrange street access fairly soon to allow for your travel, as well as supply and construction vehicles and inspectors to enter your site. Verify if street-cut permits are required and that street drainage is not adversely affected. Be sure to post a large sign with your address clearly marked near the street access.

Tractor excavating earth.

Bearing Capacity

Soil bearing capacity can be altered by chemical admixtures to improve the soil's strength or cohesiveness. Excavation of entire sub-base and in-place compaction of sub-base, in six inch lifts, may be a way to remedy a poor soil stability. A solution may be to build a footing into the soil deeper than a normal footing depth of 12 inches and build a foundation wall to the finish floor level. There are many fixes available to compensate for a poor soil bearing capacity. Additionally, bearing capacity is adversely affected by the freeze-thaw cycle of soil, which extends below grade to a distance known as the frost depth. Therefore, a footing must extend below this depth to prevent heaving action, which may cause significant damage to the building. Check with your local building department for the required depth.

Utilities

Other major improvements that you should consider at this point are the public utilities such as water service, sewage disposal (or private septic tank), electrical service, gas service, cable television, and telephone. If you plan to live on the property while building, now is certainly the appropriate opportunity to arrange for these services. In some cases, a licensed contractor is required by the jurisdiction to perform the connection to utilities such as a sewer tap or electrical service equipment connection. However, there may be some work which you can do for these contractors. Discuss this with each utility company.

Begin by contacting each utility and requesting service. They will need some pertinent information and copies of your deed and building plans. They will probably need to have signatures from all of the owners on applications. All the owners of the land may need to agree in writing to grant access for utilities. Therefore, plan ahead and avoid delays. Sewer connection is authorized by the local purveyor, usually a city or county government. Depending upon the height of the sewer tap, a basement with plumbing may need to have a sump pump or similar sewage ejector. A backwater valve may be necessary if your plumbing fixtures are below the next manhole cover of the sewer. Check on the height of the sewer line invert elevations before you design the house. There will be a sewer connection fee that may seem excessive. This is to defray the costs of providing the service to your area. In any case, you will need a permit and inspections to connect to the sewer.

If no sewer is available, you may be permitted to install a private sewage disposal system, such as a septic tank and leaching field. Septic tanks are allowed on tracts of land large enough to disperse sewage effluent from a single home. Usually land as small as a one acre tract is large enough to accommodate a private sewage system such as a septic tank. The principle is simple. Sewage is held in a holding tank where microbial bacteria decompose the sewage and cause chemical action that purifies the waste water. The remaining sanitary liquid enters the second stage of the tank and is expelled into one or more leach lines. The treated water then drains through aggregate into the leach field for further purification until it reaches the water table. The easiest way to participate in this project yourself is to offer to dig the holes but to purchase the tank from a qualified dealer. A septic tank design will be discussed in a later chapter. If you decide to install any of the system yourself, I recommend that you read that chapter thoroughly.

Electrical service will be available from the local public service utility. When you ask for electric service, the service provider will want to know what size service is needed. Use the electrical service calculation to ensure the adequate size. They will want a list of all the equipment in your house, including water heater, dryer, ranges and ovens, HVAC, welder, as well as pumps and other motors. They will want to review your plot plan with scaled distances of the desired route to your home site. The cost to install electric service to your site will be charged to you unless you can convince the utility service provider that your future energy consumption will offset their expense in building your service.

In addition, they will want to install the electric service from the street to your site. Some utility companies will install services underground upon request. If your electric service provider allows this option, take it! You can hide those unsightly power lines. You may be able to use the same trench to carry water and telephone service. You just have to place your service entrance location in the same proximity as your water shutoff valve. Be sure to coordinate between all utilities affected by their mutual use of the trench. For instance, the electrical service provider may insist on being at the very bottom to avoid a calamity if the water line breaks and must be dug up and replaced.

Temporary On-Site Power

If you plan to live on the property in a mobile home or RV, tell the utility of your plans. You can either use the primary service equipment or a temporary electrical pedestal. Just locate the service equipment near its final location. If the service is underground, ask for an extra *loop* of service cable at the base of the pedestal to allow for shifting it around to its final location on the finished house. The temporary electrical service equipment will be required to have at least one utility outlet that is protected by a ground fault circuit interrupter (GFCI) with a weather-protected cover. In addition, if you are connected to a mobile home or RV, you will need to have a breaker sufficient to power your mobile home or RV. Usually a 50 amp circuit is adequate for mobile homes. For an RV, the power requirements will dictate the breaker size. A smaller RV may have only a 15 or 20 amp breaker; others may demand as much as 50 amps.

Typical electrical utility connection.

Water Service

Water service is delivered by either public or private water companies or a private well. Water supply lines may already be installed to your property. If not, the cost to deliver water to your site may be your responsibility. If a water connection is available, the cost will be minimal; usually just the cost of a water meter. Sometimes a water company may offset the expense of extending the street water main to your property through your monthly bill or even a one-time charge.

You will be responsible for routing water service to your building site. A trench at least two feet deep is recommended. Connect to the meter and use PVC with proper solvent and glue. Extend the line to where your homesite will be and transition above grade with a copper water line and a shut-off valve. The pipe size was determined during the design stage.

Temporary Living Quarters

If you decide to live on the land while building your dream home, there is some general preparation that will make your life much easier. Living under these conditions is not for the weak or timid. I strongly suggest you not enter into this decision lightly. For those who think camping out is spending a weekend at a discount motel, this will be a very long construction period.

The good thing about living on the property is that your project is always nearby and convenient. The bad thing about living on the property is that your project is always nearby and convenient. If you're the kind of person who has trouble knowing when to take a break, you could be very frustrated about always confronting an incomplete project. However there are many reasons for living on the project site.

You save on a mortgage payment that includes interest. Also, many taxing authorities reduce the tax rate for land if it is inhabited, even by a travel trailer. Living near a site allows for you to do more work and still be with your family. Better yet, your family can participate with you in a shared dream. You can establish family roots in the area quicker. Children can adjust to school better, and at a younger age.

You can control the safety and security of your project better if you live on-site. Your neighbors are more likely to drop by and offer their help if your family is present. You can be available to your neighbors to return help if you're around when they need you, thereby helping to create a real community. You can grow a garden on-site and help out with grocery expenses.

This approach toward building, by paying as you go, can be erratic. For instance, you may save all month for $1,000 to order concrete to pour in one weekend; then wait another month for another $1,000 to buy wood. You can use the other weekends to perform supplementary jobs such as pulling weeds, erecting a fence, moving dirt, landscaping, planting a garden or similar tasks, which will cost very little, but be labor intensive.

For those of you who are willing to endure this brief Spartan lifestyle, here are some tips that will be useful. Whether you plan to live in a mobile home or RV, you will need basic utilities, including electricity, water, and sewer facilities.

Other desirable facilities will include a washing machine, clothesline and a storage area for supplies. Get a big dog with a loud bark for security. Start mail service with a mailbox. Get a telephone and television. Build a small weather porch to expand your living area. Create an outdoor game or sports area for activities such as volleyball or basketball and other things your family can do for free.

There are other things to consider when you decide to live on a building site. Follow the checklists below using your choice of temporary site quarters. There may be some concerns unique to where you live. Check with others who have done this in your area. Remember, ask questions, listen and learn!

Garage as a Dwelling

Consider building your home in stages. If you design a dwelling with a garage, you could build the garage first, move in and finish the dwelling later. Your garage design could include a bathroom and a kitchen sink to serve sanitation needs. Living in the garage gives you a sense of fulfillment toward achieving your goals. If your local building department will permit this option, you may have to agree to revert the temporary dwelling area back to the garage after the dwelling is complete.

Guest House

Consider building a very small guest house or ancillary out-building such as a bunkhouse or pool house in which to live while building your main house. A small house such as this will come in handy for out of town guests when your main home is complete. Also, when your children grow up and want to move away, the transition is less traumatic if they're within 30 feet for the first few years.

Mobile Home

If your family is too large, or these options seem too sparse, and if your zoning jurisdiction will approve, consider moving a mobile home onto your site while you build your home. An older mobile home will provide more of a real home feeling during the construction. Most zoning jurisdictions will limit the time period that a mobile home may be installed on a zone approved for site-built houses.

Recreational Vehicle

Living in an RV, even a large one, is an adventure. As my wife put it, so is going over Niagara Falls in a barrel. Memory has a way of tempering objectionable times, except for some! It'll take some time for a family with grown kids (over 10 years old), to adjust to living in a 200 square foot room. Just begin by acknowledging that there will be less privacy. You'll learn that togetherness is what really makes a family!

Checklist for Living On-Site in an RV

___ Level site for an RV

___ Level site for future porch off of the RV (extended living area)

___ Sewage disposal and plumbing drain line for house so that RV can temporarily connect easily

___ Allow for washing machine connection inside the porch

___ Position RV to maximize protection from wind, dust, and rain

___ Position RV away from trucks and excavation vehicles

___ Position RV to ease in its removal when that times comes

___ Position RV within the length of its cord to electrical service

___ Block RV to stabilize and prevent movement

___ Use pallets as floor for porch. Find broken, discarded pallets for this use from building material stores

___ If RV has a built-in awning, use it

___ Otherwise erect a small temporary porch

___ Check for an electrical outlet matching RV cord and plug

___ Tie down RV with anchor strapping material

___ If unit has a gas-fired water heater, park RV out of air drafts to avoid pilot light blowouts

___ Keep RV's wheels off ground to avoid decay

___ Do not bury RV electrical cords (unless they are rated to be underground)

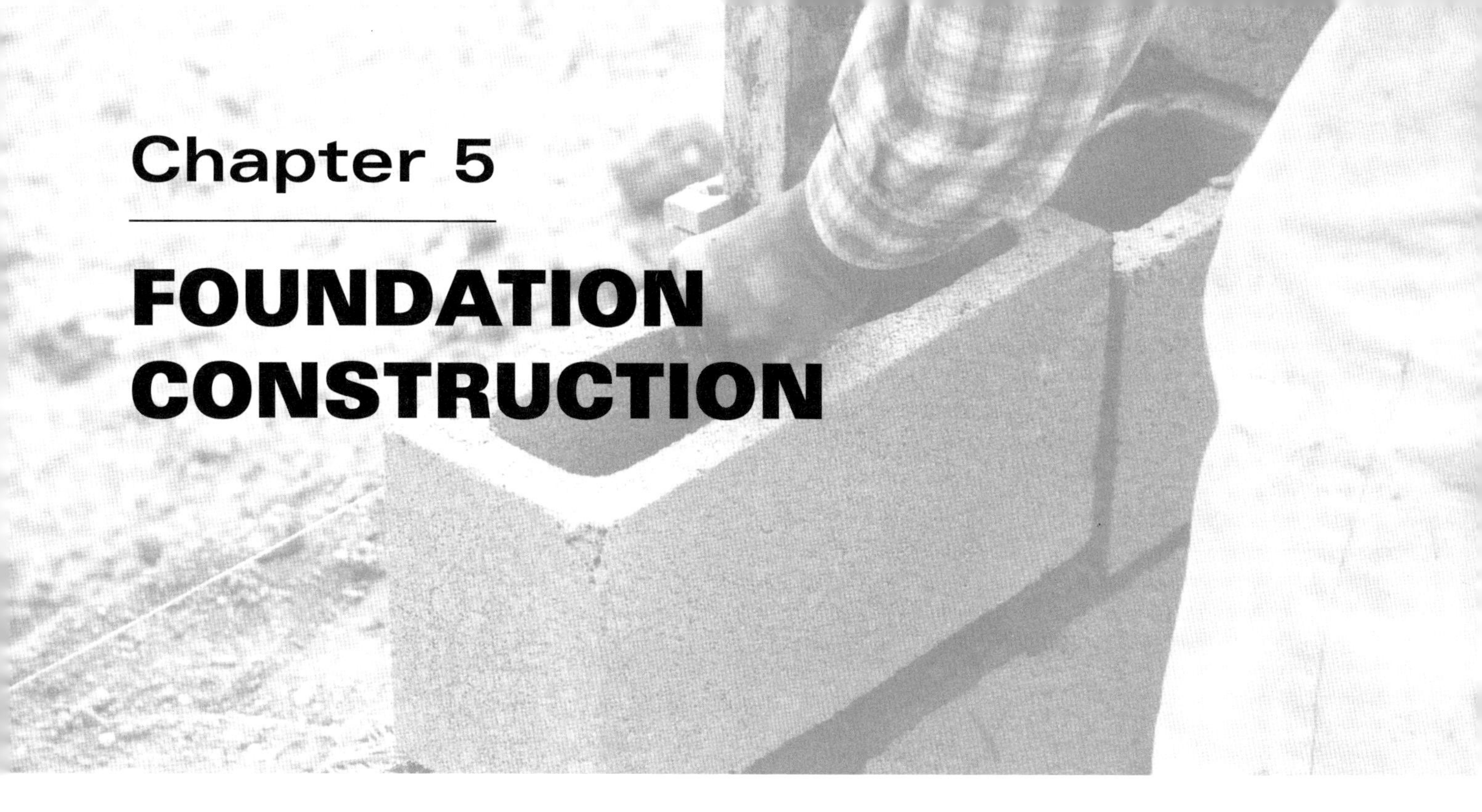

Chapter 5

FOUNDATION CONSTRUCTION

Setup

With the excavation complete, the building foundation is the next step. The foundation supports the load of the entire house and transfers it into the earth. Additionally, it stabilizes the house from moving laterally, due to wind and earthquake forces. That's why it is important for stability that the footing be cast into and against native soil, or at least firmly compacted fill. Before you dig the footing trenches according to your design depth and width, you must first define the exact corners and boundaries of your house.

Locating the House on Your Site

While understanding how to position your house on your lot is not complex, a successful outcome will depend upon the precision and care in measurements. You have decided, on paper, where your house will sit. It has been the result of decision making by numerous people. Zoning code and building code requirements affect your site selection. Your family knows what they want. Maybe even a subdivision architectural committee had to approve your proposal. Your work here will ensure that all of these parties will be satisfied. If you make even the slightest mistake, it will haunt your project forever. That is why most builders will use a licensed, professional land surveyor to install building corners. Consult with a surveyor and consider the costs and value. Perhaps, your participation will bring the cost down a bit.

Clearing and Grubbing

Work continues by extracting native vegetation, including roots and branches, which must be removed from the site of your new home. Rental of a tractor may be the best method to do this, especially if the vegetation is large or has deep root systems. Before you begin any excavation, post your permit and verify that there are no buried wires or pipes that you might disturb. There are agencies that do this at your request. They will mark the surface of the ground where surprises are buried below. Use them to keep you out of trouble. Also, be careful not to disturb any stakes that the surveyor has placed to define the building corner.

The best time for excavation is daylight, and the best weather conditions are dry and cool. If you plan to perform concrete or masonry work soon after excavation, do so when temperatures are above freezing. Excavations less than two feet can reasonably be done by hand. For deeper work use heavy equipment. If you plan a basement, renting a backhoe or renting a bulldozer and digging a hole is the next step. While a bulldozer would be a good choice for a full basement excavation, a smaller backhoe is more accurate and appropriate for single story foundations with rugged soil conditions. A hand shovel is adequate for shallow footings in softer soil.

You have already removed all trees, stumps, roots, branches, underbrush and large boulders that would have prevented you from establishing a clean layout. You might find it necessary to rent a roll-off dumpster to remove the debris. Do not over-excavate the soil at and near your building site. Unnecessary disturbance of the building pad may cause problems. Your footing must extend into natural, undisturbed soil. If you disturb the soil, a deeper footing may be required.

Pile surplus dirt far enough away, but near the excavation if it will be used as backfill before the project is complete. Earth that has been excavated is subject to sudden, unpredictable shifting and can bury whatever is under it in an instant. Rain or wind can set off a landslide. Do not let kids play on this dirt pile. Be sure to mark any excavation with safety tape to prevent accidental falls.

Batter Boards

At this point, you have roughly located the building corners with stakes. Since digging the footing will inevitably disturb those stakes that serve as the building corners, you must develop a method of maintaining the corner marker. Batter boards are used to allow for precise corner-control without risking the loss of corner markers. Batter boards are simply a series of temporary frames, much like a short fence corner, used at each building corner to adjust and maintain the exact building corner placement. The batter boards are erected in each direction behind the anticipated corner. They must be out of the way of all construction work and placed very securely into the ground. They can be discarded after the foundation wall is complete.

Batter boards serve to establish and maintain exact corners, but they have another purpose as well. Batter boards also serve to establish the height of the foundation wall. Begin by selecting the highest building corner among the batter boards. Usually the finish floor elevation is around 12 inches above the highest grade, even though most code requirements set the minimum at 6 inches. Install the cross member on the first batter board at this height. This will serve as the base for finish elevation height. Then adjust the other batter boards to match that elevation height. You will need to use a builder's level to establish this elevation height.

With the elevation height established, locate the first building corner by driving a small stake or nail into the ground at the corner. From this stake or nail, using string line across the batter boards and measuring tape, you can establish the remaining corners with pinpoint accuracy. Drive a nail into the batter board on each end of the plane of one building line to begin precise layout. Use a string line tied to each nail to define the building line. Using triangulation methods and careful measurements, establish the third corner and add another nail and string line. Continue to the fourth corner, then continue to the last corner, and finally, return to the first corner. Your string line should be close enough to the ground to use spray paint to *draw* a mark of the building line onto the ground. When you have established the exact corners, mark the nail very well—and then never disturb it again. These nails represent the control for your corner.

Of course, before you have arrived at this point, you have used careful measurement techniques and trigonometry to verify the *square* of your building. You may have adjusted the nails several times to locate the precise location of the building corner. Follow the description below for batter board erection.

Example:

Step 1 With approximate corners marked, site from behind first corner and build board about 6 to 8 feet back.
Step 2 Site adjacent wall and build second batter board the same distance away from corner.
Step 3 Repeat for other three corners with two batter boards per building corner.

The next step is to verify that the layout is indeed square. Do this by using your high school trigonometry. Remember that the Pythagorean Theorem stated, in part, that the square of the hypotenuse is equal to the sums of the squares of the two legs of a right triangle. Here is a brief refresher course.

Given a rectangle (house), 30 feet by 40 feet. If you connect the four corners with lines, you see a box with an X inside. How do you know what the length of the legs of that X are supposed to be? The reason you need to know is that they need to be equal, to create a perfect shape. Here is how to verify:

$(30^2) + (40^2) = 2500$
The square root of 2500 = 50
X legs should be 50 feet long
If legs are 49 feet 11.5 inches and 50 feet 1.2 inches, you are out of square, so try again.
This is known as the 3 x 4 x 5 triangle method

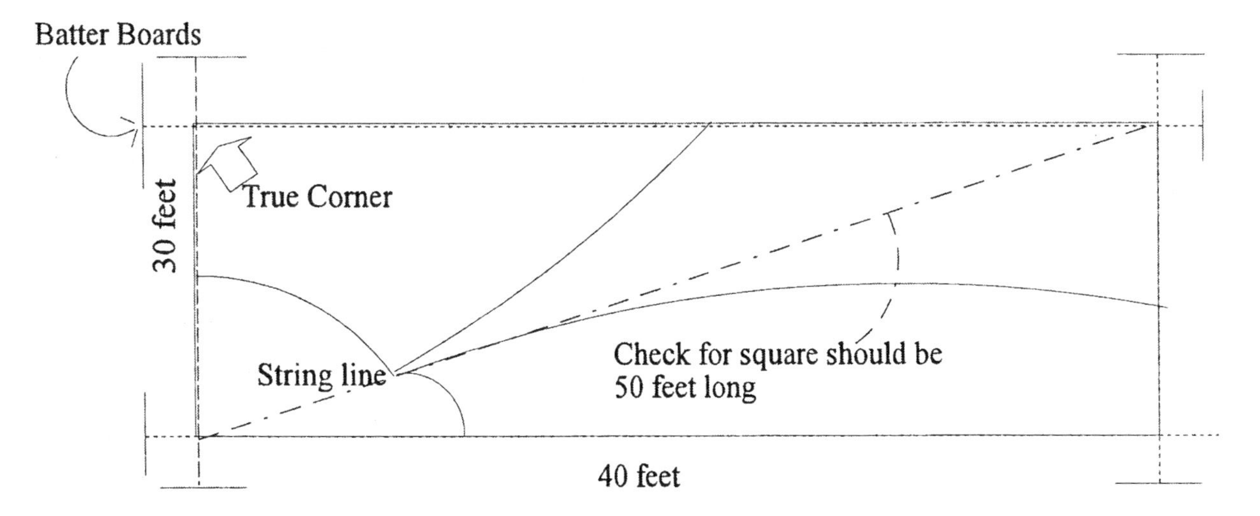

Batter board installation.

Utility Trenches

No matter what kind of footing/foundation system you intend, certain utilities such as plumbing waste and water supply pipe will enter the building from under the finish floor. The location and height of the plumbing waste pipe is dependent upon the distance the sewage pipe must travel to either a septic tank or sewer connection and its relative elevation to the house connection. There is a minimum slope that drainage pipe must have, in order to effectively carry sewage away from your home. The pipe must slope at least ¼ inch per running foot. This may affect the entry height of the drain pipe. The location of these utilities must be identified so that when excavation begins, they are accommodated. Mark the location on the ground where they must enter.

Footing and Foundation Choices

The foundation type will depend upon several things: the most common are a monolithic slab foundation, combination footing/block wall foundation, and combination footing/concrete wall foundation. Each of these has distinct advantages. Some methods are predominant and popular in a specific region. Available experienced help may be a factor in your choice. Some are more labor intensive. Others may be easier to install, but more time consuming. Some may be more useful to your particular design. You must research and determine the best type for you.

A monolithic slab is desirable due to its excellent strength. This type of foundation system is a single pour of concrete for both the footing, foundation wall and the floor slab. The benefits of this type of foundation include strength in a monolithic material and time savings for completion of the project. However, it will require lots of experienced help for most of an entire day. Additionally, weather may be a factor. Depending upon soil and weather conditions, this may not be cost effective.

A concrete footing/foundation system is a multiple component design. First, a footing is poured in a trench. Then either a block or concrete wall foundation wall is erected above the footing. The advantages to this type of foundation system include ease of installation and staged construction most commonly associated with a do-it-yourselfer. It is easier for the beginner to complete by themselves. The different stages of work can be separated in time. However, a drawback to the block foundation design is that it may require block laying skills, which you will either have to learn or hire. Another advantage that both of these types of designs have is that they allow for the possibility of creating a basement with minimal difficulty.

While most foundation walls are less than four feet in height, an extra four feet deeper will create an eight foot high basement wall that will double your square footage. A poured reinforced concrete wall foundation will provide significant structural integrity but may require expensive forms that are cost prohibitive to buy for a one-time pour. You may be able to either borrow or rent these forms if you choose this design.

Insulated Concrete Forms (ICFs) are manufactured, engineered products, which serve as forms for poured reinforced, concrete walls. The advantage to these forms is that there is substantial engineered design and manufacturing control behind the product, which is easy for a novice to install. They solve lots of problems inherent to masonry or concrete construction, including structural design, ease of assembly, energy efficiency, and cosmetic appearance. There are dozens of different product manufacturers for ICFs nationwide. They are becoming popular to those who want a solid masonry or concrete house that will meet energy code requirements. Since there are different product manufacturers, I recommend that you research the concept and determine if it is right for your project.

Footing Design

Footing design width and thickness is based on the load of the structure and the soil base on which it is resting. The code prescribes the minimum width, thickness and depth below native soil based on number of floors supported. Standard footing sizes are listed in Chapter 3. A specific design may require that a footing be wider or thicker. A cause for this might be heavier than normal wall section, such as a solid grouted double wythe block wall, or building on a weak, collapsible or expansive soil. Generally, the following footing shapes will be adequate for the conventional structure with solid soil.

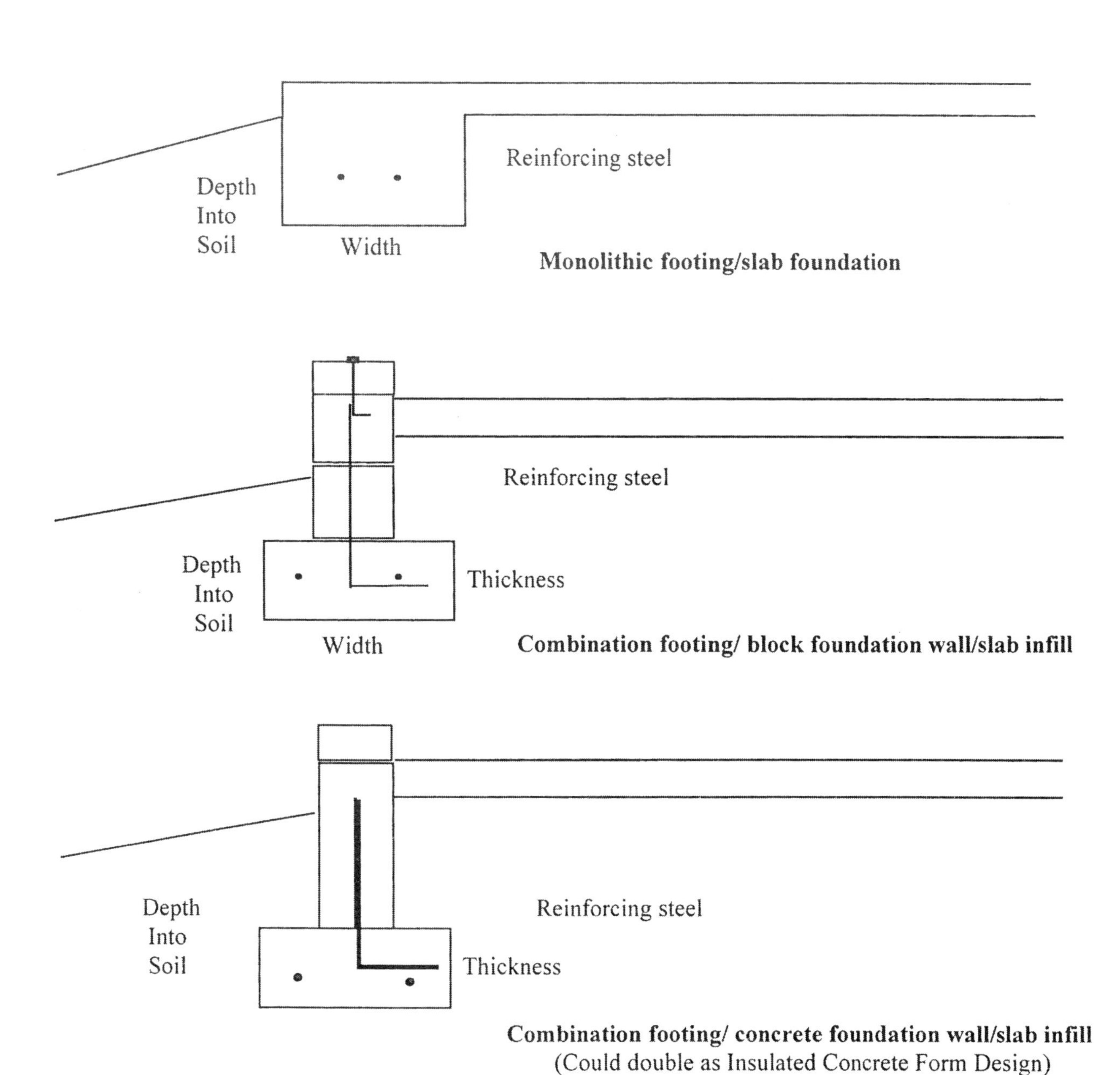

Footing diagrams.

Footings are located not only around the perimeter of the building. Sometimes an interior footing is required to be installed to support an intermediate bearing wall. These interior footings must abide by the same requirements, including required width, depth into natural soil (below frost depth), steel reinforcement and thickness. The interior bearing walls supported by these footings support more load than do the exterior walls. Therefore, they may need to be even wider or deeper.

Reinforcing Steel

Deformed steel bars in a concrete footing allow the concrete to develop more strength. This steel adds significant strength and value for very little expense. If some jurisdictions allow you to exempt reinforcing, you might be tempted to save costs by not installing it. DON'T! Reinforcing steel will cost you a little over $100 for a small house. It will be worth it, especially if you never notice a problem. Believe me. Follow the specifications determined in Chapter 3 for any required steel and placement methods. Additionally, at this point, the electrical ground wire should connect to a 20 foot length of reinforcing steel to ensure an adequate means of electrical ground.

Monolithic Footing-Slab

A monolithic concrete footing-slab is one that is poured all at one time. The monolithic footing-slab has several advantages. First, consider the added strength of having one solid piece of concrete under your structure. Additionally, a single visit from concrete mixing trucks makes life less com-

Backhoe digging footings.

Steel reinforcement bars must be bent around corners.

plicated. You also can finish a major step relatively quickly. Drawbacks include the need for lots of experienced help during the pour and finishing, up-front costs, and the need to be nimble during concrete placement.

Forms are erected to create the boundary for your exterior wall and floor. They are usually built from long and very straight framing lumber. A wood plate material can be used as forms for monolithic footing/slab outside perimeter or foundation wall design if used on both sides of proposed footing. These forms can be laid on edge, directly on the ground, just along the building line, and connected to each other at butt joints and corners. Steel stakes must be installed so as to brace these forms every few feet to prevent a form board from bending or falling due to the weight of the concrete. Forms for a concrete footing/slab will be erected only after the exact location for corners are marked. The tops of the two opposite and adjacent forms must be level and stable.

The monolithic concrete footing/slab pour is very popular in some areas. However, there are some drawbacks. Because of the heat, concrete slabs will dry quickly and be ready to support a load such as a bearing wall in a matter of weeks. To achieve a strength adequate to support construction loads, a 28 day strength is usually best.

Because of the chemical nature of concrete, you should expect some small surface cracking in larger slabs. There are chemicals that purport to reduce cracking. Some work and some don't. Weather and the saturation of the chemicals can affect the results. Any additives must be approved prior to using in your concrete mixture. To improve strength during the curing process keep the slab wet for a week or two. This will help reduce cracking by delaying the drying time. Another

Digging a footing by hand for monolithic footing.

Forms for monolithic footing in place and held together with stakes.

method for preventing cracks in large slabs is to place some woven wire mesh (wwm) into the middle of the slab during the pour. The woven wire mesh is made from 10 gauge wire woven together so that strands are 6 inches apart. This helps to hold the concrete within the slab during its curing process. Lay it on the ground beforehand, then pour the concrete mix. Embed the wire mesh into the middle portion of the slab by lifting it up into the concrete mix during the pour. Another method to limit cracking in a concrete slab is the control joint. Since cracking travels through the top surface of the slab, the control joint is a break in the continuity of the upper portion of the slab, which stops the meandering crack. A control joint can be installed with a fiberboard between two sections of concrete. You can look for examples of this control joint in large slabs. It looks like a straight line, dividing two sections of concrete. A trowel joint can provide some similar degree of protection from preventing a meandering crack from migrating across the slab.

In order to build a monolithic slab, it is assumed that you have already installed any underground plumbing or mechanical work and received an approval on that inspection. Rough-in plumbing installation is discussed elsewhere in the construction section. Be sure to wrap any plastic plumbing vent or drainage pipe material prior to pouring concrete. This will protect the pipe from damage due to abrasion and chemical action from concrete. Additionally, the building's electrical ground is usually connected to the reinforcing steel in the footing. This should be installed and connected prior to pouring. Connect a number four (#4) bare copper wire by means of an electrical ground clamp to one piece of ½ inch deformed reinforcing steel bar that is 20 feet long. Allow the wire to extend out of the footing in a location very close to the electrical service panel. Allow a sufficient length of wire to connect to the ground buss bar of the electrical service panel. If the service panel is to be mounted within 36 inches above the finish floor, allow 5 or 6 feet of copper ground wire to ensure an adequate length.

Picture of control joints tooled in top of slab to prevent meandering cracks.

Pad Preparation

You must begin with a level building pad. Establish exactly how flat the building pad is with a building level. The directions are simple. Level the survey instrument and take various readings around the pad in order to make a topographic map of your site. Some sections will be higher than others. Establish the average elevation as the design elevation and proceed to make cuts and fills as needed to produce a level building pad. A crude method used to determine the rudimentary contour of the site, is to run water over the pad and determine the necessary cut and fill needed. Your finish floor should be 3½ inches higher than this design elevation, so the top of the form will be the height of your finish floor.

Now you are ready to build the form for the monolithic slab. Earlier you learned to build batter boards to establish the building corner. This taut string line establishes the exact building corner (suspended from a taut string line). Use a plumb bob to find the building corner and drive a nail into the dirt directly under the string line. Repeat for each corner of the building. Draw lines in the dirt with spray paint or sprinkle powdered chalk. Along the lines, begin digging a trench the width and depth of the required design. Make an effort to keep the trench straight; double check regularly.

After completing the trench excavation around the entire perimeter of the building, dig any interior footings needed in the same manner. Measure the distance precisely from the reference corners to verify the exact location of all interior footings. These footings will hold up your interior bearing walls. Footing support is crucial, even for interior walls. When exterior trenches are completed, use very straight wood plate material and begin placing forms along the building line by driving 24 inch steel stakes into the ground against the form. The steel stakes have holes in them for driving nails through, into the wood form. Place steel stakes as often as possible to avoid a failure in the form during the pour, at least every 4 or 5 feet. Double-check the building line and the square of the building.

The earth is compacted sufficiently to prevent a heel mark from a boot.

Adjust as needed for accuracy. The corners, as established by the forms, should be adjusted to match the string line from the batter boards.

Prepare the interior of the pad by compacting the dirt with a pneumatic tamper. Apply lots of water in a fine mist spray fashion. Let the water compact the soil through repeated wettings. Remove all stones and rocks and extraneous debris larger than a few inches. The finished surface should be very level and hard. Install reinforcing steel in all footings as illustrated. I recommend using #4 (½ inch), grade 40, deformed steel bars. They usually come in 20 foot lengths. If you don't have a steel bender, bend in the middle for installing steel around the corners. Provide a ten foot lap at each corner. It's excessive, but doing so may cost less then renting a steel bender. Remember that reinforcing steel should be lapped in straight runs too. A lap of two feet is adequate. Tie the lapped steel together with tie wire. Be sure not to allow any steel to contact the ground. Accomplish this by placing masonry block pieces under the steel every few feet for support, or use plastic rebar chairs that are made for this purpose. Moisture from the ground will eventually cause rust and destroy the steel.

Additional steel can be used within the slab section to prevent shrinkage cracking. Welded wire steel mesh embedded in the slab section can be used to reinforce the slab area. Place on the ground just prior to pouring and then raise into concrete mix during the pour so that the mesh is thoroughly embedded in the wet concrete mixture of the slab. Another way of building a strong slab is to install some deformed rebar in the slab in a crosshatch pattern every 24 or 48 inches each way. As with the welded wire mesh, it must be raised into the middle section of the slab. Now you're almost ready to pour concrete.

Termite Protection

Termites can destroy a home. Protection is cheap compared to the cost of repairing any damage. Now is the time to protect your future investment. There are many ways to do this. The IRC establishes the requirement to use either pressure-treated wood or foundation Redwood to resist decay for wood used as sole plates touching concrete in wood wall frames. Chemicals are available that have been proven effective over the last several years. Some of these you can install yourself. Others must be installed by a qualified party, trained in safety procedures. In many cases these chemicals have a definitive life, and their value will diminish with time. There is a long-term protection method to renew the application of poison safely. During your backfill procedure, and near the top of the wall but still below grade, install a continuous perforated pipe all around the perimeter of your home with one or two ends that extend through the finish grade. The size of the pipe could be just larger than an inch in diameter to allow sufficient flow. The pipe will be used later to add a fresh supply of liquefied poison to the foundation area underground every so often.

Concrete Preparations

Before you order from the ready-mix plant, be sure that you have all the tools you need for placement and finish work. Also be sure you have enough qualified and experienced help. A slab will usually require more than a couple of neighbors. Be sure that at least one person has had extensive experience with finishing concrete. Ask them to supervise the pour. Tools will include shovels, hoes, jitterbugs, trowels, water supply, and water hose. A power trowel will be useful if someone is experienced using one (dangerous otherwise). Concrete delivery trucks are usually big enough to bring all the concrete you need during a single load. Arrange your concrete deliveries to be staggered apart to allow for time to work the wet concrete delivered from the previous truck.

Concrete Strength

Concrete is a blended mixture of Portland cement, sand, gravel and water in precise proportions. Concrete derives its strength based on these proportions. The code establishes certain minimum strengths required for various uses of concrete and its weathering potential. These are specified in Chapter 3. Footings and foundations, including slab on grade, should be at least 2500 pounds per square inch (psi). If you specify 3000 psi concrete, you will have a better strength concrete that will exceed the minimum requirements at very little extra cost. Discuss this with your concrete supplier. Water mixture affects a concrete mixture. More water may improve the workability of a concrete mixture, but it decreases the strength. So don't be too hasty to add lots of water in the mixer just before the pour.

Concrete Finishing

Shovels and hoes are used to move the fresh concrete into the general area. A jitterbug is a device that drives aggregate down below the very top surface of the finish floor. In some cases, too much jitterbug is not a good thing. The aggregate gives strength to the concrete mixture. If the entire top half of the slab has no gravel, its strength will be diminished. A concrete float is used to create a semi-smooth finish, and concrete trowels are used to create a more smooth surface. Use the trowel to push down the sand granules and bring up the cement, laden with moisture, to create a smooth hard finish. It takes lots of elbow grease. Tooling and finishing the slab is the most important job. This will be your finish floor. You will want it smooth and free of rough surfaces. Rely on an experienced concrete finisher to develop a smooth finish.

A jitterbug will help create a smoother finish by embedding aggregate under top surface.

Anchor Bolts

Before concrete sets up and becomes firm, anchor bolts must be placed into position. The purpose of the anchor bolts is to secure or "anchor" the wall frame to the footing and foundation. The most common size is ½ inch diameter bolt, 10 inches long. Each bolt must be embedded at least 7 inches into concrete. This allows the bolt to project 3 inches out of the floor or foundation. This allows for a 1½inch plate thickness and 1½ inches for threading of washer and nut.

Unless required for shear wall anchors, the spacing must be no greater than 6 feet between anchor bolts. Be sure to avoid placing anchor bolts where doors will be installed. You will just have to cut them and then install two *retrofit* anchors on either side to compensate for the one removed. Be sure to carefully measure the distance from the outside edge of the slab or foundation wall. For using 2 x 6 foot plates allow for 2¾ inches from outside edge. For 2 x 4 foot plates use 1¾ inches from edge. This will cause the bolt to emerge in the exact center of the wall frame plate. Don't worry about trying to avoid placing a bolt according to predicted stud spacing. Some will fall directly under studs despite the law of averages. Just make a "V" notch in the bottom of the stud before erecting to allow for washer and nut to be attached. While finishing the concrete, take special care not to disturb the placement of the bolts. Additionally, before placing bolts, place a little bit of grease or oil on threads to prevent corrosion or rust while waiting for concrete to cure.

Additives

I have heard of some artistic types who have used certain additives to enhance the appearance of their slabs. Some additives are approved to use and others are not. Color additives are a very cheap

Anchor bolts must be installed in wet concrete mixture at precise locations under bearing walls.

way to enhance your floor. I have heard of those who have added animal blood to concrete mixtures for color. Once, I saw metal shavings added to the surface of wet, colored concrete slab. The result was very exotic! Other ideas I have seen include scoring the exposed surface into squares to achieve a simulated tile effect. One creative concept included intentionally *spilling* some artistic paint colors in a random manner onto the wet concrete during finishing. The results of troweling over the paint gave a very different look to an otherwise drab concrete finish. But keep in mind that certain additives, which adversely affect the strength of the concrete, are not permitted. If you have any questions, ask your inspector before you pour!

Concrete or Block Foundation Wall

First, a footing is poured in a trench dug around the perimeter of the building. This footing is reinforced with deformed steel bars, which are placed so that they are embedded within the footing. Additionally, vertical steel reinforcement bars extend from the footing to a few feet above the top of the finish height of the footing. These reinforcing boards will serve to create a structural connection between the footing and the foundation wall. The foundation wall design relies on this structural connection. The quantity or amount of steel in thickness or center spacing of vertical reinforcement is subject to lateral pressure against the wall from the dirt side. If you are in an area subject to significant rainfall or where the water table is excessively high, the more steel the better. In some cases, a basement wall may require a design from a structural engineer. This will ensure the proper steel strength, size and placement.

A footing is dug and reinforcement installed for the foundation wall.

Concrete block makes an excellent wall with very substantial structural integrity. The wall below grade, as in a basement, should be designed to resist the lateral pressures from wet earth as discussed earlier in the design section. Additionally, the concrete block, which is a hollow or cavity wall system, should be filled with a solid concrete grout mix to complete the finished product. This grout engages the steel reinforcing and helps develop its strength. However, before grouting, wait long enough to allow the mortar joint to set and the block to dry. Before laying the first course of block, clean all the dirt from the top of the footing. This ensures a good bond between the concrete footing and both the block and grout, which will be poured into the cavities of the block.

After the concrete in the footing has hardened, the steel must not be bent.

Each concrete block must be connected to another on the top, bottom and ends with mortar cement. The mixing, proportions, and placement of this mortar is specified by industry standards and is outlined in the IRC. The variations of mixture proportions relate to the strength design requirements needed for specific use such as basement or foundation wall. Mortar is a mixture of Portland cement, lime, sand and water. Most inspectors will not perform a specific inspection during your placement of masonry units. The assembly of masonry units into a wall and the quality control of their placement is usually verified by testing of masonry prisms, which are assembled at the same time as your wall. These masonry prisms are an assembly of a miniature wall of a few masonry units using the same mortar and grout mix used on your job. A mason or job superintendent will have an independent testing lab verify the compressive and tensile strength developed in these prisms by breaking them and determining their tested strength. If you are not familiar with masonry construction, your home may not be the place to learn. Consider asking for help if you choose this design.

A good alternative for building a foundation wall if you lack experience with block laying, is the concrete foundation wall. In order to build this wall you must have forms for the poured-in-place concrete wall. You can rent these from concrete contractors supply stores. It is easier to establish a level foundation and more forgiving of mistakes. However, it takes some technical preparation to erect and stabilize the forms in a safe and proper manner.

Do not begin installing forms for the concrete foundation wall until after the footing is completely cured. Forms are spaced a distance apart, equal to the design width of the wall. The required width is specified by the IRC. Vertical steel will be embedded in and protruding from the footing as

Concrete blocks for a foundation wall are installed with mortar on both bed and head joints.

This owner-builder installed an Insulated Concrete Form (ICF) wall and is applying water proofing.

in the block foundation wall design. First, clean all the dirt from the top of footing. Add vertical and horizontal steel within the wall form as needed to satisfy the requirements for the structural design. The reinforcing steel requirements are specified in the IRC.

Both vertical and horizontal reinforcement must be tied together with wire ties. Reinforced ½ inch steel bars need to overlap 20 inches. Then place plywood forms on either side of the center of the footing to allow for the wall thickness planned. The wall and steel must be centered on top of the footing. These plywood forms must be held in place with an anchoring mechanism that will sustain a significant lateral load. Wet concrete is very heavy. It weighs at least 130 pounds per cubic foot. When concrete is poured from 8 feet high, a 4 foot wide plywood wall form will be restraining a significant weight per running foot. This applies pressure on each side of the wall. Wire ties are required to hold plywood forms together in numerous locations within the form-work. Metal cleats are used to restrain the plywood forms at the top and bottom. They also serve as spacers and can be placed on the bottom and top of forms to create a firm support.

Place forms so that steel is centrally positioned, and so as to establish a straight wall line. Forms are secured in place by steel stakes on either side of forms and additionally supported by stakes or other restraining members anchored into the earth. Verify that wall forms are true and plumb. When the forms are complete, double-check the *square* in the walls to be created by the forms. Adjust as needed to bring your building line into accurate alignment.

After your forms are complete, set the elevation for the finish height of foundation wall by using a builder's level set in the center of the house. This is a two-person job. One will use the builder's level and the other will hold a leveling rod and make a mark within the concrete form at the prescribed height around the perimeter of the wall. Select the highest point along the footing and add the intended height of the foundation wall. Then use the level to establish the new finish height around the entire house by making a mark on the inside of the form. Connect these lines with a colored chalk snap line. Now add one or two pieces of additional horizontal reinforcing steel about 6 inches below the top of wall. Secure into place by tying to vertical steel.

Checking the steel installation before a concrete wall is poured.

Electrical Ground Connection

At this time, before you pour concrete, you must plan for the future electrical equipment. As you learned earlier, electrical equipment must be grounded to allow an electrical fault a sure and safe path to ground. This ground connection is referred to as a *Ufer ground.* A #4 copper wire serves as the electrical grounding conductor between the electrical service equipment and the grounding electrode for services of 200 amp or less. It is connected to the reinforcing steel in the footing with an approved clamp connector. For foundation walls, be sure to extend the ufer (electrical ground) through wall prior to pouring. Concrete is then poured all around the ground wire, encasing it in the concrete. Later, the other end of the copper wire must be connected to the ground clamp inside the electrical service panel.

Pouring Concrete

Now you are ready to pour. Order the concrete mix and carefully place inside forms. If you have a tall foundation wall, consider renting a concrete pump. This pump will allow you to control the placement of concrete at a more leisurely pace. A gradual placement will help avoid failure in the forms due to pressure sustained from rapid placement. To produce a stronger wall, rent a mechanical vibrator to help consolidate the concrete mix and evacuate air or other voids. The job of pouring concrete is by no means a one or two person job. You will need numerous, experienced helpers. Pick the most experienced member of your crew to supervise the pour and finishing work. He will direct and control the job assignments and general work flow.

When the pour reaches completion, the top of the foundation wall must be precisely level throughout its perimeter. You have planned for this with careful measurements with a builder's level before the pour. Your marks have instructed the concrete finisher, the level at which to raise the concrete within the wall form. Install any required anchor bolts at this time. These anchor bolts will anchor a sill plate to the top of the concrete wall.

Anchors are important because the sill plate holds your wall and roof system secure to the foundation system. Use trowels to smooth the concrete at the top of the wall along the expansion joint material. Let the wall cure, and strip forms after 24 hours. Maintain the dampness of the wall for several days after the forms are removed. This allows the concrete to cure evenly and achieve a greater strength.

Concrete foundation wall complete with forms stripped.

Underground Plumbing and Mechanical Installation

With either the concrete or block foundation wall installations complete, it is time for the installation of all underground plumbing and mechanical components. Please consult the chapter that discusses these installations thoroughly for the appropriate advice. When careful measurements have been taken and you are confident the plumbing layout is correct, it is time for the testing of the pipe. With the ends of the pipe capped to prevent leaks, fill the entire pipe assembly with water. The pipe and all of its fittings must not leak.

When you have successfully passed the appropriate inspections, it is time to backfill inside the perimeter, around the pipe. This must be done with the correct material, using care not to damage any pipe or equipment. Carefully apply granular sand material around pipe and compact with care. Care should be exercised in backfilling. Avoid using any rocks, broken concrete, frozen chunks of earth and other rubble until the pipe is covered with at least 12 inches of tamped earth.

Block foundation wall is complete.

Concrete Slab Infill Preparation

The most secure form work in which to pour a slab is a foundation wall of either a block or concrete. If you use this approach, you will build a stem wall to the height desired. Then you will backfill a good quality sand and soil mixture inside the foundation perimeter around the ground work plumbing and water supply pipes. Bring the backfill up in 6 inch increments and compact with water and a pneumatic tamper, as you go. To pour a slab inside a perimeter foundation system, you must ensure that you have a level floor. This is accomplished by building the foundation wall to the correct height using a masonry level and survey instruments. Follow the instructions for monolithic slab for tooling and finishing.

Remember, you must have completed all plumbing, mechanical and electrical work which are under slab prior to pouring the slab. Additionally, any plumbing pipe must be protected from abrasion due to concrete by wrapping the pipe with an approved material such as duct tape or other similar material. When finishing and tooling a concrete floor, be sure to take special care around plumbing, mechanical and electrical penetrations. Do not get too close with mechanical equipment to these pipe penetrations. Manually finish with a trowel. You could accidentally damage the pipe with power equipment.

Damp-Proofing and Water-Proofing

Whether you use poured concrete or grout-filled CMU block to serve as your foundation wall, water will penetrate into the basement or crawl space unless you do something to stop it. Backfill must be installed against the outside of both masonry and concrete foundation walls and even full height basement walls. Before you install any dirt, be sure to provide an adequate means of disposing of the water drainage, which is sure to accumulate after the first rainstorm.

After screeding concrete for a rough level, and using a jitterbug, finish concrete with a hand trowel.

Depending upon the sub-soil water conditions, you may need to install perforated drainage pipe just below the foundation wall to prevent a buildup of excessive water. The pipe should be laid on a bed of gravel either above it or embedded in larger size aggregate, to facilitate water drainage. Drainage tiles are placed end to end, with a small gap between them to allow the entry and collection of water. The gap should be covered with tar paper to prevent infiltration of dirt and debris. Both the tile and the pipe should slope slightly toward the collection point. The tile or pipe can be connected to a storm sewer of other approved means of runoff disposal. The drainage water is diverted away from the foundation wall by the pipe.

The next step is to install weatherproofing on the exterior side of the concrete or masonry walls to retard water or moisture from entering. Foundation or basement walls below grade, which are subject to excessive water drainage, should be protected to retard water penetration. Paint the exterior walls with roofing tar or pitch. Be sure to wear plastic gloves while doing this chore. The tar will not look good on your arms and hands when you go to work at your day job. Apply the tar liberally around all areas of the exterior wall. Allow this to dry somewhat, then apply the next level of protection. Use a heavy thickness of plastic sheeting to drape over the walls as further protection. More conservative measures include installation of thicker roofing material such as roll roofing in overlapped layers against the wall before backfilling with dirt.

Exterior Backfilling

After you have either poured the concrete wall or grouted the block foundation wall and then installed protection from water penetration, earth must be installed against the back side of the subterranean wall. Dirt must be added in short layers referred to as *lifts*. A good design lift will add no

Here, a typical lumber package awaits the skilled hands of the carpenter.

more than 6 inches of dirt at a time. Compact each lift of backfill material by tamping the earth until it is very firm. Be careful not to disturb the waterproofing material you have installed. You might even add small amounts of water to improve consolidation and compact further. You could even add some leftover cement to your fresh earth to create a low-grade concrete mixture. At this point, adding pesticide in the backfill area near the upper portions of the wall is cheap insurance against insects such as termites.

If you are backfilling deep holes such as a basement excavation, keep in mind that dirt is heavy and adjacent backfill is loose and subject to sliding. So, be careful. Additionally, do not backfill against masonry walls unless steel has been installed and the wall cavities have been solid-grouted and are fully cured. Casting earth against an unbraced wall will subject it to a lateral load that may cause it to fall into your basement. Do not use heavy equipment such as tractors or backhoes to speed this work up. The weight of this equipment could damage or even topple the wall you have built.

Foundation and Footing Installation Checklist

___ Approved plans on site

___ Footings dug to proper width, depth. *Section R403.1.4 and Table R403.1*

___ No extraneous dirt, debris or organic material such as roots in the cavity dug for the footing or slab

This is what your home will look like after the skeletal frame is complete.

___ Reinforcing (horizontal and vertical) installed correctly. Steel is not in contact with soil. *Section R403.1.3*

___ Concrete forms are installed in proper position to define the future foundation wall. *Section R404.1*

___ If monolithic footing/slab is intended, is ground under slab adequately compacted and level? *Section R506.2.1*

___ Stretch a line across the building corners on top of forms. Verify at least 3½ inches of concrete for slab. Section R506.1

___ Is electric grounding conductor connected to a 20 foot length of reinforcing steel in footing? *Section E3508.1.2*

Uh Oh!

Problem

Reinforcing steel in footing does not have proper lap or corners are butt-jointed together.

Solution

Add extra steel alongside the butt-jointed steel or bend an additional corner to compensate for the missing corner lap. In any case, allow for at least 20 inches of lap in steel bars.

Problem

Foundation subgrade (soil beneath what will be your footing or slab) is not compacted enough. This causes failure in the concrete (cracking) and therefore failure in your wall and roof structure.

Solution

Rent a *pneumatic tamping device*, sprinkle water intermittently, then use a tamper on soil until you can walk with your heels and leave no print in the earth. Then tamp for another hour or more. Remember, this is a really important step, which cannot be fixed after concrete is poured.

Problem

No anchor bolts ready to be installed.

Solution

Since anchor bolts must be installed in fresh, wet concrete, they must be on-site prior to the concrete truck. Be sure to mark along the form, the exact location they will be placed. When the concrete starts pouring, you lose time very fast. If you forgot them, you may be able to substitute a retrofit expansion anchor which will fit into a hole drilled into cured concrete.

Chapter 6

WOOD WALL FRAMING

Wood Basics

Everyone who works in construction finds out sooner or later that there is a particular trade that seems to come easier than others. Wood framing is mine! I love working with wood. Wood is forgiving of mistakes. This is especially useful for beginners. Before I get started on framing methods, let's review some basic properties and characteristics of wood. Wood is sized in nominal dimensions, that is, the size it was before the final surfacing at the sawmill. For example, a finished 2 x 4 is really only about 1½ inches thick by 3½ inches wide. When sizing any dimension keep this characteristic in mind.

Wood is a natural material and, as such, is subject to the variables of nature. Trees are divided into two broad categories: hardwoods, which have broad leaves, and softwoods, which have needles or scale-like leaves. The cross section of a tree has several distinct zones: the outside is sapwood, the inner zone is called heartwood, and the core is called the pith. The tree grows by adding new layers of cells in the outer surface of the sapwood. A thin ring of tissue called the cambium applies new bark cells toward the perimeter and new wood cells toward the interior causing growth rings. For a time this new layer acts to bring sap and store food. It eventually dies and serves only as support for the tree as a newer, outer concentric layer becomes the active layer. The ever-increasing, inactive layer is the inner heartwood. The active layer is the outer sapwood.

Moisture content of wood is the weight of the water in the wood, expressed as a percentage of the weight of kiln-dried wood. If used for construction, wood must not have more than 19 percent moisture content, which causes adverse twisting during shrinkage. Most wood sold as framing lumber is dried sufficiently to comply with this requirement. Be sure to select your framing lumber carefully to ensure a straight board. When you are selecting lumber, notice that once in a while, one board will weigh significantly more than others. That is usually due to the moisture content.

Mechanical properties of wood describe the various strengths it possesses. Grading of wood is important because its strength and mechanical properties descend from its rating. The visual rating of wood uses grading standards adopted by the grading agency. Most framing lumber has a visual rating of #2 or better. Normally, you can expect that a grade of #2 will be adequate for the structural requirements of building a conventional wood frame home. This classification restricts structurally defective lumber that has excessive checks, splits or knots. If you have the time to sort through a stack of lumber, do it. Being meticulous with your money is a virtue. Always get straight lumber. Try to discard all lumber with twists, splits, and checks. Next, if you have an adequate supply to sort through, be picky about large knots. Timber manuals and wood construction books may provide more information than you will reasonably need in order to build your home, but they are helpful when understanding wood design. Just remember to be highly selective about your lumber. At least, get a commitment from the lumber company to replace poor quality lumber.

Tools and Materials

In order to be a carpenter, you must have certain tools. You should have at least the following tools for assembling your dream home: 20 ounce framing hammer (or two), power circular saw, handsaw (sharp), minimum 25 foot long tape measure (or two), framing square, small framing square, 2 chalk lines (different colors), framing level, wrenches for bolts, power drill with numerous large bits, nail set, razor knife, table saw, saw horses, pry bar, hacksaw, reciprocating saw, tool belt, large pipe clamps, power miter saw. Additionally, an air compressor and a nail gun really add to the ease of building a home.

Materials

For 8 foot high walls, you must buy *stud length* studs for the wall frame. These studs are 92⅝ inches long. You will also purchase wood plates, a large box of 16d nails, and washers and nuts to connect

Wood used in construction must be marked with a grade stamp showing its appropriate use.

Electric generator and air compressor will prove useful for power tools.

anchor bolts. Header material and beams must be purchased. You will use extra studs for jack or trimmer studs, fire blocking, corner spacing, king stud, and blocking. You can use any extra plate material to stabilize wall frame during erection for temporary bracing.

Command Central

At this point in construction, you may want to build an outdoor desk or work station that will double as both a plan table and a work table. Locate it out of the way or build it so that it is portable. Gather tools that you cannot carry in your tool belt in this area. This is a good place for an outdoor telephone or radio. It is a good idea to orient the top and bottom of the plan on the table according to the layout of the building. As you look down at the plan and look up at the building, their layout should match. This will help you improve your perspective and comprehension of the anticipated layout.

Stud Size

There are two basic choices available for exterior wall framing: 2 x 4s or 2 x 6s. The 2 x 6s cost more, but you will use less of them since you can space them 24 inches on centers whereas 2 x 4s should be placed at 16 inches on centers. Codes may allow a greater spacing; however, your wall will be superior to that minimum standard. Some regions of the country have other stud sizes available such as 3 x 4s or even 2 x 5s. These are less common, but permitted by code.

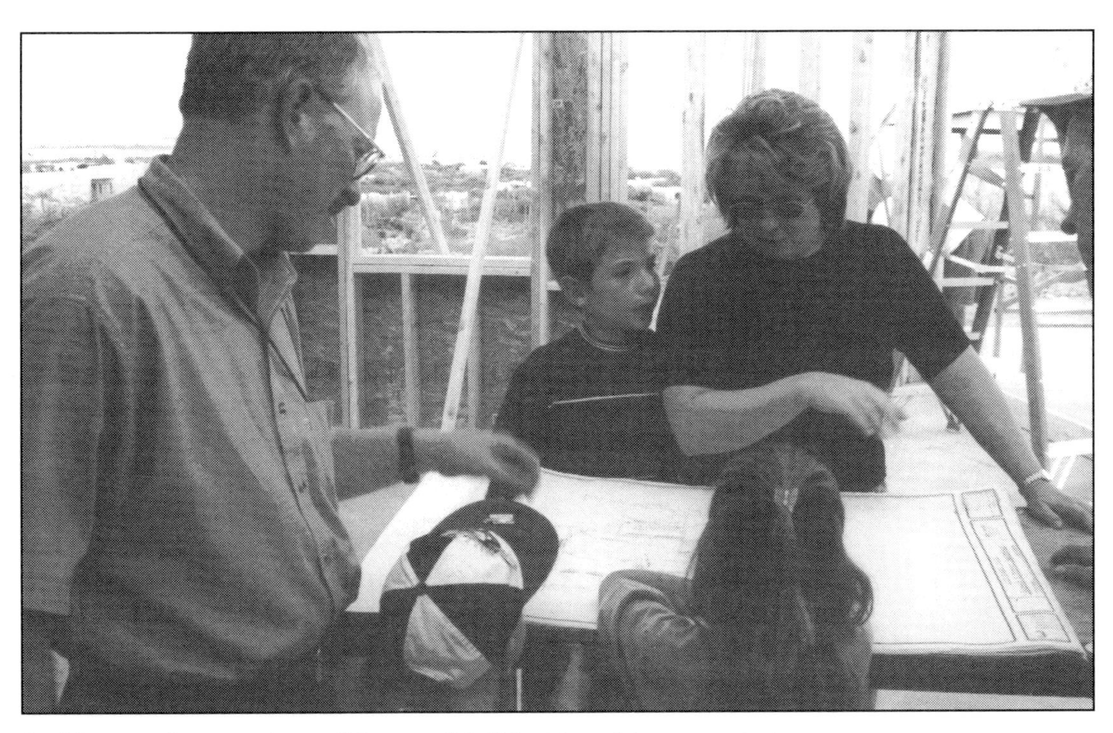

A table to use for your plans will be essential. Orient to match your project.

After selecting a stud size, determine the planned height of the wall frame. Most wall frames are 8 feet high. The codes allow for heights up to 10 feet without any special design. The higher you build, the stronger wall frame you need. For instance, you may plan to build a *rake* wall on the gable end to create strength and eliminate an additional roof truss or rafter. Depending on the pitch of the roof, you may be extending the wall between 8 feet at the end to over 16 feet in the middle. Your stud size can play an important part in the overall structural integrity of the wall and its design limits.

My recommendation is to use 2 x 6 studs at 16 inches on centers. You will gain a thicker wall cavity for insulation. Most of all, the strength of a 2 x 6 wall frame is substantially greater than a 2 x 4 wall frame. The thickness of the wall allows a greater sound barrier against outdoor noises. Although you could increase the center spacing to 24 inches o.c., some exterior siding requires a framing member support every 16 inches on center. Read the manufacturer's specifications for the siding you have selected. Plumbing walls are those walls that will have DWV and water supply pipe within them. These walls should use 2 inch by 6 inch framing to allow for the thickness of these pipes.

Laying Out Walls on Floor

Before you cut one plate, header, stud or trimmer, you must lay out where the walls will be on the floor. You do this by drawing lines onto the floor that match the proposed lines of the wall. The floor may be either a concrete slab or plywood sub-floor. Draw the lines with either a pencil or a chalk line. Plan on spending the entire day, uninterrupted, to perform this task. Do not rush this very important task or let anything distract you. This project will require the help of a passive partner. Pick someone who can think but will not bother you with discussion, conversation or questions. Kids are definitely not prone to this personality quotient. Use your plans, and establish a control corner. This will be the beginning of your layout. Start with exterior walls. Measure the width of the plate (either 3½ inches or

Carpenter begins wall layout begins with marking the top and bottom plates with locations for studs.

5½ inches) and, allowing for exterior siding, draw both lines along the length of the line of the wall. Repeat the layout for all exterior walls. The width of interior walls that are 2 inches by 4 inches are 4½ inches, (½ inch drywall on both sides of a 3½ inch wide frame).

Next, lay out the interior walls by drawing where they will sit on the floor. This will be a little harder. You must use the building corners to establish control distances to the outside corner of each wall section. You will use your geometry skills to establish a true square for each wall length. In performing this interior wall layout, you will be defining the layout of the house. For slab-on-grade floors, you will use the rough-in plumbing to establish where walls for bathrooms, kitchens are to be located. You will be using both your short and long tape measure a lot during this exercise. Double check everything!

The 3 x 4 x 5 triangle method works best to check layout. This is the method that uses the Pythagorean theorem. That is, the square of the hypotenuse of a right triangle is equal to the sums of the squares of the two adjacent sides. For example, a 3 x 4 x 5 triangle; 3 squared + 4 squared = 5 squared. Test the perpendicular angle by using this method. Remember that interior walls will eat up room space, and room size will not be exact when complete.

Mark the floor in a manner that will endure rain, wind and sun bleaching. Chalk line is okay if you will

Closeup of marks on top and bottom plate reveal the exact location for various components within wall frame.

begin wall framing fairly soon. Perhaps after the chalk line is struck, use a straight edge and a heavy felt-tip marker to indelibly mark the wall location. Keep lots of pencils on hand and recheck if you are not sure about the layout. Sometimes, you may accidentally use a previously drawn, incorrect line from your layout and be off several inches.

Basic Wall Frame

Walls are built best on a level, flat surface. The floor provides that flat surface. The wall frame consists of studs, bottom plate, two top plates, jack or trimmer studs, cripple studs, sill plate (under windows), headers over doors or windows and finally, blocking. Besides supporting a floor or roof frame, the wall frame also provides a nailing surface for exterior sheathing such as siding and interior wall coverings such as drywall.

Be sure to always build the exterior walls first. The interior walls can be built after the roof trusses are installed or even after roofing is installed. If interior walls are built and installed before trusses are erected, the trusses may be adversely affected. The engineered trusses are designed to have specific points of bearing and a certain amount of mid-span deflection. If interior walls are built too tall, before trusses are installed, the truss may not be able to deflect enough, and an interior nonbearing wall will inadvertently become a bearing wall. If you haven't planned for a bearing condition, the wall may collapse under this unanticipated weight as well as cause some damage to the truss.

The basic wall frame is composed of wall studs and three plates. There are two plates on top and one on the bottom. If you have no windows and doors within a wall section, these represent all of the components you will use. If you are building on concrete slab or foundation wall, the bottom plate will have to be treated to resist decay due to termites and weather. A pressure treatment of chemicals

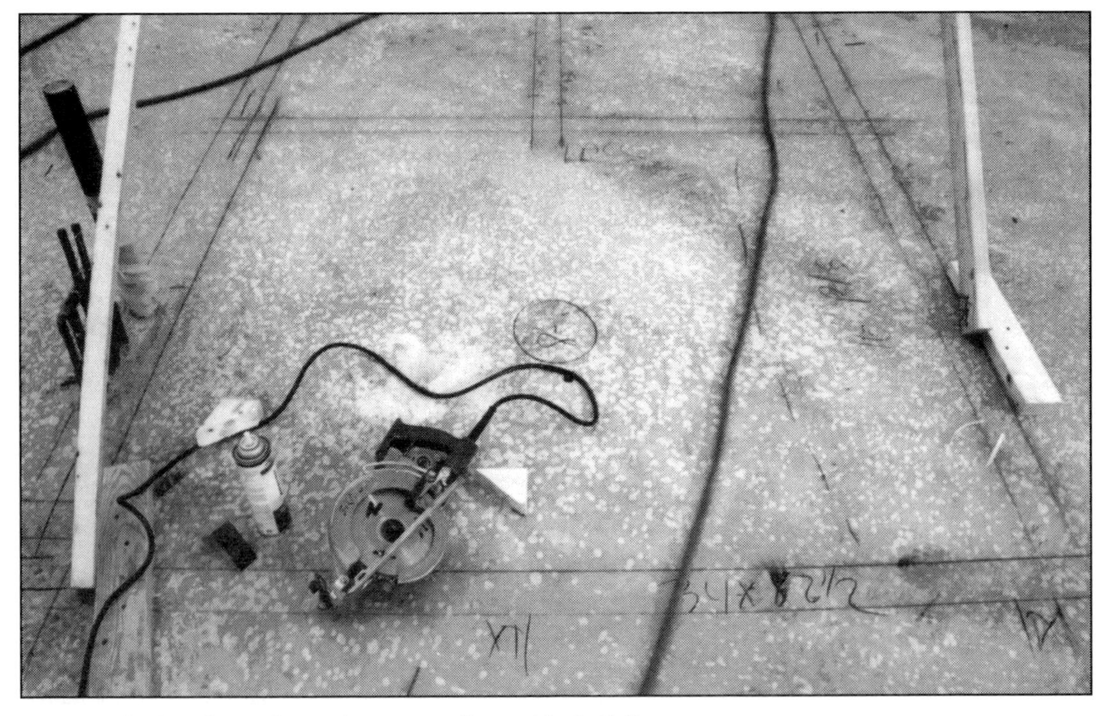

Locations for interior walls are drawn onto floor with chalk line.

is generally required to achieve this degree of protection. The wood generally looks green due to the chemical and will have cleat marks where chemicals are injected inside the wood under pressure. An approved alternative is to use foundation-grade Redwood as the bottom plate material. Remember that the overall height of the wall will be the height of the stud plus 4½ inches. That's three of the 1½ inch plate thicknesses.

Ceiling height should be at least eight feet from the finish floor (after floor covering). Precut studs are usually 92½ inches long. With 4½ inches of plate material, the overall height of a frame wall becomes 97 inches. With ½ inch drywall on ceiling and ½ inch floor covering, the finished, clear interior height is 96 inches or eight feet.

Header Construction

A wall has windows and doors that are, in effect, openings within the structural integrity of the wall frame. A *header* replaces the support removed to allow the window or door to be installed in the wall frame. The header is a site-built beam that bridges the opening across studs. Larger dimensional lumber such as 2 x 12s or 2 x 10s or even 2 x 8s should be doubled together to form this homemade, composite beam.

In order to infill the entire thickness of the framed wall, sandwich ½ inch plywood in between the two pieces of lumber. For instance, while building a 2 x 4 wall frame, the thickness of the wall is 3½ inches. The width of a 2 x 10 is 1½ inches. Doubled, it's still only 3 inches. Add a ½ inch piece of plywood the same shape of the 2 x 10s and sandwich between them. For a 2 x 6 inch stud wall frame, use three 2 x 10s with two pieces of plywood between the boards to create a thickness of 5½ inches.

Another method of building a header is the *box beam*. Here, a header is built with the same large framing members, but without the plywood sandwich. Instead, these framing elements are capped with another dimensional wood member. A 2 x 4 is nailed to the top and bottom of the 2 x 10s forming a frame assembly.

A box beam is assembled to act as a header in a bearing wall frame.

The *king stud* will be nailed to the side of the header. Inside each king stud will be a *trimmer* or *jack stud*. Headers are vertically supported in the wall frame by this *trimmer* or *jack stud*. The height of the header should be 1 inch more than needed for the rough opening for the door frame and exactly the rough opening of a window. Space above the header should be filled in with *cripple studs*. Maintain the layout for the wall frame. The king stud will be end-nailed to bottom and top plate. Use a square to verify the vertical plumb for the king stud. Next, nail the trimmer stud to the king stud and the bottom plate. For door frame openings, the trimmer will be cut ½ inch shorter than the desired rough opening height. This represents the 1½ inch thickness of the bottom plate that the trimmer rests upon minus the anticipated thickness of the floor material. The header will sit upright on top of the trimmer studs. Be sure to cut the headers 3 inches longer than desired rough opening width. This represents the 1½ inch width of the trimmer studs on *each* side of the header.

When building door frames, remember that a rough opening is usually 2 inches wider than door width. Rough opening height is usually around 82 inches. That allows 80 inches for the door, 1 inch for top door frame clearance and 1 inch for floor covering clearance. For windows, remember to account for height above floor as well as proximity along wall. Depending upon its size, a window less than 18 inches from the floor will require "safety glazing," which means added expense. The bottom of a window in a bedroom must be less than 44 inches from the floor to provide adequate egress in an emergency. In a kitchen, a window must be higher than the base cabinets and counter top. Be sure to plan and measure for furniture planned for specific rooms to avoid an "Uh, Oh." Wood framing is very forgiving of mistakes, but a good clean, professional looking job, demonstrating foresight, is what you're trying to achieve.

Wall Layout

The first thing you do before you get the nails out is to plan a layout for where the studs will be along the plates. You do this by aligning one top plate and the bottom plate together, measuring the anticipated center spacing distance between studs, and making a mark. For instance, let's say that I know that I am spacing my studs at 16 inches on center. The first stud layout is always 15¼ inches from the end. Then 16 inches apart thereafter. This is to accommodate the layout of drywall and plywood, since the first stud must be covered completely. I measure: 15¼ inches, 32 inches, 48 inches, 64 inches, 80 inches, 96 inches and so on. If I used a 24 inch on center (24 inch o.c.) spacing, I would measure 23¼ inches, 48 inches, 72 inches, 96 inches and so on. At each of these distances I make an inverted "V" mark on the wood. The exact distance is marked at the nadir (or bottom end) of the "V." Remember that all measurements should always be made from the same wall end, even if new wall sections are built onto previous ones.

Headers are above windows and doors; however, *cripple studs* are above headers and below sill plates on windows. These cripple studs must continue the pattern for the wall layout to accommodate the needed support for siding and drywall around the openings. The location of the window and door can be established by one of two ways. You can either measure to the center line of the window and then offset half of the opening width on either side, or measure to the first opening edge and add a measure beyond the full opening width. Then, identify the opening width by marking these edges with a *V* mark similar to the stud layout. Distinguish between the two marks with a *W* or a *D*, to differentiate between the stud layout and the opening. Notice that some studs are within the opening frame. These are left on the plate because a smaller stud called a *cripple* will still be installed above and below the opening in order to maintain layout for drywall and exterior sheathing. Remember that the bottom of doors will be installed at least ¾ inch to 1 inch above the floor to account for threshold height.

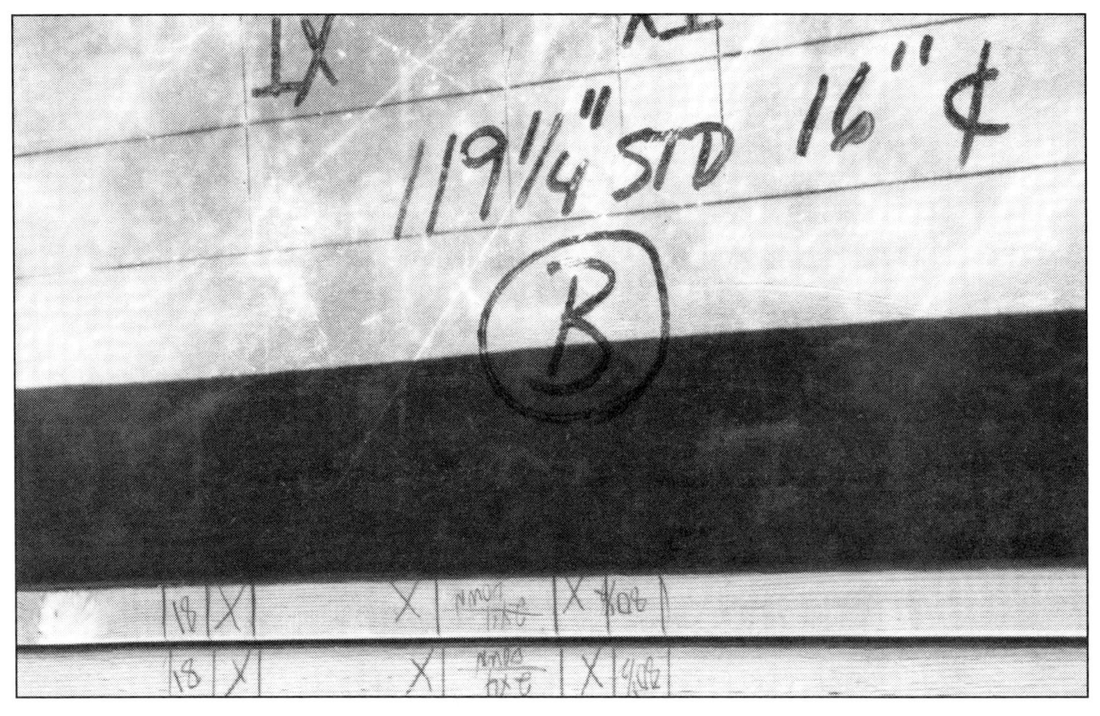

Marks on top and bottom plate account for the intended wall layout.

Next, establish where the beginning of the wall layout will begin and start from that end and actually draw an outline of the studs, cripples, king studs and trimmer studs on the two plates. Do this with a straight edge. Since the width of the stud is 1½ inches, scribe a mark with that width onto the sides of both the top and bottom plates that are temporarily sandwiched together. This ensures that where you nail the stud from each plate is exactly the same distance from the edge, thus creating a vertical alignment for the stud.

Continue to make marks on the plates. Mark the king stud, trimmer stud, cripples and studs. Each mark will be 1½ inches wide and contain an identifying letter inside this space to identify the type of member to be connected to the plate at that point. Mark the studs with an *X*, the king studs with an *X*, the trimmer studs with a *T*, and the cripples with a *C*. The space in between the trimmer studs will represent the opening, either a window or a door. Although this space will have cripple stud marks, you can add a note about the opening width required, so that a header can be cut for the space. The header length will equal the distance in between the king studs. This is two 1½ inch widths or 3 inches longer than the actual opening. This 1½ inches represents the header support on the trimmer stud.

Studs within a wall frame need to be arranged so that sections of a four foot wide wall covering such as exterior siding or interior drywall can be supported by a framing member. To do this, keep in mind that layout should begin from a corner and continue all the way to the next corner. For interior walls, the first four foot section may be short due to the corner, but that just means that the first piece of drywall at the inside corner will be shorter. Matching panel width to studs within a wall frame is known as *wall layout*. Layout for exterior shear walls is more critical than interior drywall. Just stay on layout as you build subsequent wall sections.

Wall frame in place before nailing.

Wall frame after nailing is complete.

True and Plumb

This is an out-of-place but gentle reminder that your wall must be straight and square. You get there by double checking lengths before cuts are made, using a square to mark a cut, reading your tape measure correctly, using a sharp pencil to make marks, making straight cuts and double checking your wall frame. It is a virtue to be over-attentive to the details in framing. The difference between a carpenter and a nail pounder is attention to detail. Remember the object lesson in the quip *"No matter how many times I cut this board, it's still too short."* Another popular quote you should remember right about now is *"Measure twice, cut once."* Be attentive to detail.

Assembly

Within the basic wall frame, components such as wall studs, top and bottom plates, cripples, headers, and trimmers are connected together by nails. Drive at least two 16d nails from the back side of the plate into the stud. Don't put more nails than are necessary to securely fasten the two pieces of wood together. This may damage the stud's capacity to resist a load. Drive nails perpendicular and straight into the wood. Avoid nailing into any areas of wood with knots that will cause splitting. Toe nail only when straight nailing is impractical. Drive nails straight to achieve a solid, secure connection.

Always use a builder's level to check the plumb on walls before you secure with nails.

Corners and Intersecting Walls

Wood frame walls must have a secure method of connection to prevent collapse. The way to do this is to build a means of face nailing one wall section to another. To do this adequately, three studs are required at all corners. There are numerous methods to achieve a secure connection between adjoining wall sections. If walls are being built with 2 inch by 6 inch studs, the first wall could be built with an additional stud within 2½ inches. This will allow the end stud from the intersecting

Using care, a nail gun is useful in wall assembly. Notice the studs are placed where the layout is marked.

wall to be connected to both of these studs, not just an edge connection to a single stud. For other interior, intersecting walls, the connection is similar. Two additional studs should be added in the exterior wall to allow a face-nailing connection from the interior, intersecting wall. These studs are separated so that the intersecting wall can be nailed to both and establish a secure connection.

Erecting Wall Section(s)

Try to build wall sections where breaks do not include headers (window or door openings). This will give you a sufficient amount of structural integrity to prevent a wall from coming apart during erection. A wall frame of 12 to 16 feet in length can be easily lifted into place by two healthy adults. When upright, and in close proximity to the building line, it will stand with limited bracing (unless the wind is blowing). You can nudge it into place using a framing hammer. An important aspect of segmented wall frames is that the layout must continue from the initial wall corner. Because of this, your adjoining wall frame may look peculiar for the first few inches while regaining a proper layout.

In the case of a wall frame being installed on a concrete floor, anchor bolt holes must be drilled through the bottom plate prior to lifting the wall section. Determine the location for holes by carefully measuring the distance from the edge of the building line and the distance from the end of the wall section. Where these two lines intersect, mark an "X" on the plate and drill a ⅝ inch diameter hole into the bottom plate. If possible, lift the wall section over the bolt to avoid damage to the threads on the bolt. When over the bolts, the wall frame will be more or less stable. You may find that you will need to adjust the wall section into a more secure position with a sledge hammer or large framing hammer. When finally in place, install a 2 inch by 4 inch or 2 inch by 6 inch wood plate at least 12 feet long as bracing on each side of the wall every few feet. Secure the bracing with scaffold nails to studs of the wall frame and stabilize on the ground with stakes driven into the earth at least 12 to 18 inches. Perform a final level check of wall frame to verify that it is true and plumb, then nail the end stud of the newly erected wall section to the last stud of the previously erected wall section from both sides.

At this point, be sure to install the second top plate so that splices in the two top plates are at least 48 inches apart. The way to do that is to let the top plate of the first wall section extend or *run wild* beyond its last stud for a distance of 48 inches. When erecting the next wall section, plan on starting the second top plate 48 inches in from the edge to accommodate the adjoining top plate. For adjacent wall sections at corners, at least one wall must have a stud turned sideways within the wall frame to serve as a *nailer* for the stud within the other wall frame. Be sure and verify the plumb of each wall before connecting with several 16d nails.

Erecting a wall section. Gather friends or neighbors to help.

It is essential to discuss the option of installing siding before erecting the wall section. The only argument against installing siding when the wall section is horizontally laying on the floor is that it may prove to be out of square after erection. Sometimes, to develop a true and plumb wall frame around the perimeter of a building, some give and take between wall sections and sides is necessary. Siding creates a more rigid wall frame which prevents that give and take. However, if you are meticulous with a T square, there is every reason to pre-install the siding. Half inch siding weighs approximately thirty pounds. It is very cumbersome to hold a 4 by 8 foot sheet of panel vertically and at the same time nail it to a wall frame by yourself. However, lifting a wall with siding pre-installed requires more help. It is heavy and subject to toppling over. If you decide to install the siding on the floor first, there are several things to consider. Before you begin, make sure that you have some experienced help.

First, be thorough about developing a true and plumb wall section, which involves measuring each stud and verifying the uniformity of lengths, as well as using a T square religiously more often than you need. Avoid twisted studs altogether. Remember to start your wall layout from where the siding will begin. You can still fine-adjust the exterior wall section a little by using a block and tackle or a come-along. Nail the siding with corrosion-resistant nails. Six or eight penny nails are adequate.

Connecting adjacent wall sections requires racking wall sections together while still maintaining plumb.

They should be nailed with a spacing of no more than 6 inches o.c. around the perimeter or edge of the siding. The interior connections between the siding and the studs are referred to as the field and should be nailed at no more than 12 inch o.c. spacing in this area.

If your wall section is greater than 8 feet high, your siding will be shorter than the wall frame. In that case, it is a good idea to install wood blocking at the point where there is a joint between pieces of siding. Blocking is a piece of wood framing installed between studs. This is to ensure adequate support for the edges of siding as well as shear value for the wall section. You see, the humble wall section you have just built serves more of a purpose than just a vertical support for a roof or floor. When wind hits a wall side, the load transfers through the adjacent wall sections and resists or transfers by means of the *shear panel*. This shear panel derives strength through the siding attached to the stud wall frame. The semi-rigid wall frame is said to have *shear value*. That is to say that it can resist wind and seismic forces and transfer those forces into the foundation through the anchor bolts. Blocking edges of the siding allows the panel to more thoroughly act as a shear panel.

Stabilizing the Wall Section

The initial wall section must be stabilized temporarily, while building or erecting other wall sections. Use 2 x 4 or 2 x 6 plate stock at least 12 feet long (16 feet is better). One end of the bracing kicker will be nailed to the wall section with scaffold nails, and the other end will rest on the ground, secured in place by a steel stake. If you have opted to not place siding on first, place the kicker alongside of a stud in the framed section. The other end of the kicker will rest on the ground and be held in place, nailed to a metal stake to prevent its movement. If you have installed siding first, use the window or door openings as a secure place to nail the kicker. If necessary, temporarily place a small piece of framing lumber to the side of the wall frame and place the kicker underneath to brace. Place as many kickers as necessary to completely stabilize the wall section—at least one on each end plus one every 12 feet.

Use a level to verify that the wall section is true and plumb before you nail the kicker. You can connect either end of the kicker first. If you are by yourself and have a small wall frame to work with, nail the kicker to the steel stake and let it rest in between the studs (if you have opted not to pre-install siding). If siding is pre-installed, place the end of the kicker inside the door or window frame that you plan to secure to the bracing kicker. After the wall frame is in place, double-check level to verify true and plumb, and using scaffold nails, secure the kicker to studs as high as practical. Take your time, because as you progress, it gets harder and harder to correct an out of square wall section.

If your dream home is rectangular, the shorter wall section would be the first one to be raised. It is easier and establishes a corner for adjacent walls to brace against. Ends of subsequently erected wall frames can be nailed with 16d nails into previously erected wall frames, since they already have a kicker and true and plumb have been established.

When a subsequent wall section is erected, a 4 foot section of the second top plate will either be running wild or will be absent. After initial true and plumb is established and the kicker is in place, connect the *wild* 4 foot section of the second top plate into the first top plate of the adjacent wall frame. Use 16d nails every few inches and always at stud locations. This will help firm up and tie the wall frame sections together.

Bracing is installed to temporarily stabilize the wall.

Erect all of the exterior walls at one time. Do not leave the job site before all walls are erected and adequately braced. After all of the walls are erected, check the true and plumb on each stud all the way around the perimeter again. You will probably need to adjust or *rack* the wall frame to achieve exact true and plumb. Get a feel for where you are out of plumb and how much in relation to an opposite side. Keep in mind that now, when you adjust one side you are affecting the other side simultaneously—sometimes adversely.

With that in mind, use the kickers and a hand winch to rack the wall frame into its proper place, true and plumb. If you are off a great deal, you may need to apply lots of pressure. Do so by bending the kickers in the center. This applies horizontal pressure to the top of the wall frame where the kickers are connected. You may need to use a come-along or similar pulley assembly to pull an entire wall section. Connect one end to a pickup truck or tractor, then ratchet the wall into position. Double check the true and plumb of the wall often. When you are satisfied with the true and plumb of the entire perimeter, install the wall corner bracing and then reset the kickers. Maintain these kickers until the roof decking is installed.

Anchor Bolt Connection

Now that the wall is safely stabilized, the washer and nut must be secured to the anchor bolt. Clean any rust or debris off the bolt, and apply the washer over the bolt. Then begin screwing the nut onto the bolt. Use a ratchet or box-end wrench to tighten enough to prevent the wall from moving, but not so tight as to damage the bottom plate.

Bolts are secured to bottom plate with washer and nut.

Exterior plywood is used for conventional brace wall sections. They must be at least four wide and installed every 25 feet. Notice blocking is installed at narrow end of plywood where walls exceed eight feet high.

Corner Bracing

Building codes establish the need for lateral bracing of a building to resist wind forces as explained earlier. The general requirement is that every corner and 25 foot interval around a building be equipped with at least one 4 foot panel to serve as a shear panel. The panel must be at least ½ inch thick and must extend the full height of the frame wall section. Several alternatives are specifically referenced by the code. Among them are various types of 4 foot by 8 foot panels such as plywood and composite sheathing. These 4 foot by 8 foot panels are required to be nailed with 6d nails at 6 inch intervals around the perimeter and 12 inches on center in the field. Additionally, let-in bracing and metal bracing are approved alternates.

Let-in bracing is framing lumber, at least 1 inch in nominal thickness and four inches in width, which runs at a 45 degree diagonal from the top plate to the bottom plate. Each stud along the path traced by the let-in brace is notched and the notched portion removed to accommodate the let-in brace. The let-in brace is then nailed to the stud at the notch with at least two 16d nails. Another alternative is the metal brace. The type of metal brace that is designed to resist lateral forces has a cross section that appears like the letter, "T." The bottom of the "T" is "let-in" to each stud just like the wood version, although with a much smaller notch. The metal is nailed into each stud, bottom plate and both top plates with at least one 16d nail. I have found that these metal braces are the best alternative to panel sheathing. They help create a more rigid wall even if you have installed sheathing. Be sure to check with your inspector about the specific kind of metal bracing you intend to use as permanent wall bracing. Some types are limited to use as temporary bracing only.

Metal bracing is installed from top to bottom plate at an angle.

The well built wall frame.

During the design phase of your project, you determined where you would need shear walls. You should have marked these locations very clearly on your plan. Certain shear walls require hold-downs as anchors to the foundation. These should have been installed at these locations during foundation construction. These hold-downs may be standard ½ inch anchor bolts or as complicated as a manufactured specialty device with high strength bolts, nuts and washers. Be sure to install these exactly as your approved plan has indicated.

The Well-Built Wall Frame

If all goes well, your wall assembly should be secure, stable and plumb. The headers should be supported by jack or trimmer studs that are attached to king studs. The cripple studs should maintain layout for exterior siding and interior drywall. Breaks between the top plates should be offset at least four feet.

Chapter 7

FLOOR AND ROOF FRAMING

Floor Joists

If you have opted to build a multistory building or a foundation wall over a crawl space or basement, you will need to install floor joists or floor trusses. You have worked out the design criteria in Part One of this book. The floor joist or truss span is limited by the size, shape and strength of the structural member. Floor joists are supported by bearing walls, both exterior and interior. The joists may be secured to the bearing wall by either bearing on top of a wall with toe-nailing into the wall or with joist hangers supported by ledgers connected on the side of the wall frame near the top of the wall.

Floor joists serve as the structural frame for a wood floor system. This type of framing system is appropriate in at least two common installations. A short foundation wall can support a wood floor frame, creating a crawl space underneath. Extending the foundation wall beneath grade just a few feet can create a full or partial basement. Floor joists can rest on top of either of these walls and serve as a platform for an additional story. If designed to do so, a frame wall can be built on top of this floor system.

Without ledgers, the floor joists should rest on top of the masonry or concrete wall. However, before they are installed, a wood plate must be installed all around the perimeter of the foundation wall. The base plate should be treated to resist decay. The plate will be anchored to the foundation wall with bolts embedded in concrete. The foundation anchor bolts will be spaced no more than 6 feet apart and be at least ½ inch in diameter, 10 inches long and embedded at least 7 inches into concrete. Drill holes at the appropriate locations so that the plate matches the exterior line of the foundation wall. Install the wood plate over the anchor bolts, then add a washer and tighten the nut on the anchor bolt to secure the plate.

The top of interior walls must be equal in height to the bottom of the joist or truss as it rests on the exterior bearing wall. Verify this first, before installing any floor joists. Then, using the same technique as wall layout, install floor joists, one at a time, on top of the plate. Lay the initial floor joist along the edge of the perimeter of the wall parallel to the direction of the floor joists. Even though it is resting on top of a bearing wall, anchor it down with toe nails and blocking. Then, while maintaining layout, install subsequent joists. Be sure to constantly double-check your layout. Between joists, solid blocking must be installed to prevent rotation. One method of maintaining layout is to precut solid blocking the precise length between joists. For example, if your floor joists are 1½ inches thick and on 24 inch center spacing, the distance between joists should be 22½ inches. If joists are designed to be 16 inch o.c. the blocking should be 14½ inches long. In any case, as with wall framing, the second joist in the layout should be ¾ inch less than the planned layout. For instance, for a center spacing of 24 inches, the layout would be 23¼ inches, 48, 72, 96 and so on. Simply precut blocking to this dimension and install as you are installing joists.

Engineered Trusses

I cannot emphasize enough that a first time home builder should stick to using an engineered truss for a roof framing system. The problems are few. The costs are minimal. Erection time is short. Structural design is backed up by a Registered Professional Engineer. The truss or lumber company will usually stand behind any flaw or irregularity. Difficult roof design can be accomplished much more easily.

Cut roofs or *stick-framed roofs* are the stock and trade of the carpenter who has had years of experience in a variety of complex and difficult roof design challenges. With time and quality lumber, an experienced carpenter can create a magnificent design. The roof of a house will help define its appearance. However, you can still achieve the look of a fancy house by working closely with a truss design artisan at your lumber company. There is usually one person who specializes in designing truss layout. Seek them out and benefit from their talent and experience. For the most part, they are as qualified and experienced in roof design as some architects. They may be capable of providing you with different elevation views from which to select. Choose a roof design and let them design the individual trusses.

Roof design will involve several considerations including the span across the building, required roof slope for water drainage or snow load, the proposed roofing material, the desired elevation appearance or shape as well as the structural aspects such as wind speed and seismic zone. Coordinate with the engineer at the truss plant and create the design you need for proper structural integrity as well as the shape and general appearance you desire. Among the options available to you are *pitched roof truss, open web truss, low slope roof truss, attic truss, low slope composite roof joist,* and *open web flat roof truss.*

Pitched Roof Truss Erection

Your trusses will be delivered to your job site banded together to prevent twisting. When you are ready to install, cut the band and lift the trusses individually in place. It is a good idea to get some help with this chore. Two people will be needed to move the truss in place and secure it to the wall frame while you and a friend secure the top in place with temporary scaffolds. At the base of the truss on the wall frame, blocking will be required. If you place blocking now, you can ensure a perfect layout as well. If your trusses are designed to be placed on 24 inch centers and they are constructed of 1½ inch wide lumber, then cut blocking material 22½ inches long. Place this blocking material in front of each truss to ensure a correct layout. Then place the next truss against this blocking and repeat the process. If your attic is to be ventilated at the eaves, a product is available that will provide both blocking and ventilation. These screened vents can serve as required blocking.

It's a good idea to double-check your layout at each succeeding truss placement. Measure back three or four trusses to verify multiples of 24 inches. Two trusses would read 48 inches, three trusses would read 72 inches, four would read 96 inches, etc. Being off layout is a problem for both the roof decking on top as well as the drywall underneath, so be precise. Both support ends of the truss should be secured by toe nailing two 16d nails from opposite sides of the bottom chord of the truss into the bearing wall's top plate. Stagger each nail a couple of inches away from each other. Also, nail from the side of the bottom *chord* of the truss into blocking with two 16d nails. The top members of the truss are called top chords; the bottom member of the truss is called a bottom chord.

A rake wall is a wall in which studs of different heights define the desired pitch of the roof. For instance, if you intend to build a simple pitch roof, the studs will peak in the middle. With siding applied, the wall frame becomes very stable. When the base of the first truss ends are secured to a rake wall, the top may lean against the frame. You must move the top of the first truss into position and secure it to the rake wall frame. Use a tall ladder to place temporary scaffolding material to the top or bottom of the top chord, near the apex of the truss. (The value of placing the scaffold nailing material underneath is that it can remain even after decking is installed. However, it is harder to install correctly since you have to nail into the bottom of the top chord of the truss, while upside down.)

With the first truss secured in place, continue to install and secure one truss at a time in the same manner. Be sure to maintain the proper layout by using your tape measure regularly. The scaffolding will connect each successive truss. In fact, a scaffold may be 6 or 8 feet long and connect 3 or 4 trusses. In this fashion, each truss will be braced temporarily to the previous truss. Removal of this scaffolding occurs when the final roof decking is installed. This decking will create a diaphragm across the top of the trusses that permanently braces them as a frame assembly.

Pitched roof trusses are hoisted into place with a crane.

Trusses are designed to resist loads perpendicular to the plane of the earth. Because of this, all trusses must be thoroughly and effectively blocked and braced to prevent rotation or other truss failure. This is accomplished by blocking at the point of support on the ends and sometimes at the highest point of the truss. The truss manufacturer will tell you exactly how to block their truss. Be sure to follow their specifications as to location and nailing pattern. Connect the truss to the bearing wall by toe nailing four nails into the top plate. Additionally, install a metal connector between the truss and the top plate to ensure restraint.

Flat Roof Truss Erection

Weather and rainfall in your area will help determine the type of roofing materials you need. Where rainfall is regular and intense, a pitched roof with shingles makes sense. Where rainfall is rare, you could design a pueblo style house and use a (nearly) flat built-up roof system. In some areas of the desert Southwest this style is more prolific than a pitched roof. Of course, the flat roof is not entirely flat.

There must be a minimum slope to allow for drainage when it does rain. The normal minimum slope is ¼ inch per foot. In some cases, local jurisdictions have increased the minimum pitch to ⅜ inch or ½ inch per foot or even more. Trusses can be made to accommodate this design.

Erection of the open web, flat wood truss is just like the pitched roof truss, except it is usually lighter and easier to lift into place. There are a variety of flat roof truss types. Open web trusses are made from wood framing members connected by truss plate connectors just like the pitched roof variety. There is a proper orientation to these trusses. Be sure to verify with the truss manufacturer which side of the truss is designed to be installed on top. Connecting the truss to the bearing wall

A metal connector at the apex of the roof truss serves to stabilize the truss. Notice the diagonal bracing that serves to stabilize the end trusses.

Carpenter installs metal connector between exterior bearing wall and truss.

is usually done by toe nailing four 16d nails through the bottom chord of the truss into the top plate of the wall. Additionally, you must install a metal connector between the truss and the top plate to ensure restraint.

Composite Low Slope Wood Joists

There are a variety of composite wood joists made from wood frame material and either plywood or oriented strand board, (OSB). Wood plate material "sandwiches" plywood or OSB by bonding through the use of glue. The result is a composite wood joist which is usually stronger and straighter than an equivalent size wood joist. The cross sectional shape of the composite wood joist is a little like a steel I Beam.

These composite wood joists are easy to install. Though manufacturers' installation instructions may vary, the installation is similar to that of solid wood joists. The manufacturer will establish the maximum allowable span based on size, shape, center spacing and strength. End and mid-span blocking may be required to maintain the proper vertical orientation. Stiffeners at the end of the mid-span of the joist are sometimes recommended by the manufacturer. These are pieces of the same composite material, cut to infill the space between the flange and the wood web members. This effectively prevents the ends from collapsing from abnormal stresses imposed upon the flange at the point of support.

Composite low slope wood joists are a manufactured product with an engineered design. As such, they must be installed according to the manufacturer's recommendations.

The method of connecting the composite wood joist to a bearing wall depends upon the method of support and the manufacturer's specifications. If the composite wood joist is setting directly upon the bearing wall, nail through the bottom flange of the joist into the top plate with at least two 16d nails. If the composite joist is supported in a ledger fashion to the bearing wall, an approved joist hanger may be required. The hanger must be an adequate size and be specified by its manufacturer's recommendations.

Interior Bearing Wall Support

Solid wood joists or composite wood joists may not be long enough to span the entire length of your house. Interior bearing walls may be required to support these members. The finish height of the bottom of the joist at the exterior walls must be the same as the height of these interior bearing walls to create a level ceiling. Interior frame walls resting on a concrete floor must have a bottom plate that is pressure treated to resist damage from termites. The interior wall frame must be built using all of the framing members of the exterior wall. The framing methods must be the same as well. Mechanical connections between wood framing members such as nails and framing ties and clips must be installed in the same manner. Layout for studs within the wall frame must follow the same procedures as the exterior bearing wall frame. When complete, the wall frame must support the floor load by supporting as much of the load as an exterior bearing wall. So do not skimp just because it is an interior wall.

Chapter 8

FLOOR AND ROOF DECKING

Material Selection

There are a variety of basic materials to choose from in order to cover your roof or floor frame. Generally, a sheet of decking material will be 4 feet wide and 8 feet long. Composition of the panel could vary—from a multiple layer plywood to oriented-strand board to wafer board. The APA: The Engineered Wood Association, is one association that establishes standards for engineered wood panels. They will validate a product through exhaustive testing and continuous inspection of both the product as well as the plant. Their stamp on each sheet of material will serve as your assurance that the product will perform as designed. Each sheet is identified with a panel span rating. This rating will indicate the maximum permissible span based upon both the use and the center spacing of the supporting structural frame.

For example, a certain plywood may have a panel span rating of 32/16. The first number represents the maximum center spacing in inches allowed for a roof application. The supporting roof trusses or rafters may be at or less than 32 inches on center for the plywood to be adequately supported. The second number reflects the center spacing in inches allowed for this particular panel to act as floor decking material. The supporting floor joists may be no more than 16 inches apart to safely support the plywood floor.

The same is true for the other types of manufactured decking products. Each of these products should be tested by an approved agency and listed according to the testing criteria. The listing will describe its approved method of installation. Each panel is identified by its grade and certificate of inspection. There are similar stamps on each panel that certify the approved span and use of each product.

General Requirements

First, all sheet-type paneling installed as floor or roof decking should have staggered joints. In other words, the edge of one panel must adjoin the middle of the edge of another panel in the next row. You achieve this staggered effect by saw-cutting the first piece of paneling in half at every other row. The roof or floor decking acts to transfer wind forces from a windward wall to the adjacent wall and ultimately to the foundation. A staggered panel roof or floor deck will more effectively transfer these shear forces, which come from many different directions, into the plane of the deck.

Second, certain roof decking materials must be spaced a certain distance apart in order to accommodate expansion in the panels and maintain the alignment of each panel. Metal spacing clips may be used to achieve this purpose. Simply place the required number of clips on the ends of the first panel and install subsequent panels into this metal clip.

Third, the long dimension of all panels should be placed perpendicular to the framing members. Any section of panel must be supported by at least three structural members (rafters or joists). The panels must be nailed with 8d nails at not more than 6 inches on center around the outside edges and no more than 12 inches on center in the field (interior) of the panel. In certain special conditions such as a high wind area, nail sizes must be greater and nailing patterns must be closer together.

Fourth, with proper center spacing, each panel will extend from the center of the first joist to the center of the last joist. Be careful to verify proper center spacing. You may need to adjust one or more joists or trusses to maintain proper layout. Otherwise, you will be wasting lots of paneling.

Plywood must be installed with staggered joints.

Attic Ventilation

Along about here, you need to decide how you will ventilate your attic. If you use gable end vents, they are already installed in the wall frame. If you planned to install soffit vents, now is the time. At this time, your trusses or rafters are resting on and overhanging exterior bearing walls. At the ends, and in between these trusses or rafters is where your soffit vents will be placed. They are sold by most large hardware stores, or you can make them very easily. Measure the distance between each truss. If your trusses were constructed of 2 inch nominal thickness and your layout was 24 inches on center, the distance between them should be exactly 22½ inches long. The height will be from the top plate to the top of the truss at the outside edge of the top plate. Cut a piece of thick plywood to these dimensions. Using a hole saw or bore, cut two 3 inch circles in the plywood at equal points along the length. Then, staple some wire screen on one side of the board over the hole. This assembly is referred to as bird-blocking since it provides attic ventilation at the soffit, yet prevents birds from using your home as their nursery.

Under-Floor Foundation Ventilation

As in attics, crawl spaces under a floor are subject to a build up of moisture. Natural ventilation can prevent this moisture build up. In order to achieve this ventilation, openings in foundation walls must be provided to permit the drying action of air. The size of openings are regulated by the IRC. The openings

Wooden louvers in front of insect netting is an aesthetic solution for providing attic ventilation.

must be at least 1/150 of floor area, and they must be located so that cross ventilation will be achieved. Screens may be installed over these openings to prevent the entry of vermin.

Floor Decking

For a solid floor, I recommend at least a 3/4 inch thick tongue-in-groove panel rated for not less than the spacing of your floor joists. The method of installing the tongue-in-groove is to glue the bottom of the panel to the wood joist below. Use wood glue and squirt out a 3/8 inch wide ribbon along the top of each joist. When enough glue for one panel has been placed, ease it into place. Verify that its placement is true and square. Then connect the four corners with nails. After its initial position has been fixed, complete nailing as outlined earlier.

Roof Decking

Roof decking could include any of a variety of materials, including sheet products such as wood structural sheathing. Use of one of these materials represents a very economical method for decking a roof. The sheathing also provides the required shear transfer for lateral loads caused by wind or seismic forces. I recommend that you wear skid-resistant shoes when working on a surface as slick as sheet panels.

To avoid repeated trips up and down a ladder, build scaffolding that will store a few sheets at a time. You can't do this until after the first or second row. Before installing panels, check for a level surface on top of rafters or joists. This can be done with a string line or a long piece of *straight* lumber at least 10 feet long. If the surface is not level, use shim stock as necessary to create a level surface. If rafters or trusses are warped or bowed, use either blocking or bar clamps to straighten them to fit the panel edges. If trusses are excessively warped, contact the truss manufacturer and discuss the problem before you install decking.

Begin installing the decking with a starter row along the bottom end of the roof line. Then stagger the joints and work your way up one row at a time to the apex of the roof. The requirement to stagger the joints means that you must cut a full sheet in half every other row to accommodate the correct layout. The overhang portions that are exposed to weather will need to be rated for exposure. Nailing requirements are that a 6d nail be placed every 6 inches on the perimeter of the panel and attached to a framing member every 12 inches in the field. In high wind areas it is recommended that you increase the required nails.

Roof Drainage

Into every life, some rain must fall. The same is true for all roofs. Because of the nature of a low slope design, parapet walls are sometimes built on some or all sides of a flat roof to hide the roof from view. When this is done, holes in the parapet wall above the roof line must be made to allow rain to drain off the roof.

These holes are channels for rain to be discharged from the roof. They usually protrude out beyond the exterior wall. In some locations they are called *canales*, which is a Spanish derivation of the word canals or channels. Several of these canales are spread along the lowest slope of the edge of the roof.

These canales can be clay, masonry or metal and are installed into the roofing system. On a low-slope roof, there is a need for diverting rain water toward these canales. The diversion method used most commonly is a simple application of wood framing on top of the roof decking under the built up roof assembly. Plywood is used to create the effect by using triangular pieces abutting each other on the common high end. This method of water diversion is known as a *cricket*. The cricket is covered with flashing and then roofing material, just like the rest of the roof.

Different Roof Lines

Depending upon your carpentry skill, you may have selected a complex roof design with different roof lines. In that case, these roof lines will intersect and must be connected together. Where these connections occur, plywood must be cut to fit the different slopes with unusual shapes. Care must be taken to measure and make the proper cut for each of these different cuts. You may need help in making the cuts. Rely on experienced advice or assistance when making cuts that exceed your skill level.

Careful measurement and cutting is necessary to assemble roof decking over different roof lines.

Intersecting rooflines.

Window Installation

Windows can be installed after walls are erected and before siding is installed. However, it is best to wait until after the roof and floor framing members and deck material are installed to avoid accidental breakage. Each window manufacturer will specify the dimensions of the rough opening required for proper window installation. Build the rough opening in the wall frame according to these dimensions. The lower portion of the header should be at the height of the window. Trimmer studs should align with the window's sides and a sill plate should be located at the bottom end of the window. It is very important that you verify the squareness of the opening. Nothing looks as noticeable as a crooked window. Use a building level, framing square and even a plumb bob to determine a perfectly square opening.

Building paper installed around the wood frame opening will help seal out weather infiltration. When the window is placed over the opening and carefully moved into place, the paper will serve to seal the joint between the window and the wood frame. Use nails to secure the window to the frame. Drive nails through the metal flange into the header, trimmer studs, and sill plate. The type of nail and center spacing should be specified by the product manufacturer. Be very careful with driving nails close to a glass surface. Consider using a trim hammer for better control.

A window is installed in a wall frame over building paper.

Roof or Floor Deck Nailing Inspection

___ Are roof or floor plywood panels installed perpendicular to roof joists or trusses?

___ Are plywood panels nailed to trusses with an 8d nail @ 6 inches o.c. on plywood edges?

___ Are plywood panels nailed to trusses with an 8d nail @ 12 inches o.c. on plywood interior?

___ Is blocking installed above the bearing wall between joists or trusses?

___ Is blocking installed at apex of roof system if required by truss manufacturer?

___ Are metal framing connectors installed between wood top plate and each joist or truss?

___ Are nails galvanized type and resistant to weather decay?

___ Is there a support under each edge of plywood decking? You may need to add blocking to provide support.

___ Are the plywood sheets staggered so that they do not have a common end joint?

___ Are the plywood sheets oriented so that they are perpendicular to the layout of rafters?

___ Are canales installed in parapet walls on the low or drainage end of the flat roof system with a parapet?

___ Are crickets installed properly to divert rain water toward canales in the same application?

___ Is flashing material installed in areas of the roof subject to potential water penetration?

Uh Oh!

Problem

The grade of wood products is different than plans call for or is not stamped at all.

Solution

Don't buy wood without a grade stamp. If you do, return to the lumber yard and get a letter from the mill attesting to the design strength of the wood. Try to get a letter of testimony for the wood from an approved grading agency if possible.

Problem

Trusses do not bear a manufacturer's label stamp, which attests to their origin and design. Engineered design for trusses is in question.

Solution

Don't accept any trusses on job site without a stamp identifying their origin. Always acquire a copy of the engineered, sealed calculations and design of each of the trusses you purchase. After all, would you buy tires without a warranty? In the event you purchased trusses from a factory that does not employ a registered professional engineer, you can still have an independent engineer analyze your trusses and certify their design. It will be a costly lesson!

Problem

Excessive cutting or notching of framing members.

Solution

Along the side of the stud, joist, or other framing member that is over-cut or notched, add another full-size stud or significantly long member along the side of the affected structural member, and remember not to do so again.

Problem

Wall intersections and corner construction are inadequate.

Solution

You must add an extra full-size stud to make the proper corner. Remember, at least three studs are required to make a corner. You can do it a number of ways, just be sure it's secure.

Problem

Lateral bracing, such as shear walls, are inadequate or are nailed improperly.

Solution

Remember that the shear wall is an integral part of the structural integrity of your house. Do not skimp or try to stretch the limits of this vital part of your wall frame. The rule is that every corner and every 25 foot interval must have at least a 48 inch plywood panel (most wall sheathing counts). The panel must be applied vertically with at least three studs and nailed with 8d nails at 6 inches o.c. on the perimeter and 12 inches o.c. inside the perimeter. If these are missing, try using an alternate shear wall from where it is required. Any shear wall panel farther than 12 feet away would not meet the requirements of the IRC for a shear wall.

Problem

Columns and posts are installed differently than specified on the plan.

Solution

You may find that the connector that you specified when plans were drawn is not available when you start building and you are forced to use an alternate connection. Have the manufacturer's details and specification on hand indicating that the new hardware is equal to that specified on the plans.

Problem

Sizes of headers and beams are not per plans and are too small, but are currently bearing roof or floor loads and cannot be replaced.

Solution

You should have brought the plans into the building department prior to the inspection in order to verify that the new beams or headers were adequate prior to their installation. If you haven't done this and they are substandard, try installing a steel *C* channel or a double steel angle below the header supported by the wood trimmer stud. Check with your building department before you install a substitute.

Problem

Overall height of building above finish grade is greater than plans permit.

Solution

Some locations limit the height of a building for zoning or aesthetic reasons. If you have inadvertently done this, there are some remedies available. Since height is measured from grade, the best method is to add soil to your finish grade to make the overall height lower. This would look like a dirt berm within 6 feet of the building wall, creating a channel next to the building. If the berm is within 6 feet of the building, it establishes a new reference datum for the finish grade. But, surface drainage may be a replacement problem. Water must be diverted from the building.

Problem

Attic access is not installed or is not the minimum of 30 inches by 24 inches.

Solution

While the inspector is on-site, enlarge the opening to the required size by moving a piece of framing lumber. Remember that the attic access should be accessible, as the name implies. Locate the access in a room that permits rapid access, like a laundry or hallway, not a small clothes closet.

Problem

Sill plates are not pressure treated or foundation grade redwood.

Solution

If you forget to install foundation grade redwood or treated sill plates, shame on you! These plates are designed to resist decay and ward off termites and other vermin. Try using an after-market poison product, which sprays onto the plate much like insecticide. This is not the best solution, but it may satisfy the inspector. Show the details of manufacturer, warranty and product listing.

Problem

Anchor bolts for sill plates are not installed or are installed incorrectly.

Solution

Remember, when the wind blows, the shear wall can only do its work if these bolts are installed into the footing no more than six feet apart with at least two bolts per plate segment. If this is not the case, you must buy an after-market expansion bolt and drill holes through the plate into the foundation at least 7 inches. The holes must be sized according to the expansion bolt diameter. Read the instructions for drill bit size. When the hole is drilled, remove all of the dust and debris with an air compressor before setting the bolt. In certain instances, you may need to use an epoxy grout to secure other types of bolts. Check with the manufacturer for requirements.

Problem

Metal connector clips are not installed between walls and roof trusses with proper nails.

Solution

This is an easy fix. You can install them at this stage. In fact, that's when it's done normally. Just be sure and use the proper nails for the connector. Read the manufacturer's installation instructions for proper methods and nail type.

Problem

Missing fire blocking throughout wood wall frame.

Solution

All wood frame walls must retard or attempt to block the progression of smoke and fire from entering the roof and attic. This is achieved in frame walls with fire-blocking. It is simply any solid material such as 1½ inch wood framing, 1 inch thick plywood assembly or similar material that can be installed inside the wall frame. Normally, the top plate serves as the fire-blocking. However, if the roof trusses are supported by a ledger, then the top plate extends above the truss support, and the wall must be blocked at or below that level in addition to the top plate.

Problem

Trusses or roof framing are missing structural blocking or solid blocking at points of bearing.

Solution

At the point of bearing, all trusses and roof or floor joists must be blocked with solid blocking to prevent rotation. That means that a 2 by 12 serving as a roof rafter must be blocked with a 2 by 12 block. Trusses may use a variety of blocking methods approved by the truss manufacturer. Be sure to ask the truss manufacturer exactly how to block their trusses, then have those instructions on-site during inspections to validate your methods.

Problem

Plywood sheathing is not installed correctly. It runs with the framing members instead of perpendicular to the framing members.

Solution

By far, the most common mistake made by an amateur is installing plywood vertically in relation to the framing members. This is commonly called *railroading*. The listing on most engineered wood products requires the panel to be installed horizontally with respect to the structural load carrying members. Its design strength anticipates this method of installation. Fixes for this are limited to installing a framing member between each joist or truss every few feet. This improves, but does not meet, the intent of the product design. The panels would probably lose their warranty. Be sure to read the manufacturer's listing for installation methods and procedures, since every product is slightly different.

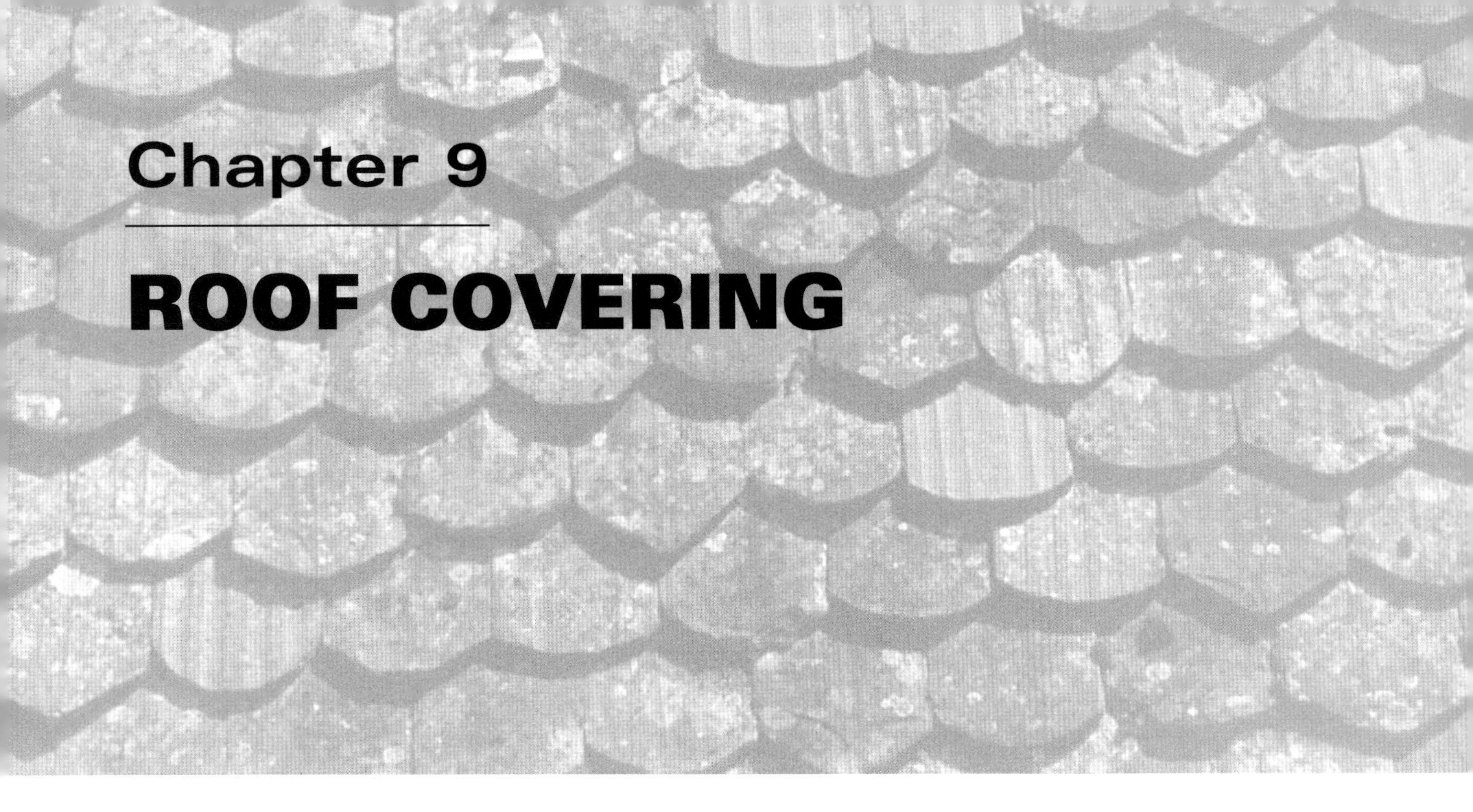

Chapter 9

ROOF COVERING

Choices

Slope considerations will dictate the most appropriate roof covering material. The type of roofing materials necessary will depend upon the configuration and slope of your roof structure. Generally, the greater the slope, the less water-repellent the roof materials must be. In the extreme, a nearly vertical roof (as in an A-Frame) may use materials appropriate for siding with no traditional roofing materials applied. Conversely, a flat roof may even be designed to hold or *pond* rainfall for domestic use (as in a cistern). In this case, the roof cavity must be designed to be waterproofed with materials equivalent to a swimming pool, and the roof superstructure must be designed to support the weight of the water.

Most roofing material and application decisions will be based on the slope of the roof. For slopes of 3:12 or greater, a 3-tab composition shingle over layers of an approved underlayment is satisfactory. Between 1:12 and 3:12 slopes, a more substantial roll roofing over double layers of heavier underlayment may be needed. For near flat to 1:12 slopes, a built-up roof application, sealed on top with aggregate, may be needed with melted tar applied between coats.

Materials

Shingles are a thin layer material applied in overlapping rows and can be composed of many different types of material including wood, metal, slate, concrete or water repellent fabric. Composition asphalt fiberglass shingles are a very common shingle material. Concrete or slate tile is more expensive, but lasts longer. Wood shingles or shakes are very commonly used in timbered regions. Built-up roofing is successive layers of roll roofing joined and connected by nails and roofing tar. Sometimes the finish surface will be sealed with gravel over hot tar to form a stronger, more protected surface. Liquid-applied membranes such as elastomeric roof coatings are available for certain roof slope applications. Metal roofing panels are both long lasting and attractive.

Asphalt or Composition Fiberglass Roofing Products

There are three types of asphalt roofing products: underlayment, roll roofing and shingles. Underlayment and roll roofing are both asphalt-impregnated and are sold by weight such as 15 pound, 20 pound, or 110 pound. This weight designation represents the weight of the particular material it takes to cover 100 square feet of roof decking. Underlayment is used only as an initial roofing base and is not intended to be exposed to weather. Roll roofing is strengthened with additional asphalt and mineral surface and is intended to be an exposed weather barrier over an underlayment. Most asphalt/fiberglass shingles have a surface coating of mineral granules over emulsified fiberglass and asphalt. Consider having the material supply company actually stock your shingles (which are very heavy after 10 trips up a ladder) on the roof for you. The stocking is usually included in the price anyway because of the public demand.

There is an alternative product, which is superior to the traditional three-tab shingle: the laminated asphalt shingle. This shingle is the same size and shape but has no slits, thus no tabs. This type of shingle is known as an architectural shingle, due to its appearance. It is a heavier grade and looks magnificent. The warranty is much longer than on traditional shingles, which are lighter and less substantial.

Flashing

Where different portions of roof lines come together, they will create a valley. This valley scribes the bottom of each separate roof assembly's section. Because it is subject to significant drainage, it must be equipped with metal flashing to better support corner shingles and prevent leakage. This particular flashing is called valley flashing. It should extend at least 8 inches on either side of the joint being covered.

Whenever different materials are encountered, metal flashing is also installed to help seal the joint before roofing material is installed. A common example is where a masonry chimney breaches a wood roof deck. Because of the potential of a fire, wood must be separated from the chimney walls. Metal flashing will be installed in a cove fashion around the perimeter of the chimney at the joint. Skylights are another example of material that may need flashing. For plumbing vent penetration, there are special boots or covers available that install over the vent pipe and lie under the roofing material.

Metal Drip Edge

A galvanized metal drip edge should be installed on the edge of the roof deck. Nail the edge into the roof deck from the top by using galvanized nails. This drip edge prevents premature decay in the wood roof facia by diverting water away from the wood, allowing it to drip off.

Underlayment

The more weight used for the underlayment, the better. The choices are 15 pound or 20 pound felt. I recommend a 20 pound felt underlayment. It's easier to manage on a roof subject to wind and its water repellent properties are better. The underlayment is installed first directly onto the roof decking. The underlayment is nailed with very small roofing nails. For roof slopes between 2:12 to 4:12, begin at the eave and apply a 19 inch strip of underlayment felt parallel with and starting at the eave. Then install the 36 inch wide sheets overlapping successive sheets 19 inches. Roof slopes exceeding 4:12 are installed in a similar manner, except the overlap is reduced to at least 2 inches. Any horizontal joint in the underlayment should be lapped at least 6 feet, though manufacturers' requirements may supercede these requirements.

Composition Shingles

The most common type of roofing material installed on medium slope roofs is the composition shingle. There are several reasons for its popularity, including cost, value, durability, ease of installation, and appearance. The cost of the composition asphalt-fiberglass shingle makes it most popular. Colors and grade weight make this type of roof very attractive and affordable. It's easy to install, and depending on the quality of the product, it provides a long-lasting roof. The shingles come in a variety of colors and textures. These composition shingles are relatively easy to install.

They are described as 3-tab shingles because they have two slits that create three tabs for half of the width of the shingle. They are 36 inches long by 12 inches wide and are sold in packages that are one third of a square (a square is 100 square feet), which means that you need three packages to cover 100 square feet. This coverage assumes that you provide the proper lap over the previous shingle. The lap is well-defined on the shingle by a variation in the color. The exposed texture and color will be pronounced along the line of lap on ascending rows of shingles.

Getting Started

Getting a proper layout is essential to ending up with a straight roof. You do that by squaring your roof section. Begin by measuring each side and comparing the two lengths. If there is some difference in length where there is not supposed to be (and it will only be an inch or two), spread out the difference over the entire width of the roof section, and it will be barely perceptible along the course of shingle application. It would be a good idea to mark the plywood every twelve inches (1 foot) with a pencil mark to help you in your subsequent layout.

Next, use a chalk line and lay out the line for the underlayment sheeting. In the most simple application, a pitched roof, underlayment is laid on the roof deck. This underlayment has printed guidelines on which the line of shingles may align. Continue laying the underlayment so that it is never more than two layers ahead of the line of shingles as you progress upward on the roof slope.

Installing Shingles

Shingles are placed over this underlayment a row at a time. Begin on the base of the roof line with two layers of shingles that are staggered to maintain over-lap coverage. The first row of shingles, known as the *starter strip*, is usually doubled (with the bottom row inverted) due to potential damage by wind. It should overhang the metal drip edge slightly.

The first shingle will be installed in a corner of the roof. Continue until you reach the other end of the roof. You may have to cut off a portion of one shingle. When you start the second row, begin by cutting the shingle. The size of the offset with which you start will dictate the appearance for the rest of the roof. You will notice a unique herringbone pattern develop, which is created by the offset. The offset could be as little as six inches or as much as 18 inches. Decide the style you want by looking at other shingled roofs and measuring their offset. Read the back of the shingle package for a more detailed explanation.

The first line of shingles should extend slightly beyond the metal drip edge. The first course establishes the slots in successive rows of shingles. You can see that the top of the slots in the shingle is covered by the shingle in the next row. Successive rows are laid overlapping previous rows. The overlap dimension is usually specified by the manufacturer. The exposed portion of the shingle is normally between 4 and 6 inches. Generally, shingles are nailed to the roof deck with galvanized steel roofing nails which are long enough to penetrate the decking at least ¾ inch. The nails must be at least 12 gauge shank type and have a minimum ⅜ inch diameter head. The nails are placed on the *top* of the shingle in the area that will be covered by the next ascending shingle. There should be a minimum of four nails in each shingle or more according to the manufacturer's instructions on the shingle package. If the slope exceeds 20:12, special methods of fastening are required.

It is important to keep the rows parallel to each other and not to meander above or below the true line of reference, which is parallel to the bottom and top of the roof line. Maintain your layout for the row of shingles. Do this by regularly measuring from the ends of the roof line upward along the rake of the roof a distance equal to the next shingle. Then, striking a line from side to side, create a border to follow for the next row. A reference string line stretched across the roof can help as a guide.

The ridge line of the roof can be a source of leaks, so do not skimp at the top. The row closest to the top should ideally extend to touch the row of shingles from the other side (on a simple pitched roof). Fold or cut the shingles so that they just meet each other. Then add the ridge cap by folding a shingle at one end of the roof and nailing on either side of the apex of the roof into the roof decking. Follow this pattern until you reach the other end of the roof along the ridge.

What you have learned so far is valid for all simple sections of roof decks that are a simple pitch and rectangular in nature. If you have built a valley or hip roof structure, you must deal with the geometry created by that variation. In addition, a metal valley flashing must be installed prior to the underlayment. Nail this metal flashing into the decking, then repeat the steps for underlayment and shingle application. When you reach the centerline of the flashing with each shingle, you can either cut along the shingle along the line of the metal flashing or weave the shingles together from either side. If you cut, be sure to cut straight. The straighter you cut, the better the job will look.

Nailing

Nails for shingles are usually galvanized to avoid rust. They are formed with annular or spiral thread to resist withdrawal. The location for nailing is clearly identified on each shingle package. Do not nail too near the edge or slot, which can cause tearing. Shingles stay in place by sealing themselves onto the lower shingle. There is a strip of mastic on the bottom of the upper shingle covered with plastic. The mastic is activated by heat from the sun. Strip off the plastic and let solar energy do its job. It will seal the upper shingle onto the lower shingle and prevent wind from blowing it off.

Flat or Low Slope Roofs

In order to prevent rain from leaking into your new home, roofing materials must be appropriate to the slope. For example, a medium sloped roof may use shingles, but a lower slope roof must use a more weather-resistant barrier such as roll roofing material or even a built-up roof system. Most codes require that you provide a slope to discharge water safely off a roof.

If you have designed one of these low-slope or flat roof systems, you may need to provide a built-up roof. This roof includes successive layers of underlayment, each mopped with hot asphalt to seal that layer. The final layer could be covered with gravel to prevent erosion of the asphalt. Another option for a flat roof is to provide underlayment without hot tar mopping. After two or three layers of underlayment, install a heavy (110 pound) weight roll roofing. This roll roofing will be sealed to the previous layer with mastic along the edge to prevent wind uplift. As with other roofing methods, an overlap is required between layers of material according to the manufacturer's specifications.

Metal Roof

The use of a metal roof used to be limited to barns and pre-twentieth century houses. Now, more and more houses of all values are turning to this type of roof. Colors are available that dramatically improve a home's appearance or curb appeal. They are really beautiful, easy to install and very long-lasting. These roofs usually are made in 24 inch wide segments, and since they are manufactured uniquely for you, they can stretch the full pitch of your roof.

Alternatives

There are a variety of other new roofing products on the market, which have a good reputation for reliability and a short learning curve for the amateur applicator. There are synthetic chemical polymers and urethane-based materials, which are particularly beneficial for flat or near flat surfaces. Some roofs may be designed to pond or channel water to cisterns. These roofs must have water-resistant lining material. Roofs designed to pond water must have structural members designed to hold that additional weight as well. Metal shingles are a manufactured product that look very much like wood shingles but do not erode as quickly.

Among the prettiest of all roofs is the concrete tile roof. This is a roof to last a lifetime. It is very popular and very heavy. If you decide to use this roof, design the roof trusses for the added weight. However, there are numerous brands of light-weight alternates that look very much like concrete tile, but are material substitutes. The concrete tile and its lightweight clones are only intended for slopes exceeding 3:12.

Inspections

Even though most jurisdictions do not inspect the installation of roof covering materials, it is still important to do the job correctly. Weather penetrates the roof that is not installed correctly. In most cases, the correct method for installing a particular roof covering material will be based on the manufacturer's installation instructions. While the code will specify certain conditions for roofing installations, an inspector will not necessarily be available to guide you in this aspect of construction. You may need to seek advice from other reliable sources such as the manufacturer, product vendor or even a roofing subcontractor.

Roof Covering Installation

___ Is roofing material appropriate for the slope?

___ Is underlayment installed correctly, with a proper lap?

___ Is metal drip edge installed properly?

___ Are shingles sealing themselves together? Is plastic tab removed?

___ For low-slope roofs, are canales, crickets and cant strip installed correctly?

Uh Oh!

Problem

Roof covering is not appropriate for the slope.

Solution

The toughest question is how to fool rain into thinking that an inferior roofing material is adequate. Nature doesn't usually compromise. Unfortunately, there are only two options. You can take the old material off and start again, or, in some cases, you might be able to cover the existing roofing with an adequate roofing material and regard the initial roofing material as the underlayment. Check with each product manufacturer to verify if this is an acceptable option. Look for adequate connection of the second layer by using correctly sized roofing nails that penetrate through the first layer and into the deck.

Chapter 10

WALL SIDING

Timing

Wall siding can be installed either before or after stud wall erection. It depends upon whether you think you can lift a wall section with siding installed. Lifting a frame wall without siding is heavy enough, depending upon the length and height of the frame wall section. Siding could add over 30 pounds for every four feet of wall frame length. Test yourself first.

Of course, it is easier to install siding horizontally on the floor, and the degree of precision you can achieve by laying the siding onto the wall frame is significant. It's also much easier to achieve a true square in the frame wall. Then, when the wall is erected, you can avoid the need to rack the wall into true square.

Materials

The easiest exterior siding for a novice to install is an exterior grade panel siding. You could install this exterior grade panel siding even if you intend to cover the exterior walls later with stucco or masonry veneer such as brick. The siding attached to the wood wall frame acts as a stabilizing shear panel, which helps resist wind forces. There are other approved methods, but for the money, siding is the best. It serves as a finish exterior siding as well as a backing for future veneer or even stucco.

Siding comes in several qualities and price ranges. Obviously the best is what you should install, if you can afford it. Look for a solid tongue and groove plywood siding. This exterior grade plywood siding has a textured appearance that looks beautiful when stained or painted. Nailing requirements for this siding are subject to the requirements of exterior plywood.

Building Paper

Exterior siding performs many significant tasks. It acts as a shear wall to prevent the building from turning over. It acts in harmony with the wall frame to achieve structural integrity. It supports windows and doors within the exterior wall. It holds insulation in place. It supports interior drywall applied to the interior side of the wall. It also serves to prevent weather from entering into the home. Its ability to do this is enhanced by applying building paper under the exterior siding. There are a variety of different products available that serve this purpose. Some building wrap products are very strong and durable.

Grade D paper is an asphalt-based building paper that is available in rolls 40 inches wide and 150 feet long. The paper unrolls easily and can be attached to the studs along the exterior of the wall frame. Attach with staples or small nails. Begin the installation from the bottom. Circumnavigate the perimeter of the building before adding a second row above the first row. Allow the second row of paper to overlap the first row at least 3 inches. Continue around the building and add successive rows until you reach the top of the wall. Cover openings such as window and door frames as if they were not there. Then, install siding as outlined in this chapter. Later, carefully cut the paper around these openings from the inside to ensure a tight seal at the frame openings.

Installation

If you built the wall frame, including top and bottom plates, to be exactly 8 feet high, the 4 foot by 8 foot panel should fit exactly. If you designed a slightly higher wall, be sure to block the studs in the wall frame at the panel height (8 feet). Depending upon how careful you were on your stud layout, the edge of a four feet panel should fall directly in the middle of a stud. If not, you must either cut the panel or add blocking to provide support for the panel and accommodate the overlap.

However you install the siding, before you erect a wall section or after, be sure to use corrosion-resistant nails. Nails should be at least 8d in size and nailed at least 6 inches on center on the outer perimeter or boundary of the siding and should firmly engage the stud supporting the siding. The interior of the siding should be nailed to a framing member with the same nail every 12 inches. Be sure to nail the siding to any blocking installed in the wall frame.

Openings such as window and door frames can be difficult to install without careful measurement and a perfectly square opening. This is a task that should never be rushed. Use a steady, level surface when measuring and cutting. Strike the proposed opening with a chalk line to ensure a straight cut. Use a circular saw with the appropriate blade to make the cut. Stop just short of the corner with the circular saw blade. Because it's circular, if you continued, it would cause an over-cut. Then finish the corner cut with a hand saw or power jigsaw.

Exterior Horizontal Lap Siding

Another possible method that is easy for the beginner is to install horizontal lap siding. Plywood sheathing must still be applied to exterior wall frame as specified for shear walls. This plywood sheathing is required to provide rigidity and a substantial backing for the siding. The primed hardboard siding is made from composition wood particles and is treated with materials to resist decay. The siding is usually 12 inches wide by 12 feet long. The thickness for this material is usually $\frac{7}{16}$ inch.

The siding is easy to install. Begin at the bottom of the wall and complete each row with butt joints between 12 foot sections. Then, on each successive row, overlap the siding in a vertical manner and connect at least two corrosion resistant nails to each stud. The overlap is usually at least a couple of inches but is always specified by the product manufacturer. Be sure to use a tape measure to maintain the exact lap between rows. Use the same pattern you used for layout of roof shingles to establish and maintain proper layout.

This type of exterior siding costs a bit more than panel sheathing because you still have to provide plywood underneath. Although this type of siding is slightly more expensive, it is extremely easy to install and achieve professional looking results the first time. As with panel siding, you must paint or cover the lap siding to prevent long term decay. An exterior grade latex house paint is most appropriate.

Wall Siding Installation

___ Is siding properly installed according to listing and manufacturer's specifications?

___ Is siding labeled with manufacturer's name? Listing requirements?

___ Are the installation instructions for wall siding available? Are the listing requirements available from manufacturer?

___ Are galvanized nails used? *Section R703.4 and Table R703.4*

___ Are enough nails used? In the right locations? *Table R703.4*

___ Is blocking installed at top of panel edges where studs are longer than 8 feet? *Section R703.3.1*

___ Is siding nailed to this blocking? *Table R703.4*

___ Is siding damaged by nailing? Does this void the manufacturer's specifications?

___ Is a weather barrier required to be installed under siding? *Section R703.2*

___ Do lap joints fit tightly together? *Section R703.3.1*

___ Does the stud spacing match the manufacturer's specifications for siding support? *Table R703.4 footnote (a)*

___ Do construction joints between dissimilar building materials fit tightly? *Section N1102.1.10*

Uh Oh!

Problem

Siding is not installed according to its manufacturer's listing for stud and edge support spacing.

Solution

If the wall siding manufacturer requires the studs to be spaced no more than 16 inches o.c., and your stud layout is 24 inches on center, you have a problem. If you leave it, the siding will warp in an outward manner when exposed to time and weather. The right thing to do is to add an extra stud in between every two studs to make the layout 12 inches on center. Short of that, ask the manufacturer for consideration to add a couple of pieces of blocking in the stud cavity, perpendicular to studs, as an extra support. This will be time consuming and make insulating harder, but it will keep your siding straight.

Problem

Galvanized nails were not used.

Solution

Add galvanized nails in between every other nail.

Chapter 11

PLUMBING: ROUGH-IN AND TOP-OUT

Which First?

Always plan for how you get water or waste out of your house before you plan for bringing it in. This means that you will install the drain, waste and vent system before you install water supply pipe. If you install water supply pipe under a slab, this means that your DWV system will be installed and covered with sand to provide a comfortable bed on which it may rest.

The Drainage, Waste, and Vent System

The purpose of the DWV system is to remove wastes while venting away sewage gases. The DWV system includes vertical and horizontal drain pipe, vertical and horizontal vent pipe, fixture traps, trap arms, and waste clean-outs. Waste from the fixture, such as a sink or shower, drains into a trap (which blocks gases from entering the house), then travels through the trap arm into a drain pipe, then into larger drain pipes until it ultimately empties into the building sewer tap or septic tank.

The material most commonly used in residences for DWV pipe is a combination of Monomers Acrylonitrile-Butadiene-Styrene, abbreviated ABS, and is available in standard trade sizes such as 1½ inch, 2 inch, 3 inch, and 4 inch. This plastic pipe is a manufactured product and must be installed according to the listed manufacturer's instructions. It has significant toughness and corrosion resistance properties. The material has been reliable for use as drainage piping for over 35 years. It is resistant to rodents, termites, and other vermin. The joint connection integrity with approved solvents has met the test of time. The pipe material has a smooth interior surface, which resists deposit formation. The pipe may be cut by traditional methods such as a saw and connected with fittings only with approved ABS cement.

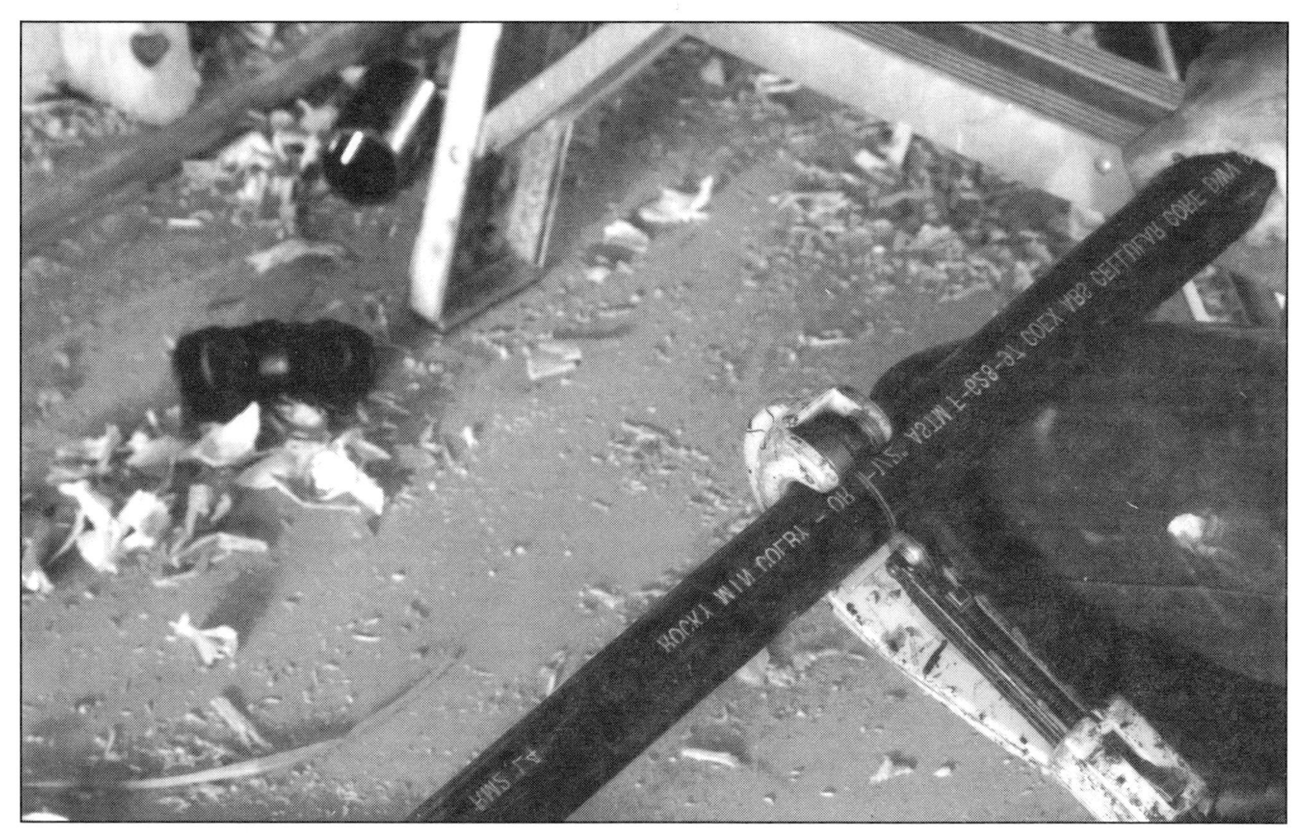

A plumber makes a cut of a section of ABS pipe with an approved pipe cutter.

Polyvinylchloride (PVC) pipe is acceptable for use as drain, waste and vent pipe as well as exterior water supply pipe if it is buried to a proper depth. There are three types of PVC pipe available; I, II and IV. Type I is the most commonly used pipe. Type II is less resistant to damage by abuse. Type IV is commonly known as CPVC and has a higher resistance to heat, thus making it suitable for hot water. It is also approved for use within the house to serve as water supply pipe for both hot and cold water. Sections of PVC pipe can be cut using saws equipped with a carbide-tipped circular blade. There are a few different methods of joining PVC pipe together. Solvent-cemented joints are the most common method of connecting PVC pipe and fittings. Special glue is used to connect joints and fittings together. Threaded connections are available in Schedule 40 and 80 PVC fittings. Transition joints are fittings of materials different than PVC. In other words, PVC pipe or fittings can be connected to a copper pipe or fitting with a transition fitting. Be sure to follow the manufacturer's instructions for connecting pipe.

Roughing In

Installing plumbing water supply or drainage pipe under floor, inside wall frames, and above ceiling is called *roughing in*. If you plan a slab-on-grade foundation, the rough-in occurs prior to the concrete pour, and you must be extremely precise in your layout, since a concrete floor slab is not very forgiving. If you plan a wood floor, or have a two story building, the rough-in occurs after the frame is complete and usually after the roof material is partially installed and the house is dried-in. For these wood floors with a basement or a crawl space, the drainage system may be installed under the floor in the joist space.

Note that the wood studs within a wall are only capable of encasing a certain size pipe. This requires cutting and boring holes or notching wood frame members. Be sure to avoid cutting more wood than necessary. There are structural limitations that limit how much cutting, notching and boring is permitted. Normally, holes in an interior, nonbearing wall may not exceed 60 percent of the depth of the stud. Holes inside bearing walls are limited to 40 percent of stud depth. For example, a 2 inch by 4 inch stud in a nonbearing wall frame is only 3½ inches wide and can only enclose a 2 inch pipe with fittings. It is necessary to build a 2 inch by 6 inch stud wall to enclose a 3 inch or 4 inch pipe. Therefore, bathroom walls are normally nonbearing and at least 6 inches thick.

Install your DWV piping at the point where the pipe will exit the building. This should be the side that is closest to the sewer tap or septic tank. You can consider that this is the trunk of your plumbing tree. It is also the lowest point in the DWV system, unless you are installing a sump pump. Begin by mentally clustering your plumbing groups. These clusters will represent terminals for branch lines. Let's say that you plan two standard bathrooms that are separated from each other, a kitchen sink and a clothes washer. If the kitchen and clothes washer are separated from the bathrooms and each other, you will have four clusters. Your main drain can be a 3 inch pipe and would *branch* away from the horizontal drain line to each of these clusters. The branch lines to each of the bathroom clusters would be a 3 inch pipe and the branch line to the clothes washer and the kitchen sink would be a 2 inch pipe. Each of these branch lines would angle away and slope upward toward the cluster at the minimum rate of ¼ inch per foot.

A plumbing waste drain pipe transitions through a nonbearing wall section. Perhaps too much wood has been removed for this size stud wall.

Three Rules of Plumbing

Now that you're ready to install the interior waste plumbing, it's time to recall that old adage about the three rules of plumbing. Waste flows downhill, hot water's on the left and payday's Friday! The lucid point behind this adage is that the most important thing to remember about rules of plumbing is that sewage must flow downhill based on gravity drainage. Not much else matters, but then there are many ways of getting there from here.

To do this, you must consider all of the sources of waste collection, the size of each fixture trap, the size of each vent, the distance of each fixture trap to its vent, the drainage branches, the collection of traps on a single vent. How you get there from here is through the effective use of plumbing fittings! Let's start with the most simple of all; a single bathroom with a tub, water closet (toilet) and a lavatory (hand sink to you and me).

First, the drain pipe must be large enough to accommodate the largest plumbing fixture. In this case it is the water closet. The minimum size of drain pipe for a water closet is 3 inch diameter pipe. Tables P3004.1 and P3005.4.2 in Chapter 3 will allow you to determine the minimum size of this and other fixtures.

Branches

Clusters of neighboring fixture drains can be designed as branches. For example, the most common drainage branch is the single bathroom. From the main building drain pipe, a 3 inch pipe runs horizontally and then veers upward toward the water closet in the bathroom at a minimum slope of ¼ inches per foot. The 3 inch pipe is connected to a 90 degree bend, which rises upward toward a plumbing wall with a low slope 90 degree bend. Three fixtures will feed effluent into the single 3 inch drain: the water closet, the tub or shower, and the lavatory.

Two fixtures may be served with a fitting called a *cottage tee*. The bottom of the cottage tee fixture has four openings: a three inch sanitary tee, that will serve the water closet and a two inch sanitary tee that will serve the tub or shower. The top of the cottage tee will extend upward to become the vent for this fixture or group of fixtures. The length of pipe between the top of the cottage tee and the fixture trap is referred to as the *trap arm*. There are limits to the length of the trap arm based on the diameter of the pipe.

On top of the cottage tee, a 2 inch pipe is connected to the opening and rises upward to act both as a vent for the water closet and shower and as a vertical drain for the lavatory, which is served by installing a *sanitary tee* on this pipe. This is referred to as *vertical wet venting* and is limited to certain fixtures such as a lavatory. This vent will either extend through the roof or connect to a vent that extends through the roof. Connecting to another venting system is commonly referred to as *re-venting* or *branch venting*.

If you have an isolated water closet, there is another effective use of fittings. Notice that you can connect a sanitary tee to a closet bend. The top of the sanitary tee extends straight up and can become a vent. The closet bend is measured carefully, then cut to fit into the sweep portion of the sanitary tee. The center line of the closet bend will be exactly 12 inches from the edge of the finish wall (which includes drywall).

Branches can include many fixtures or just a few. Each branch line has limitations on the angle at which it may turn before a separate clean-out is required. The IRC limits the turning angle to a maximum of 135 degrees without installing a drainage clean-out fitting.

A bathroom group is installed along this plumbing branch. Notice the fitting, which permits the use of multiple traps. Only approved fittings are permitted for such use. Check with your inspector if in doubt.

Traps and Trap Arms

Every fixture must be provided with a trap to prevent odors and sewage gas from entering the home. Some fixtures, like the water closet, have integral traps. By definition, an integral trap is a trap within the fixture. The sizes of traps are regulated according to the fixture they serve. Table P3201.7 in Chapter 3 lists the trap sizes you may expect to install in your home.

Turns and Angles

Any change in direction of drain or vent pipe, either horizontal or vertical, must be made with the use of approved fittings. These fittings have specific uses and limitations. They also are identified by trade name, resulting from their most prominent use. Table P3005.1 in Chapter 3 tabulates the limitations on fittings and their respective allowable changes in direction. The following list is abbreviated to those fittings you will normally encounter in building a standard house. Keep in mind that there are other, more exotic fittings with highly specific uses.

The sanitary tee is only permitted to convey waste from a horizontal direction to a vertical direction of flow. A long sweep is permitted to be installed horizontally or rather to convey waste from a vertical to a horizontal direction of flow. Check at your plumbing supply outlet for the proper use of each fitting.

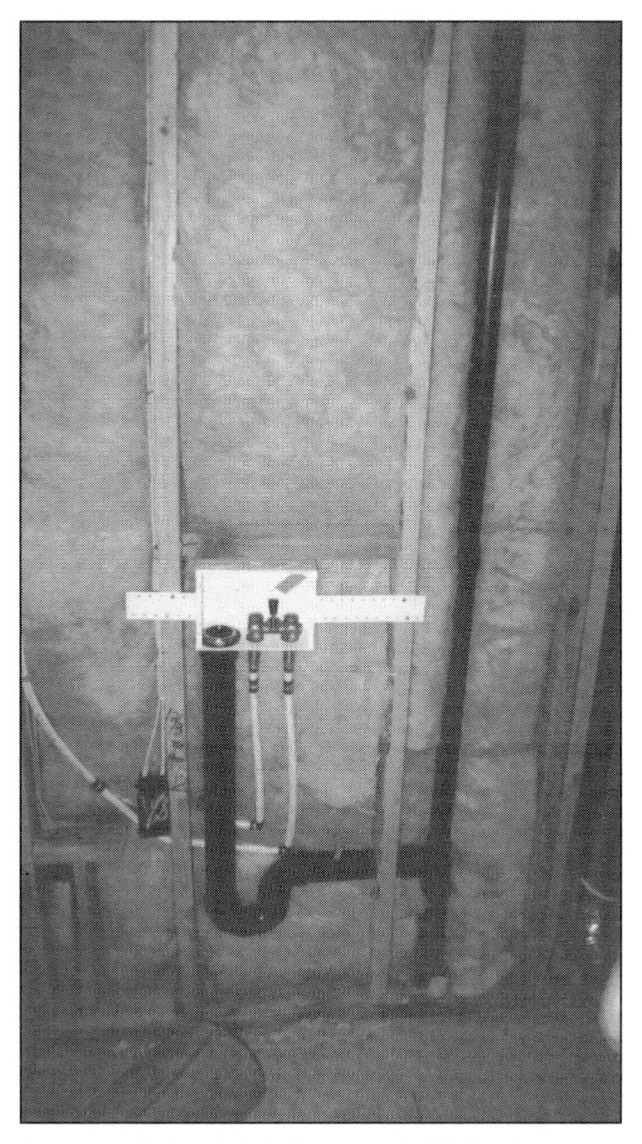

Laundry drain trap is installed correctly. Notice the hot and cold water supply are within the same outlet box.

Turns in drain pipe must be made with approved fittings.

Popular Fittings and Their Uses

Sanitary Tee

This fitting allows effluent to change direction of flow from horizontal to vertical. Most commonly, it is a fitting between a trap arm and a drain at the vent. Sizes:

- 1½ inch
- 2 inch
- 3 inch
- 4 inch

Wye

Used to join drain or vent pipes when they intersect at 45 degrees. When used as a vent fitting, they must be inverted. Sizes:

- 1½ inch
- 2 inch
- 3 inch

Check the height of plumbing using a string line across the concrete forms before and after backfill is installed.

Reducing Wye

Used similar to Wye except allows a different size branch pipe to join. Sizes:

- 2 x 2 x 1½
- 3 x 3 x 2
- 3 x 3 x 1½
- 4 x 4 x 3
- 4 x 4 x 2

Coupling

Used to join two pipes together. Sizes:

- 1½ inch
- 2 inch
- 3 inch
- 4 inch

Long Radius Tee/Wye

Used to connect a vertical drain flow into a horizontal flow. Also used to join horizontal drains at the same level when they intersect at 90 degrees. Sizes:

- 1½ inch
- 2 inch
- 3 inch
- 2 x 2 x 1½
- 3 x 3 x 2
- 4 x 4 x 3

Cottage Tee

Used to join traps from two fixtures together. Sizes:

- 3 x 2
- 3 x 3

String line establishes level of finish floor for checking drain line height.

Venting

Vents serve to prevent dangerous and noxious sewage gases from accumulating in the drainage system and to prevent siphoning of waste into the fixture. Vents should avoid receiving any water or waste products. This is called *wet venting* and is only permitted in certain conditions because of its propensity to cause siphoning of waste into the trap or fixture. Vent sizes are limited by the number and size of the fixtures served.

Building Vent

A good design practice is that the drainage system of a building be vented with one or more vent pipes, whose total aggregate size must equal that of the building drain. For instance, if the building drain is 4 inches, or 12.56 square inches, the venting system may be either a 4 inch pipe (12.56 square inches), two three inch pipes (14.13 square inches) or four two inch pipes (12.56 square inches) to meet this requirement.

Individual Fixture Vent

Each fixture must have its own vent that may branch into the larger building vent described above. The minimum size vent for each fixture is a function of the required drain size for that particular fixture or group of fixtures. The requirement from the IRC is that the diameter of the vent be at least one half of the diameter of the drain that serves that fixture.

Under certain conditions, two or more fixtures may share a vent, if the size of the vent is large enough for both fixtures. This sharing of vents is achieved with the proper use of fittings and is standard practice. The fixture derives air from the vent through the trap arm. There are rules that regulate the use of these fittings. One such approved fitting for fixture sharing is called a cottage tee fitting described previously.

An individual fixture vent is installed inside a wall frame.

Fixture Installation

Generally, plumbing fixtures are installed after drywall has been painted. There are two fixtures that must be installed during the top-out and before the drywall is installed. The bathtub and shower cannot reasonably be installed after walls are covered with drywall. They are difficult enough to install without any obstruction.

Bathtub

If you have installed the ground-work plumbing under a concrete slab floor, you will have followed directions and boxed out a space around the bathtub or shower drain, which will provide access to install a trap for these fixtures. If you are building a wood floor on a foundation wall, you must install the bathtub connections from the opposite side of the wall and beneath the floor. If you have open and unobstructed access to the front end of the tub, the fittings are easy to connect.

When you have installed the vent and trap, move the tub into its place. Position the tub approximately where it will rest and mark the height where the tub's upper lip intersects each stud. While the tub is in position, check the level of the tub to ensure adequate drainage of water. Use shims to provide a proper drainage surface. Now, mark the location where the tub's drain will sit on the floor. Remove the tub and install solid blocking between the studs at the height of the lip of the tub.

A bathtub equipped with an overflow apparatus (most are), must allow overflow into the drainage system. A bathtub installation kit will usually include a waste and overflow pipe kit, a drain stopper and several washers. Both plastic and metal (brass or stainless steel) waste and overflow pipe are available. Plastic is the most common material due to price.

The proper use of fittings will allow for venting multiple fixtures and offsets within wall frame.

Now, assemble the waste and overflow drain and connect to the tub. You may have to cut into some wall framing and floor decking. Be careful not to cut any crucial structural members. A tub is heavy when filled with water. Filled with water, a 25 gallon tub and a 200 pound person will weigh over 400 pounds. Install the tub drain stopper and the face piece of the overflow.

After installing the waste and overflow, connect them to the drain stopper. You will notice that the trap is capable of swiveling to accommodate the location of the waste and overflow tailpiece. The connections of the plastic pipe will be cemented with a special glue unique to that pipe. ABS pipe uses ABS glue and PVC pipe uses PVC glue. If you have elected to use stainless steel or brass fittings, they are joined with a screwed connection. In all cases, you must connect the tailpiece of the waste and overflow to the trap. Connect the trap to the trap arm, then to the drain.

The installation of a shower trap is much simpler. The drain pipe rises vertically to connect to a sanitary tee fitting, which rises to a vent and extends laterally with the pipe to become a trap arm. The trap is positioned at the desired location and connected to the trap arm. Since these fittings are ABS, they are cut and glued together with ABS solvent. Be sure to note that the proposed height of the trap's weir must be above the inlet at the vent. This is to prevent any siphoning effect, which will draw water back into the shower basin.

An alternative to the manufactured shower pan is a tile floor shower. This shower requires an impermeable pan on top of the floor and under the tile. The shower floor must slope toward the drain.

A manufactured shower pan is installed after framing and before drywall is installed.

After concrete is poured around a shower trap, an impermeable shower pan is installed.

Testing the Drain, Waste, and Vent Pipe Integrity

Test the water tightness of the DWV system. Plug the sewer tap with a test ball and install caps on all vents and traps. Fill water into the highest trap in your system. There should be no leaks. The test should hold without losing any water for at least an hour. Look for wet spots at all joints between DWV pipe.

Water Supply

Now that your drainage system is installed, you can turn your attention to the domestic water supply piping installation. Every dwelling is required to have an adequate supply of potable water for each fixture. This can be a well, private water system, public water system, or even a sanitary cistern. Of course, you must acquire separate permits for access to each of these water services. Wells are usually authorized and inspected by the state government. A public or private water company will usually provide connections.

The aspect of sanitation is the primary reason for the regulation of water sources. Again, usually the state will regulate water purveyors and adjudicate disputes. They review the sanitation of the water supply before it gets to your water meter. After the water meter, the local administrative authority will normally regulate water sanitation. The administrative authority is usually regarded as the plumbing inspector.

Plug the sewer tap with a test ball designed to restrict water flow in order to test the water tightness of the DWV system.

There is an entire section in the IRC that relates to the prevention of a cross-connection control between the water supply and sewage disposal. The basic consideration is that sewage disposal pipe or equipment must be separated from any potential water supply pipe or equipment. The most effective separation is an *air gap*. This means that after fresh water leaves the tap from your faucet, it must travel through the air to reach a basin or drain line. You can visually see how this works if you look carefully at either a kitchen sink or a bathroom lavatory. Notice that the spout is higher than the flood rim of the basin, thereby preventing water in the bowl from flowing into the spout and immersing the faucet with water from the bowl.

In most fixtures, there is an overflow drain that prevents the water from rising too high in the bowl. This would only be a problem when the faucet is turned off and backflow or back-siphonage causes some of that bowl-water to be sucked into the potable water system. The good news for you is that most fixtures such as sinks, lavatory bowls, and appliances such as faucets are designed with the faucet high enough to avoid these safety issues.

Yard Line Water Pipe Sizing

As discussed in Chapter 3, the size of water supply pipe is dependent on three variables: water pressure, the length of pipe, and the number and type of water supply fixtures served by the pipe. Your water company will tell you what the prevailing water pressure is in your area. You must count the total fixture units (not just fixtures) within your home and then measure the distance between the water meter and the most remote fixture. Once these variables have been determined, use Table P2903.6 in the IRC to determine the minimum size water pipe required to extend from the water meter to the water heater. Remember that you must provide a main shut-off valve at the house before the water pipe enters.

Pipe Material Choices

The material originally specified by the plans you developed should be the same as that selected for use as water supply pipe. Plastic pipe is approved in some areas for water supply. Even if plastic pipe is approved for use within your jurisdictions, not all plastic pipe may be approved for use. For example, PVC pipe may be approved for yard line uses only in exterior locations and only with proper burial depth and adequate cover. It is not approved for use as interior water supply pipe. On the other hand, CPVC may be approved for use in concealed locations inside wall frames or attic locations for interior water pipe but may not be approved for use within concrete or block wall systems. Polyethylene water pipe is another water supply pipe gaining in popularity. Be sure and check with your local jurisdiction for the approved material and locations for each type of pipe.

CPVC is the easiest pipe material for the novice to install successfully the first time. If you have ever put together a plastic model car or airplane, this should be no problem for you. CPVC is approved for use as both hot and cold domestic water pipe. The pipe is rigid and capable of being cut with a plumber's saw and joined together with plastic CPVC solvent. There are fittings for every possible turn, joint, assembly, connection, or combination you can imagine. These fittings are made from the same material and must not be substituted for a dissimilar material. Polyethylene (PE) cold water building supply and yard pipe is approved in most jurisdictions. In some instances, variations of these products are even approved for interior water supply pipe.

Product Standards and Limitations for Use

All materials used for water supply pipe are regulated by the IRC codes. The manufacturer is usually the best source for determining the exact methods of assembly or limitations on its use. Since the manufacturer must gain the approvals from various code enforcement organizations and jurisdictions, they are intimately knowledgeable as to the approved application of their product and the limitations on its use. If a plumbing supply store seems less than qualified to assist you, contact the manufacturer directly via phone or even the Internet for advice on installation of their product in your project.

The manufacturer has a certain liability for their product. They limit this liability by establishing the use or installation of their product. If you violate the installation method, you void the warranty and gain the liability for the installation. The manufacturer has paid for exhaustive testing and listing of their product. It is ensured to perform according to the product listing if installed properly. Be sure and read every aspect of the installation standards that apply to the product.

Water Pipe Branches and Supply Routing Methods

After you have reached the water heater with your main water pipe, you must extend the lines to every fixture. There are innumerable ways of achieving this. Manifolding is a method of adding multiple branch lines that reach individual fixtures. You can also connect the main water line, in series, to individual fixtures. There are even manufactured products that distribute water to fixture branches with a manifold.

Since most of the fixtures in your home have a hot and cold water supply, you must provide two pipes to each fixture. Before the main water pipe enters the water heater, extend a branch that will be the source for all of the cold water supply in the building. Then, for the branch that leads toward

Underground water supply pipe approved for such use is installed in a manifold.

Both drain and water supply pipe must be protected from accidental damage. Inside a wall frame, a pipe can be damaged from nails. Here nail plates are installed to protect the inner surface of drain pipe.

the water heater, add a cut-off valve with a ¾ inch threaded connection, which will eventually connect to a flexible copper connector. This connector will feed cold water to the water heater. The location of this cut-off valve is critical and dependent upon the height above the floor of the water heater and its distance from the cold water inlet. Therefore, shop for a water heater before you install this valve.

From both the hot and cold water branches, extend a manifold connection that will lead to all of the hot and cold water locations in your home. Assemble this manifold by using tee fittings in a series configuration. Each of these tee fittings can lead to a general location instead of a specific fixture. Then, in the general location, such as a cluster of bathrooms, you can manifold again to direct the water pipe to each specific fixture. This will avoid running a dedicated water line to each fixture for both hot and cold water.

Individual Branch Water Pipe Sizing

Before you manifold water pipe to provide distribution to every fixture, you must determine what size each branch line needs to be. The common assumption is that the water pressure is maintained at the water heater. Measure the distance to the most remote fixture in each branch line as you did with the yard line calculation, and use the fixture unit calculation for the cluster of fixtures to determine that specific branch line pipe size.

Protection of Pipe

Water pipe must be protected from physical damage or breakage. Piping must be installed in a manner that does not cause undue strains or stresses. This means that pipe cannot be bent or contorted beyond that prescribed by the manufacturer. It also means that holes in wood framing must be large enough for pipe to pass through without causing abrasion to the pipe. Pipe passing through concrete or block walls must be protected with sleeves of nonabrasive material. The face of studs where holes are bored for pipe that leave less than 1½ inches thickness of wood material must be protected with metal plates. This prevents damage to pipe when you want to hang that family picture on the wall with a 16d nail.

Underground or buried pipe must be protected as well. Adequate cover and proper burial depth ensures the protection required. Plastic pipe must be buried at least 18 inches below grade. Sand or soft gravel must be placed on the bottom of the trench to prevent rocks from damaging pipe. Load from footings must be kept away from plumbing trenches. An angle of repose of 45 degrees away from the bottom of the footing must not contain any water or drainage pipe.

Fixture Cut-Off Valves

Water pipe leading to every plumbing fixture should be provided with a water cut-off valve. There are specific types and sizes available for each fixture. For example, an angle stop can be used for lavatories, toilets, and kitchen sinks. A laundry tray comes equipped with hot and cold water faucets inside a combination supply/drainage box. An ice maker tray allows for a smaller ¼ inch or ⅜ inch water line to extend to the freezer. Each of these valves must be accessible for shutting off in case of a leak or fixture replacement. These cut-off valves may be installed after the drywall goes up. Until that time, use end caps to seal the charged water pipe, so as to perform water pipe pressure testing.

Bathtub

The water connections to the bathtub or shower faucet must be made in advance of the drywall installation. Faucets for bathtubs with showers are usually combined as one. Bathtub/shower combination faucets have two outlets (up for shower and down for tub). Shower-only valves are required to have a temperature limiting device installed, which prevents accidental scalding. Install the bath tub and/or shower faucet onto a solid wood blocking above the tub surface. Then connect the hot water supply pipe to the left faucet and the cold water supply pipe to the right faucet. Remove the handles and store in a safe location where you can locate after drywall is installed.

Water Heater Selection (Size, Height, and Location)

An electric water heater is recommended for the novice installer. Installation of the water heater is among the last and simplest parts of building a home. During the final trim, after the drywall is installed, the water heater is positioned where it will set, then hot and cold water are connected and the electrical connection is made. The cold water supply valve is then turned on and, after filling, the electrical power is turned on.

The water heater loop is an appropriate location to tie the water supply together to perform a water test.

Before this process occurs, some simple steps will ensure ease of installation. The height of the water heater you select is important when you determine the location of the cut-off valve for the cold water supply leading to the water heater. A flexible copper pipe will extend from the cutoff valve to the water heater. If the cut-off valve is in the wrong place, you'll have a problem connecting to the cold water inlet for the water heater. Position the cold water inlet high enough to clear the water heater, but not so high as to exceed the length of the flexible copper pipe connector. Also, when you install the water inlet pipe, mark it as such to avoid a reverse connection to the water heater. Use a blue marker to color code the cold inlet pipe and a red marker to identify the hot water outlet pipe. Make sure a shut-off valve is installed on the cold water inlet pipe.

The size of the water heater you need is a function of the recovery time for the water heater and the number of teenagers living in your home. No aspersions cast on any of our children, they just tend to shower longer than the rest of us. For a family of four, a fifty gallon electric water heater should be adequate.

Water Supply Pipe Pressure Test

Before you call for an inspection, and when all of your water pipe is installed and the ends are capped, it is time to test your skill at plumbing. Temporarily connect the hot and cold inlets to the water heater. Install caps or valves on the terminals of all fixture outlets. Then, gradually turn on the building water supply and test for leaks. Have members of your family at strategic locations looking and listening for gushing water. Be prepared for the possibility of some leaks. Mark those very clearly. Then turn the water off and remove and replace that entire section of pipe and test again. This test must be provided for the inspector to witness in order to verify the integrity of the pipe.

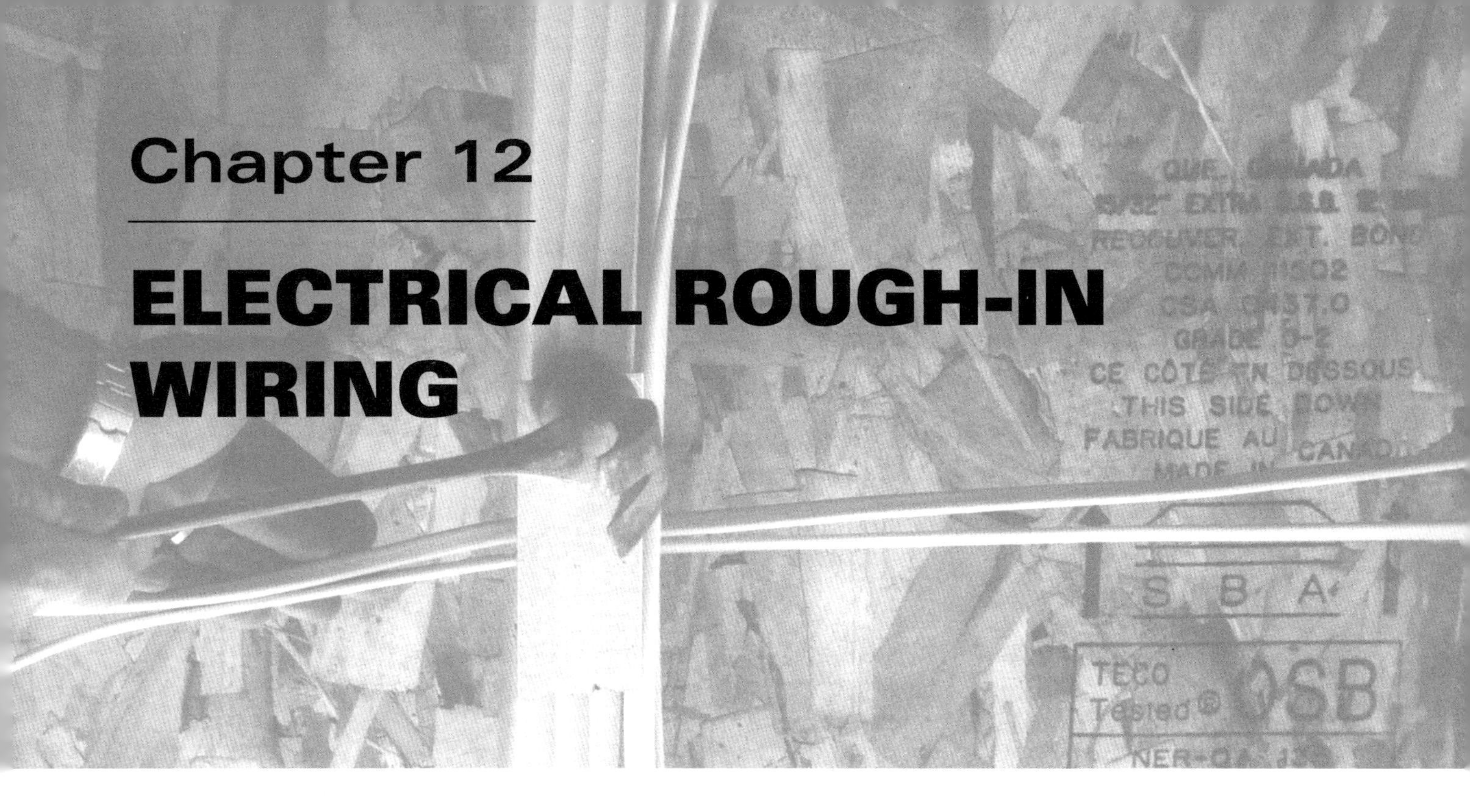

Chapter 12

ELECTRICAL ROUGH-IN WIRING

Electrical Service Main Disconnect

The power center for your dwelling is a distribution panel that channels electricity through various circuits that are located inside the main panel. These circuits are controlled with circuit breakers of various sizes. Circuit breakers limit the maximum amperage available to a specific circuit. To begin with, electricity is normally fed to your home from an electrical transformer near the street. This transformer belongs to the electric company and converts electricity into a voltage that is usable inside your home. This voltage at the transformer is very high. It is transformed into a usable power by stepping down to 120/240 volts at the transformer. The power is fed to the main panel with three very large wires. Two of them are phase conductors, and the third is a neutral conductor. The two phase conductors each feed one half of the power in your main panel. Each of these phase conductors carries 120 volts. When the two are connected with an approved circuit breaker in your service panel, you have 240 volts available to use for equipment such as a water heater, electric range or clothes dryer. The third wire, the neutral conductor, will provide a return path for electric current back to the transformer. It will also carry any electrical ground fault or unbalanced current back to the transformer.

The main disconnect must be sized properly according to the proposed demand load of the dwelling. Shop for that size service disconnect equipment that best suits your needs for size and cost. Be sure to purchase a panel that accepts commonly available circuit breakers.

Service Drop

The service main disconnect will probably mount onto the side of your home. Electrical current is supplied from the transformer via three wires, two phase conductors, and a neutral conductor. These three wires can be installed either underground or overhead according to the standards of the electrical supplier. When buried, they improve the aesthetics of your neighborhood. If buried, the trench must meet certain depth and width requirements. These are generally left up to the electric purveyor. Other utilities may be installed in the same trench if the other utilities all agree. Therefore, you could dig just one trench to your home.

If the electrical service is installed overhead, it must meet minimum height requirements above any roads, driveways, roofs, and surrounding walkway. Additionally, the services are restricted to specific distances away from openings in walls. These clearances are portrayed in drawings in Chapter 35 of the IRC.

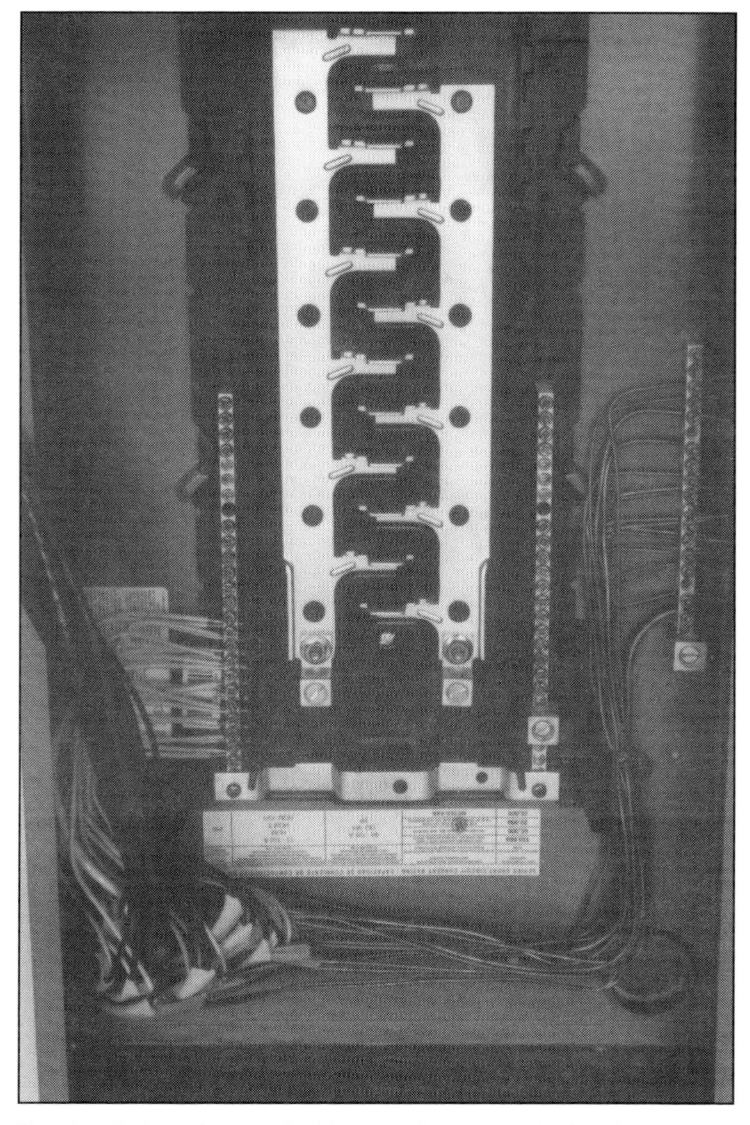

The electrical service panel with ground and neutral wires installed.

Grounding and Bonding

The main electrical service panel must be effectively grounded to the earth in order to provide a safe path for electric current to seek ground for any high current surges, such as lightning strikes. It will also assure an equipotential grounding plane. The size of this grounding conductor is based on the size of the service panel. A nominal size for a grounding conductor for a 200 amp panel is #4 solid Copper wire. This wire must be connected to the earth in one of several ways. Grounding electrodes are ⅝ inch in diameter, 8 feet long, copper clad steel rods that may be driven into the soil near your service panel. Two of these, separated by six feet, are regarded as an effective grounding method for a smaller residential service panel. The easiest and most effective grounding method is to connect the copper wire to a single 20 foot long piece of reinforced steel that is embedded in the concrete footing. This ubiquitous grounding method is simple, effective, safe, and cheap. A #4 copper ground wire is clamped onto the reinforcing steel with a proper clamp. Extend enough wire out of the footing trench to allow a connection to the electrical service panel's grounding buss bar.

The electric service provider needs to know your plans for electrical usage and proposed electrical service panel size. They will send an engineering team to your site and measure the distance between the transformer and your building. They may ask you to agree to an easement to install their power line. This means that you forever give away a certain portion of your property for the exclusive use of the electrical service provider. This allows the electric company to legally replace or repair their cable or other equipment without your permission. Check with a lawyer if you're worried, but it is a common practice.

Circuits and Overcurrent Devices

Power from your main service panel is distributed throughout your home with the use of circuits. The size of circuits are based on the size of the overcurrent device (more commonly known as a circuit breaker). Each circuit is capable of supplying only a limited quantity of outlets and lights. The most common circuit sizes for residential use are 15 amp, 20 amp, 30 amp, 40 amp, and 50 amp.

The use of 15 and 20 amp circuits are the most common and are used for receptacle outlets and general lighting. The use of 30, 40, and 50 amp circuits are limited to appliances that use 240 volts, such as a water heater, electric range and oven, electric dryer, and electrical heating and cooling equipment.

These circuits are an electrical wire that feed the device(s). The wire may not be one continuous wire, but a series of wires that are linked together by outlets, switches or junction boxes. Circuits may be 120 volt or 240 volt type. The 120 volt circuit will be a cable with three wires.

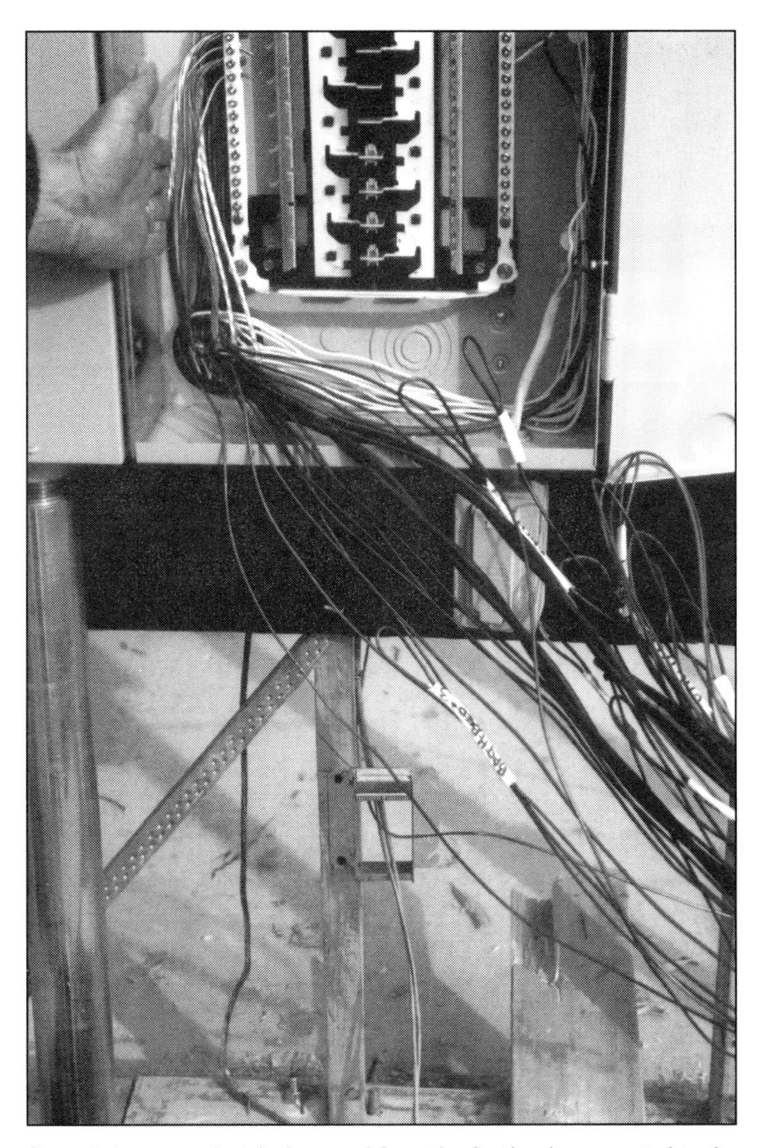

Stranded copper electrical ground from the footing is connected to the grounding buss barr within the service panel.

The black or red wire is the hot or phase conductor. This connects to the circuit breaker in the service panel. The white or natural gray wire is the neutral or common conductor. It connects to the neutral buss bar in the service panel. The third wire is a bare copper wire which is the ground. It connects to the grounding buss bar in the service panel. Note that both the neutral and the ground wire are electrically connected in the service panel enclosure, unless you feed the circuit from a secondary electrical panel or subpanel.

Normally, a room or two is dedicated to a single circuit, which will serve outlets and switched lights. The number of outlets that may be on a single circuit can be calculated with simple mathematics. For general purpose receptacles and lighting outlets, a 15 amp circuit can serve approximately 600 square feet of floor space, and a 20 amp circuit can serve up to 800 square feet of floor space.

Please note that these are safety minimum requirements. As a rule of thumb, 10 or 11 duplex receptacles or lights are permitted to be connected on every 15 or 20 amp circuit.

Wiring Methods

Methods of electrical wiring are usually synonymous with standards of installation quality. Other codes usually let the inspector enforce such standards. However, the NEC specifies the minimum standards for workmanship of some wiring techniques. These standards of quality include more issues than are presented here. However, I have included some of the most common mistakes in workmanship by a novice.

Installing breakers to individual branch circuits within the service panel.

Protection from Damage

Wire must be protected from physical damage. Sounds reasonable, but what is adequate protection? What qualifies as physical damage? One example is protection from nails. Wire in wood stud wall frames can be damaged by nails and screws penetrating from drywall and exterior siding. If the siding is already installed, you can still damage the wiring with stucco lath nails. What constitutes protection is at least 1¼ inches distance between the edge of the stud or wood joist and the beginning of the hole drilled for the wire. The theory is that you will not use a nail or screw longer than 1¾ inches for drywall or siding. What this rule means is that if you have a hole larger than 1 inch in a 2 x 4 or 2 inches in a 2 x 6, you must install a metal plate on both surfaces to prevent nail penetration. There is a similar rule for wiring attached to the sides of wood joists or rafters.

If wiring is not approved for direct burial, it may not be installed underground, unless it is inside a conduit approved for such use. The particular size and material for the conduit and the required depth of burial is also regulated. Generally, direct burial cable will be identified with the nomenclature: "U" in the identification of the cable. Example "UF" cable is approved as Underground Feeder cable. "USE" cable is known as Underground Service Entrance cable. These cables rated for direct burial may be buried not less that 24 inches below grade. If cable is installed in approved conduit, the required depth decreases depending upon the type of conduit. Rigid metal or intermediate metal conduit can be as little as six inches below grade. Nonmetallic raceways must be buried 18 inches deep.

Wiring Support

Wiring must be supported from framing members to prevent sagging that will result in mechanical damage to the wire. Support between studs and across trusses is adequate for the most part. How-

Wiring must be protected from accidental damage. Nails into outside stud wall can damage wire and cause a hazard. Here a nail plate is installed on each side of the stud to protect the wire within from accidental damage.

ever, the type of connectors and the method used for support is important. Incorrect installation could cause damage. Raceways or conduits may not be used as a means of support for other cables, wires or other raceways or conduits. That means that you cannot tie a dangling wire to a firmly supported conduit. When wire reaches an outlet or switch box, the conductors must be pulled an additional six inches through the box. This is called free conductor length and enables safe connection to a switch or outlet.

Wiring Connections

All joints and splices or connections in a wiring system must be made inside a junction box, outlet or switch box. A connection of one wire section to another must be accessible at all times. If your wire section is not long enough and you have no place to install a junction box, use a longer wire.

Cable Location

You may not use ductwork, plenums, or other air-handling spaces as a route for installing wiring. There are several reasons to avoid this. Blown, heater air tends to decay the insulating qualities of the cable. Additionally, a fire can start in a duct from faulty wiring and remain concealed while spreading throughout the duct system. The fire can spread rapidly throughout the house unimpeded by interior partitions or wood framing.

Appropriate Use of Wire

Certain cable types are required for wet locations. Generally, cable approved for use in wet locations will be identified with the nomenclature: "W" in the identification of the cable. Example "THWN" cable is approved as Wet Location cable.

There are limitations on the permitted ambient temperature for each cable type. The allowable temperature is printed on the outside of each cable. The normal operating temperatures are 60 degrees, 75 degrees, and 90 degrees centigrade. This will be important if you are in a region of the country that gets hot in the summer. The attic may exceed 60 degrees centigrade easily, and a 90 degree wire rating will be required for that installation.

Wire type must be approved and bear an identification from the manufacturer. This means that you may not use that bargain wire you bought at the swap meet if it does not have proper labeling. Stick with the electrical hardware outlets for approved electrical wire and hardware.

Outlet and Light Boxes

Boxes are made of plastic material as well as metal. The plastic boxes are approved for use within houses and they are easy to install. They usually come equipped with approved attachment nails. Remember that when you install the boxes, they should project ½ inch in front of the edge of the wood stud. This ½ inch projection is to allow for a flush finish surface after drywall is installed. If you feel carefully, there are small marks exactly ½ inch away from the edge of the box. Simply align these marks with the edge of the stud, and your finish wall surface should be flush. If you intend to use thicker drywall, or add siding or paneling, the box face must be flush with the outside surface of the wall. This means that you must adjust the box location to accommodate this difference in dimension.

The minimum size of the outlet or switch box is regulated by the number of wires inside. There are specific rules for allowable box fill. For 3 or 4 #14 wires and a single outlet or switch, a standard 16 cubic inch outlet or switch box will easily be sufficient.

Required Outlets and Switched Lights

Outlets must be installed along a wall in habitable rooms, so that an electrical appliance with a six foot long cord may be located anywhere along the wall while plugged into an outlet. That means that you must install an outlet every 12 feet around a wall. Even a small wall section 24 inches long must have at least one outlet. The height for the outlet placement above the floor is not regulated by the code but is normally 14 inches. In kitchen countertop areas the outlet spacing must be such that no point is more than 24 inches away from an outlet. The requirement for kitchen countertop outlets includes an island or peninsular space. At least one outlet must be provided in any island used as a counter space. An outlet should be installed for any anticipated fixture such as a dishwasher or waste food disposer. The outlet for the waste food disposer should be connected to a switch located near the sink. This switch/outlet assembly works much like a switch/light connection, except the outlet is activated only when the switch is in the *on* position.

Additionally, outlets are required in bathrooms, adjacent to the lavatory. An exterior outlet must be installed at the front and rear of the dwelling. Laundry areas must have an outlet for washing machine and dryer. Garages and basements must also have at least one outlet. Hallways that are at least 10 feet long must have an outlet. If you have heating and cooling equipment, a service outlet

Notice the small molded mounting guide on the outer portion of the electrical box. This marks the installation location on the stud to allow for ½ inch drywall and still provide a flush surface.

must be adjacent to the heating and cooling equipment. If you have equipment in the attic, an outlet and switched light must be provided.

Each habitable room, bathroom and outside entry location must be provided with a switched light. You may substitute a switched wall outlet in the bedroom, family room, den or living room in lieu of a switched light. The switch should be close to the latch side of the door. Additionally, at each level of stairs, a three-way, switched light must be installed. At least one switched light must be installed at the entry to an attic or underfloor space where utility equipment is installed, utility room and basement.

Standard Circuits

In a residence, outlets and lighting are fed with *standard circuits*. These are the most simple type of circuits, which are limited to a single phase conductor, a neutral, and a ground. The circuit is fed from a circuit breaker inside the panel board. Wire is within a cable assembly. The cable has a nonmetallic material surrounding the conductors. In addition, both the black, hot wire and the white, neutral wire are individually insulated. Cable is identified with the wire size followed by the number of insulated conductors and a suffix indicating the presence of a ground. An example is 12/2 G. Each wire inside the cable has a specific use and identification.

A cable with only three wires supplies current to a standard circuit. The black wire is a phase conductor, meaning that it carries a current or an electrical phase. Another wire has a white plastic coating that identifies it as a neutral wire, also known as the *common*. The remaining wire is a bare copper wire that serves as the grounding conductor. This circuit is normally limited to either a 15 or 20 amp circuit. The normal type of cable used for this circuit is NM cable.

The size of wire used in the cable varies according to the breaker size. For a 15 amp circuit breaker, a number 14/2 G wire size is required. If the circuit breaker provided is a 20 amp type, a 12/2 G wire is the minimum size acceptable. Electrical cable is available with multiple phase conductors for multi-wire branch circuiting. In this case, the second phase conductor is usually covered with a red insulation and the wire is specified as 14/3G or 12/3G. If you do not have extensive experience, I do not recommend that you use the multi-wire branch circuit method of wiring installation.

How Does Electricity Work?

Here is how alternating current (AC) electrical systems work: At the electrical generating plant, an electrical potential, called voltage, is generated by rotating wire through a magnetic field. The current is pushed along a wire toward your home by the voltage. In order to get the electrical current to your home over great distances, the voltage is very high. Near your home, the voltage is stepped-down to a voltage that is usable by your appliances. A transformer converts this voltage into 120/240 volts, that is the voltage at your electrical panel.

There are three wires that traverse from the transformer to your electrical panel. There are two phase conductors, each carrying the design system amperage at a rate of 120 volts. There is one neutral conductor. The third wire from the transformer connects to the neutral buss bar inside the panel board. This neutral conductor allows AC power to return or close the circuit to the transformer, thus closing the circuit or route for the current.

After traveling through the electrical meter and into your electrical panel board, the two phase conductors are connected to two different electrical phase conductor buss bars. These are nothing more than isolated metal strips that are now energized by electrical voltage. The white wire in the cable leading to a circuit is connected to a neutral buss bar. The phase conductor buss bars will allow circuit breakers to tap into the electric potential and then provide that voltage to outlets and lights in the circuit. The amperage available on the circuit is limited to the rating of the circuit breaker (or it will trip). This amperage will be pushed, to wires in a circuit, by electric force at a rate of 120 volts. At the outlet or light fixture, the current will perform its function and then return along the neutral wire to the panel board and back to the transformer.

Special Circuits

Special circuits include any 240 volt outlet or device. These include electric ranges, ovens, dryers, water heaters, or heating/cooling appliances. A 240 volt circuit is fed with a cable that carries two separate 120 volt phase conductors (when required), one neutral (called a grounded conductor) and one equipment ground wire (called a grounding conductor). Unlike standard circuits, this special circuit will supply only one device or outlet. The cable will extend from a 240 volt special circuit breaker to a special outlet box, specified for that particular appliance.

Some single phase, 120 volt circuits are included as special circuits. Some of these special circuits include appliance circuits, bathroom lavatory outlet circuits, dishwasher and disposer circuits, and other circuits dedicated to a particular outlet, device, or appliance. Kitchen and dining room spaces are required to be provided with special heavy duty wiring to accommodate large demands from cooking or food processing equipment. The wiring requirements include using a 20 amp circuit, which would be supplied with 12/2 G wire. There must be at least two appliance circuits provided in a kitchen. More may be required if you have a dining room or breakfast nook. Since a refrigerator is an appliance, it may be connected to an outlet on an appliance circuit. However, due to the loads, I recommend that you provide a dedicated 20 amp circuit just for the refrigerator, in addition to the two other appliance circuits for the countertop.

More Wiring Methods

The standard circuit will feed more than one outlet. In fact, it usually will feed a combination of 10 to 11 duplex outlets and lights. These outlets or devices are linked in the circuit in a daisy-chain fashion. The length of wire between the circuit breaker in the panel board and the first outlet is referred to as a *home run*. The links between the outlets traverse through holes in wall studs, joists, or in between openings in roof or floor trusses to the next outlet or switch. At the point of link between two outlets there should be one cable that contains three conductors: a hot (black), a neutral (white), and an equipment ground (bare). These conductors will be joined together electrically by an outlet or other mechanical means such as a wire tie. As the cable enters each box, extend it to at least six inches beyond the box before cutting. This six inches is required and referred to as free conductor length. It allows for connecting the outlet or switch so that it may be removed and replaced with ease.

Subpanels

A subpanel is a remotely located electrical panel box fed from the main panel board. It will contain additional circuit breakers that serve devices or outlets. The size of the subpanel box is based on the loads it will serve. Subpanels are used to avoid multiple homeruns from a significant distance. There may be a cluster of bedrooms located on the opposite side of the house from the main service panel. A subpanel could be installed adjacent to the bedrooms, then branch to each room with an individual circuit. You may use a subpanel to supply electrical current to heating or cooling equipment as well.

Wire Size and Type

The selection of wire type and size is based on the limitations of use for each wire or cable product. The variables include the wire size (AWG), the material used to manufacture the conductor (such as copper or aluminum), the temperature rating of the insulating cover over the conductor, the maximum overcurrent protection for the wire (circuit breaker), and the environmental conditions that exist around the cable, such as humidity, dampness, heat, etc.

Type NM Cable

The most common type of wire used in residences is NM type cable. It is available in 12 and 14 gauge thicknesses. Type NM cable is permitted to be installed inside interior and exterior wall frames as well as in attics or other concealed or exposed locations that are normally dry. It may be used inside the voids of CMU block walls if such walls are not exposed to excessive moisture or dampness. It may not be embedded in concrete or aggregate.

Ground Fault Circuit Interrupter

A ground fault circuit interrupter outlet after finish trim.

Many of the provisions of electrical safety codes serve to protect equipment, persons or prevent fire damage. However, one provision of the NED is established for the exclusive purpose of protecting personnel. This provision requires that circuits will disconnect upon detection of a ground fault. Electrical current seeks the easiest path toward a ground source. A ground fault occurs when a phase conductor accidentally comes in contact with a ground wire or other metal parts that are connected to a ground wire. You experience a ground fault if you grab a metal faucet (or any other metal) and get a shock. What you are experiencing is electrical current that has sought you as the easiest path toward a ground.

The Ground Fault Circuit Interrupter (GFCI) is either a circuit breaker or an outlet device that senses a ground fault and opens the circuit when it senses a fault that reaches 5mA (1/5000 of an amp). GFCI protection is required at specific locations within a dwelling. Outlets at exterior locations, within garages, bathrooms, unfinished accessory buildings, storage rooms, crawl spaces, kitchen countertop locations, and wet bar sinks must be protected by ground fault circuit interrupters. There are two methods of providing the required protection. GFCI circuit breakers will cause the entire circuit to open upon detecting a ground fault anywhere in the circuit. A GFCI outlet will stop current flow at that specific outlet. Which method to select depends on which device is in need of protection.

Smoke Detectors

Because of the success in saving lives, smoke detectors have become ubiquitous throughout residences. Smoke detectors emit an audible alarm when they sense smoke. In order to detect smoke, the detector uses electrical energy to power a sensor that activates when it is occluded with smoke. The reason smoke detectors work better than fire detectors is that smoke is the first product of a building fire.

Since electrical wiring may be the first casualty of a fire, most codes require that smoke detectors be provided with battery backup. Since smoke is lighter than ambient air, it will rise to the top of a room or space. This is why smoke detectors should be installed high on the wall or on the ceiling. In all cases, smoke detectors must be mounted a certain distance away from supply or return air plenums, which are better known as heating and cooling ducts. This is to avoid rapid air movement that would distort the sensors inside the detector. Usually, each smoke detector is listed by an approved agency such as Underwriters Laboratory and carries specific installation standards. Read and follow these instructions to ensure proper operation.

Locations where smoke detectors must be installed are established by the building code. Generally, they must be inside of each sleeping room. They must also be outside in the hallway, immediately adjacent to each sleeping room. Where bedrooms are clustered together, it is permissible to use one detector in the hallway. If the hallway leads to a room where the ceiling heights are higher, an additional smoke detector may be required on the surface of that higher ceiling. Consult with a code enforcement official in your area for specific installation locations.

A smoke detector must be installed in each bedroom, outside each bedroom, or in hallways leading to each bedroom, and at each floor or level within a dwelling. They work best when installed high on a wall or on the ceiling.

Balancing Circuits in Main Service

Inside the main service panel, you will add circuit breakers for various circuits. These circuit breakers will be of different sizes and for different loads. It is important to balance them in order to not overload one phase and the neutral inside the main service panel. There are methods to achieve this. It is important to understand that any 240 volt circuit breaker is tapping into both phases at the same time.

You must balance the single phase loads between phases. If you should connect all of the 15 and 20 amp circuits to one phase, all of the current would flow on that phase and the neutral, thus possibly overloading that phase and the neutral. The single phase loads are distributed equally between the phases. The 240 volt loads are self-balancing.

Testing Circuits and Wiring Continuity

After you finish the wiring, but before you install insulation or drywall, a safe way to test circuits is by using an ohm meter to check for wiring continuity, shorts, ground faults, and opens. Apply this testing procedure to one circuit at a time until you have tested them all. These tests should help you determine that the circuitry will be functional and is properly installed. Begin by terminating the ground wires in the panel, but do not install the circuit breakers, receptacles, switches or lights. Strip about an inch of insulation from the ends of the black and white wires in the panel and in each junction box throughout the entire circuit. Make sure the exposed ends of the black and white wires in the panel are not touching each other, the ground wires or the panel structure.

When you reach the end of the circuit, connect the ohmmeter probes between the black and white wires, then between the black and the ground wires, and finally between the white and the ground wires. The meter should indicate infinite impedance (you should have an *open* circuit). If the meter indicates there is a connection between any of the wires, you must retrace the circuit starting at the panel. Check every box in the circuit and make sure that the wires are only connected to their respective color and are not touching each other. If you cannot find the problem in the junction box, carefully trace each of the cables from box to box making sure that both the exterior and interior insulation of the cable is undamaged. A cable that has been damaged, exposing one or more of the interior conductors to a grounded object such as a water line, would effectively close the circuit. Once the problem is found, correct it and recheck the readings.

The next step is to confirm that the wiring is complete from the panel to the last outlet of a circuit. Begin this test by starting at the panel. Pick out a circuit in the panel and temporarily connect the stripped ends of the black, white and ground wires together. Then go to the end of the circuit and place one ohmmeter probe on the stripped end of the white and ground wire. This time the ohmmeter should show that you have almost no resistance (you should have a *closed* circuit). If the meter indicates an *open* between the black and one or both of the other wires, you have an *open* circuit condition. Retrace the circuit and make sure that all wires are connected to their respective color in each box and that the black, white and ground wires are connected to each other in the panel. When you locate the problem, connect the wires together or install additional cable as needed and repeat the testing.

Installing Receptacles and Lights

Notice that the duplex outlets have four screw posts for wire connections. Look closely and you will notice that two of the screws are silver in color and the other two are gold. They identify the electrical polarity required for wire attachment. The silver posts are for the white or neutral conductor. The gold or brass post is for the black or phase conductor. If you have these reversed, your electrical appliances will not work correctly. A polarity tester is a simple device that can verify if your outlets are installed properly. It plugs into the outlet and gives a lighted signal indication of the condition of the circuit.

Turning the Lights On

At some point you will be tempted to turn on circuits and actually test lighting and outlets. This represents a landmark event in your project. Congratulate yourself and celebrate the occasion. You are getting closer! There are some advantages to having light available for your project. Use the artificial light to increase your working hours into the evening. Test the thermal integrity of your exterior siding for light leaks by looking for them at night outside.

Chapter 13

MECHANICAL EQUIPMENT ROUGH-IN

Choices

Hopefully, in the planning process, the method of heating and cooling was selected and designed. Climate factors, energy source, expense, ease of installation, personal experience, and compatibility probably were factors in the decision. More than likely the energy source choice was electricity. Hopefully, the heat pump was your choice for heating and cooling equipment. I have found that, with the exception of a few faults, the heat pump is the very best choice for the first time owner-builder, particularly if you need both heating and cooling. If you can avoid installing natural gas or propane in your first home, you will thank me later. The degree of skill required for successfully installing gas pipe and appliances is outside the range of most first-time home builders. Safety may be compromised with a haphazard installation, and leaking gas is not forgiving of mistakes or poor work.

Heat Pump

The heat pump concept has been around for more than a generation now and has a very good track record. The heat pump is a single piece of equipment that uses a refrigeration cycle for both heating and cooling. The heat pump transfers heat in an exchange process. The most common type of heating transfer method is using air. Electricity is used to operate the refrigeration cycle, which is how the heat exchange process works. The heat is pumped from the interior spaces to the outside in the summer cooling season and from exterior spaces to inside in the winter heating season.

There are laws that may restrict your intention to purchase and install your own mechanical equipment. These laws relate to the environmental damage caused by improperly installed refrigeration equipment and the chemicals associated with this equipment. Dealers may not even consider selling equipment that requires active installations. Some dealers may consider selling a self-contained unit that has refrigerant lines factory-sealed and ready to operate. The installation and repair of refrigerant lines are restricted to a certified mechanical equipment technician. If you select this type of HVAC system, you may need to secure a licensed subcontractor to supervise the entire installation.

Heat Pump Principles

A typical heat pump will include three components that work together to transfer heat from one location to another—from outside to interior spaces during heating season, and from inside to outdoors during cooling season. These three components are the condenser, the evaporator, and the blower. During the cooling season, the blower will circulate air through the evaporator to allow the heat to be removed. The evaporator will absorb heat from the interior space. The condenser, which is located outdoors, will dissipate heat collected from the evaporator. In the heating season, the process is reversed.

Refrigerant is used to transfer heat between the evaporator and the condenser. During the cooling cycle, the condenser will draw air across coils containing the heated refrigerant and extract heat out into the atmosphere. The refrigerant will condense and return back to the evaporator. In returning to the evaporator, the refrigerant, which is now a liquid, will evaporate and absorb heat. The refrigerant then returns to the condenser to complete the cycle. During the heating cycle, the evaporator and condenser will reverse roles. This is accomplished by means of reversing the flow of the refrigerant. A *reversing valve*, which is controlled with the thermostat, will switch the direction of the refrigerant.

The heat pump uses electrical energy in a more efficient manner than baseboard or wall heaters that use resistance to create heat. For example, one kilowatt hour (KWH) of electricity will produce 3413 Btu per hour of heat in an electrical resistance heater such as a baseboard heater. However, a heat pump with a coefficient of performance or *COP* of 2.0 will provide 6826 Btu per hour. You can see that the COP indicates the efficiency of the heat pump's use of electrical energy to provide heat and exceeds 100 percent during normal operation. That means that the heat pump is super-efficient in providing heating compared to normal electric resistant heating systems.

The heat pump conditions (heats or cools) the inside air only. It does this by bringing conditioned air into the blower unit from the return air duct. The heat pump can effectively add or reduce about 20 degrees to the older conditioned air. In other words, during the heating season, if the conditioned air is 60 degrees, the heat pump can increase it to 80 degrees. However, it also recycles existing air, which may tend to get stale. Anybody who has entered a home after an absence, with no heating or cooling, knows the air tends to become stagnant after a while. Because of the deterioration of the indoor air quality, at least a small amount of outside air should be added to the system. Good design practice is to add fresh, outside air at the rate of about 10 percent of the total supply volume. The outside air is usually added to the return air duct before it recycles to the fan coil. If you install an outside air inlet to connect with the return or supply air, be sure to install a screen with ¼ inch openings. Duct size and volume dampers are good ways to balance the volume. Check with a mechanical professional on how best to add a mixture of outside air to your particular system.

Heat Strip

As outside temperatures decrease, the efficiency of the heat pump to deliver heat decreases. The heat pump can extract heat from outside air as cold as 20 degrees with normal operation. When temperatures dip below that, additional auxiliary heat may be necessary. During the heating cycle, the heat pump relies on adding heat to interior air in smaller increments. The initial heating period may be longer than desirable. In colder climates, it may be necessary to add a heat strip to the system. This heat strip is nothing more than a heating element (or elements) inside the blower unit, which instantaneously provides heat during the initial cycle.

Outlet Boxes

Flexible ducts transfer conditioned air into a room from the air handler. At the point where conditioned air leaves the duct and enters a room, it passes through a register grill. The outlet box is made of sheet metal and has a cylindrical lip on one end. This lip is used to fasten the flexible duct to the outlet box. The cylindrical lip must accommodate the diameter of the anticipated supply duct. However, before fastening to the duct, the outlet box must be installed in the room.

If a room is small enough, it will only have a single supply air duct. If there is only one outlet in the room, select a central location on the ceiling. Remember where lights or ceiling fans will be before selecting a location. All four sides of the outlet box should be supported with wood blocking. It will be necessary to frame in the opening where the boot is to be installed. The surface of the outlet box will have a small flange that is ½ inch long. This ½ inch is to accommodate drywall on the ceiling.

Flexible Duct Material

The formal name for flexible ductwork is *factory made air duct.* Because they are a manufactured product, they must be tested by an independent agency and given a product listing, which details installation instructions, approved locations, surface burning characteristics, insulation performance, limitations of use and other considerations. An independent testing agency such as Underwriter's Laboratories will test the product according to previously established standards. The standards for factory-made air ducts is Standard 181 and 214 of Underwriters Laboratories.

The duct is made with a spiraling wire that defines the cylindrical shape. The wire is covered with materials that have been tested for a maximum flame spread, surface burning characteristics, ability to retard mold growth and resistance to corrosion accumulation. Each product must pass a battery of tests to ensure structural integrity, leak resistance, and ability to avoid bending and collapse. Each duct is tested to ensure that it will withstand torsion, tension and pressurization. The ducts are insulated and are tested to verify their ability to retain thermal energy.

These factory-made air ducts must be marked with identification that shows the manufacturer's name, the rated air velocity, negative or positive pressures, and specific information relating to the materials and installation. The manufacturer's installation instructions will advise you more thoroughly how to connect ends and joints of the duct.

Installing Duct

Flexible duct is the easiest type to install for an amateur. Because the duct is manufactured with coiled wire for structural integrity, it expands and collapses much like an accordion. Flexible ducts are available in a variety of sizes and lengths. You are permitted to use most factory made flexible air ducts for supply and return air. If a length of flexible duct is too short for a particular span, there are splice sections that will extend the reach.

Flexible duct material must be securely fastened to outlet boxes.

Designing the route for ductwork will depend on where your air handler is positioned within the attic. If it is installed near the center of the home, the duct system will look like an octopus with ducts stretching in every direction. If the air handler is positioned on one end of the attic, the duct routing will look like the rays from a spotlight. Either location has merits and difficulties. Flexible ductwork must be installed so as to prevent a sag or crimp in the material. When routing the duct toward its destination, be sure to provide a level surface. Remember what happens when you bend a garden hose? The water flow is diminished and the pressure increases. The same applies to forced air within a flexible duct.

It is best to install a short section of sheet metal duct immediately adjacent to the air handler. That way, when the air handler pushes conditioned air out, it will not adversely deform the ductwork. Another advantage to installing a hard rectangular sheet metal duct at the air handler is that you can immediately connect duct branches running to adjacent rooms. It is necessary to have a sheet metal worker manufacture a distribution box that will have connection flanges for duct material. You must have the proposed routing and duct sizes already decided before you make this box. The size of openings in the distribution box must match the proposed ducts.

To apply the layout, begin at the boot or cylindrical portion of the outlet box and connect the end of the duct around the lip of the outlet box. Use a hard plastic band to secure the duct to the boot. The band is an irreversible cinching belt that continues to tighten around the lip of the boot. Then, while in the attic, extend the duct in the general direction of the proposed location of the air handler. Repeat the process for each outlet box in your home.

Return Air

Air that has been heated or cooled and sent into your home will eventually lose its conditioning. In order to maintain the desired temperature, the air must be re-conditioned. This conditioned air returns to the air handler via the return air plenum or duct. This return air duct is very large, since it must carry all of the volume of air supplied to the entire house. Just like the supply air ducts, the cross sectional area of the return air duct must be at least 6 square inches per 1000 Btu. In the design stage, you determined the necessary size. Locate the outlet for the duct in a central location such as a hallway or living room. The shorter the return air duct, the better the system will work. A large, flexible duct tends to sag and lose shape over time. This causes diminished air flow back into the air handler. In order to enhance the mixture of air throughout the house, the return air grill should be located near the air handler but away from the rooms being heated or cooled. This will cause air to thoroughly circulate throughout the house. The return air duct will be connected to a similar but much larger outlet box. It is connected in the same manner as the smaller ducts.

You cannot install the return air inlet in certain locations. It cannot be located in a bathroom, kitchen, laundry room, or any other location from where it will pick up objectionable odors, fumes, or flammable vapors. Remember not to locate a supply air or return air duct anywhere close to a smoke detector. The close proximity will diminish the effectiveness of the detector. Since the air around the return air outlet is the least conditioned, it helps to locate the return air close to where the thermostat will be located in order to maintain the desired temperature.

Return air duct may also be installed with flexible duct listed for that purpose.

HVAC Equipment Installation

For split system heat pumps, the condenser will be installed outside, and the evaporator and air handler (single unit) will be installed inside the house. If open-web wood trusses are used, flexible duct may be easily installed in the attic. The attic will also accommodate the air handler. The evaporator/air handler should be positioned near one end of the house and over bearing walls. Ask the HVAC mechanical dealer to establish the weight of the unit. Then, coordinate with the truss manufacturer to help design the open-web wood trusses, which will provide a flat support for the equipment with enough structural integrity to support the unit. If the evaporator/air handler can be installed in a vertical manner, select a location near the condenser and use as a heater closet.

The condenser, which is installed outdoors, must be on a level, solid surface. A three inch concrete pad, slightly larger than the condenser, resting on level ground, is adequate. The condenser should be near the same end of the building as the air handler. Maintain a modest space between the house and the equipment to allow for repair work. There will be several pipes and electrical cables between the two units that rise inside the wall into the attic.

Attic Access

If you install the air handler in the attic, you must provide access to service and replace the unit. The size of the attic access must be large enough to remove the equipment, providing a minimum space

A condenser for a heat pump must be installed on a substantial base above grade.

Attics must be accessible by means of a removable panel or operable door.

of 22 inches wide by 30 inches long. The attic access must be in an accessible location. Don't try to hide it in a small closet. You will be using this access regularly during the construction phase. A hallway near the mechanical equipment in the attic will be especially handy. You may consider fold-down ladder access for the attic. This will eliminate the need to drag out the ladder every time you want to get in the attic. You might even get the truss manufacturer to design some light storage in the attic.

There are some safety and electrical aspects for attic installations. If you provide mechanical equipment in the attic, you must provide a switched light for illumination. The switch must be located near the attic access. All mechanical equipment connected to electrical power must have a means of disconnecting that power (circuit breaker) which is in view of the equipment. This is to prevent someone from accidentally energizing the equipment while it is being serviced. A metal guard must be installed around exposed moving parts of mechanical equipment. An electrical service outlet must be installed within 25 feet of mechanical equipment.

Working Clearance

In order to perform installation and repairs on mechanical equipment, it is necessary to maintain a *working clearance*. The need for a reasonable working clearance around both units is not trivial. It is time consuming and cumbersome to do repair work on inaccessible mechanical equipment connected to a house. It is also dangerous. An effective working clearance for mechanical equipment is 30 inches in depth, width, and height adjacent to the mechanical equipment.

The condenser may be outside, but it is still subject to repair work. Do not place the condenser near any inside corner or near any other obstruction that would prevent installation or repair work. Also, be sure and maintain the required working clearance for other equipment, such as the main electrical service panel. They should not have to share a working space.

The evaporator/air handler, which is usually installed in the attic, must be accessible and have adequate working clearance as well. Access to this installation is more problematic. Attic space is typically congested, filled with insulation and insects, and dark as well. The same 30 inch minimum width, depth and height for access is required for service work on mechanical equipment in the attic.

Access Passageway

Equipment installed in the attic must be accessible to a repairman. You must also provide a walking surface beginning at the attic access that leads to the equipment. This access ramp must be provided to allow you to walk to the mechanical equipment and avoid ductwork, electrical cables and plumbing pipes. It cannot exceed 20 feet in overall length and must be at least 24 inches wide. In addition, a level platform must be provided around all controls of the equipment for repair work. This platform must be structurally safe for a repairman to carry tools and other equipment.

Protection of Equipment

All mechanical equipment should be thoroughly and effectively braced to resist a seismic event. It must be anchored in place with straps or other connections that are capable of supporting the unit's weight. Heating and cooling equipment must also be protected from accidental damage. Guards to prevent damage should be installed. For example, if your condenser is installed outside near the garage driveway, a guard post would prevent a collision by an automobile. Look for any situation that may cause damage and erect appropriate protection.

Electrical Thermostat

The thermostat will sense the ambient temperature within the home and engage the mechanical equipment to either provide heated or cooled air. It is best to install the thermostat near the return air inlet box. This ensures that it will sense the temperature of the least conditioned air in your home. Follow the manufacturer's instructions for connecting the thermostat.

Air Filter

Everyone knows that they should change their car engine's oil every 3000 miles to prevent damage. For the same reason, a good filter located at the return air grill will prevent dust accumulation inside the evaporator/air handler. It will also help keep your air cleaner and remove dust and debris. Without a filter or grill, the suction at the return air inlet is significant and can collect lots of dust, insects and other fragments that could cause damage or decrease efficient operation of the mechanical equipment. Once, I answered a house call to inspect the lack of efficiency of an HVAC unit. Upon removing the return air from the air handler, I found the problem. A child's basketball approximately 10 inches in diameter was obstructing air from entering the air handler. The filter and grill were not installed for a few months, and along with the ball, dust and debris had clogged the air circulation and caused significant mechanical failure of the unit. Install a filter before you turn on the system and change it regularly to prevent damage to the equipment and ensure efficient operation.

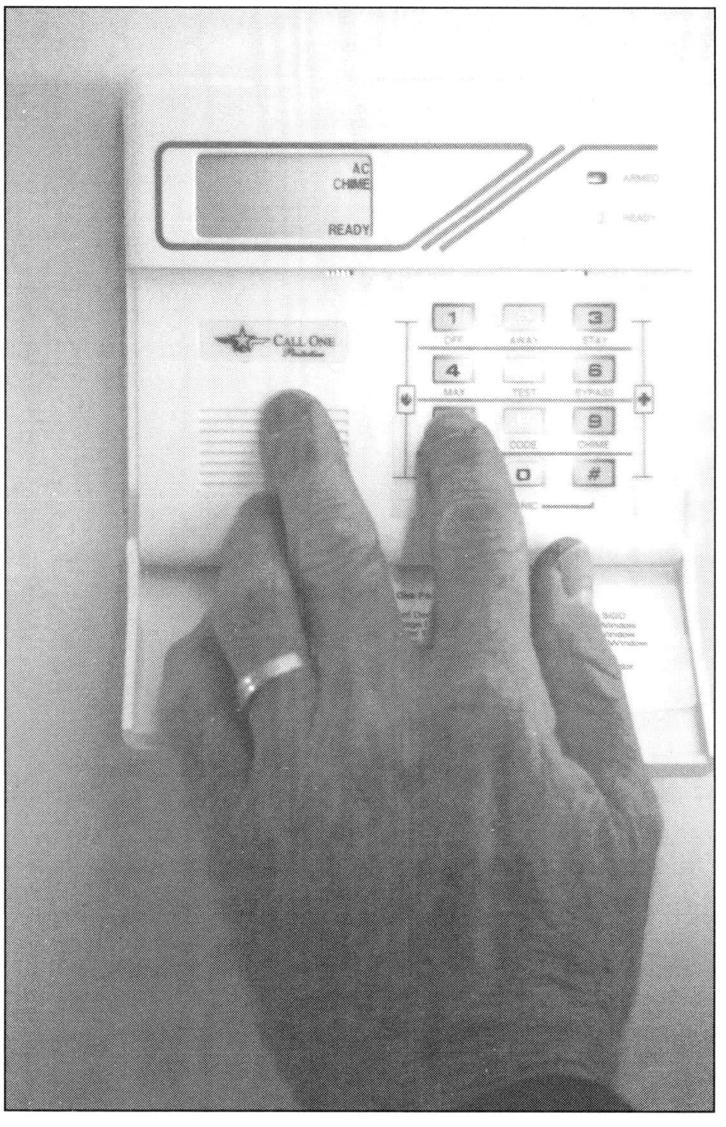

The modern electric thermostat has many improvements that add convenience to comfort.

Installation Standards

Mechanical equipment is manufactured according to standards set by the industry and adopted by the code. Heat pump standards are established by Underwriter's Laboratories. The standard designation number is UL559, which was adopted in 1975 and amended in 1984.

This manufactured equipment is tested according to this standard by an independent testing laboratory and assigned a product listing. This listing will demonstrate the conditions of the approval by the testing lab. It will validate the manufacturer's installation instructions, including the method of assembly, where it may be installed, any limitations on its use and other conditions of use. The required clearances for the equipment are part of the product listing. Even if you do not intend to participate in the installation of any mechanical equipment, you should review this product listing and compare it to your installation. If you have any questions, the manufacturer or the agency who performed the test will answer any question about the installation. It is important to the manufacturer that you be successful.

If you cannot acquire a product listing from either the HVAC equipment sales agent or the installer, you may be considering an unlisted appliance. While the unlisted appliance may be adequate for your needs, it must meet more stringent clearances than a listed appliance, and its safety may be questionable. Additionally, the building inspector may not approve it, even after you have installed it. While you may be tempted to save money on a cheaper product, I recommend that you limit your selection to listed and tested equipment.

Refrigerant and Refrigerant Lines

In the earlier part of this chapter, the use of refrigerant to transfer heat was explained. Without refrigerant, the heat pump would be nothing more than a large air handler. There are many refrigerants available. Refrigerant is installed by the mechanical equipment installer, after the units are assembled and connected. In 1990 Congress passed the Clean Air Act. The law controlled numerous sources of air pollution. Chlorofluorocarbons (CFCs), hydrochlorofluorocarbons (HCFCs) and hydrofluorocarbons (HFCs) are regarded as ozone-depleting gases. The ozone layer effectively blocks almost all of the solar radiation of very small wavelengths from reaching the earth's surface. This radiation would injure or kill most living things. Ozone is in a region of the atmosphere between 6 and 30 miles above the earth's surface. Despite the fact that the ozone layer is about 25 miles thick, it is extremely vulnerable and very dispersed. If all of the ozone in a cylindrical column through the atmosphere were brought to atmospheric pressure at sea level, it would be less than a few millimeters in height.

Because of this, certain refrigerants that were common a few years ago, are no longer available. Additionally, the installation, handling, extraction and disposal of these refrigerants are regulated. Certification by the Environmental Protection Agency is required for handling these products. For this reason, you may need to hire a qualified party to perform this task.

Condensate Drain

Heated air, filled with moisture vapor, cools to create rain or condensation. The same principal applies to the miniature ecosystem inside a mechanical heat pump. During the summer months, humid, hot air, laden with moisture is drawn from inside your home. The heat is exchanged in the air handler, and the sudden temperature drop causes condensation to form. This condensation represents the moisture that was in the air inside your home. It is extracted along with the heat. In humid conditions, the condensation from the return air is significant. It is much like having a garden hose bibb turned on to a trickle, enough to water a small tree. You have to get rid of this water to prevent damage to your new home. The condensation is routed out of your heat pump from the air/handler. Connect a ¾ inch PVC pipe section to the outlet from the air handler to your plumbing drainage system or anywhere outside the building. You might find a deserving tree or shrub nearby that would enjoy the added benefit of the condensation.

A backup plan is needed if the mechanical equipment develops a clog causing an overflow of condensation. In that case, the backup plan is to provide a tray underneath the heat pump to collect the overflow of condensation in case the regular condensate drain develops a clog. This tray must have its own separate discharge pipe that leads to a point, outside, that can be readily observed. This will serve as an alarm that your mechanical equipment is in need of repair. The condensate drain pipe is a simple yet imperative requirement.

Unit Installation

The outdoor unit does not need to be installed until you are closer to final trim. However, the interior unit should be installed before drywall is installed to ensure ease of installation. Move the interior unit in place. Install a secondary drain pan under the unit and set it into the pan. At this point, your mechanical professional can assist you in connecting refrigerant lines, condensate drains, secondary drains, electrical and thermostat wiring.

Other Mechanical Systems

A clothes dryer must be connected to an exhaust duct to expel moisture. The interior surface of the metal duct must be smooth, so that no lint will adhere and cause clogging. The dryer exhaust duct must extend to the outdoors and be provided with a backdraft damper, which will prevent weather or vermin from entering the duct. You must not install a screen on the terminal portion of the duct because it will tend to collect lint and debris. You should not use sheet metal screws when assembling the duct for the same reason.

Bathroom and laundry rooms must be provided with a moisture exhaust fan and vent. Normally this is a smaller fan capable of creating a negative pressure in a bathroom. These fans are capable of removing around 40–80 CFM (cubic feet per minute). The fan/vent assembly installs onto the ceiling joist or truss. The outlet is usually around 4 inches in diameter. This will allow a 4 inch diameter duct to carry the fumes and vapor outside when the fan is turned on. Although not required by any code, kitchen ranges may require an exhaust vent to fulfill the requirements of the range/over manufacturer's installation instructions. This exhaust fan/vent assembly is designed to create a negative pressure.

A listed chimney vent is installed during rough framing allowing for clearance to combustibles.

Wood Stoves and Fireplaces

If you plan to install either a wood stove or a fireplace, the single best piece of advice I can offer is to buy a listed and tested product. Do not purchase a used product or a home-made appliance. There is danger in any piece of equipment that is designed to contain a burning fire within your home. Unlisted appliances must meet very stringent requirements for clearance from combustibles. Additionally, the unit may be damaged in a manner of which even the seller is not aware. Do not compromise on this very important safety issue.

If you buy a new, listed, and tested wood stove or fireplace, the manufacturer will advise you of the installation methods, clearances and limitations. Normally, manufacturers require a listed double or triple wall chimney to serve as an exhaust vent for either a woodstove or a fireplace. Woodstoves are generally more energy efficient. Both offer that delightful smell and warmth which I dearly love.

Woodstoves and fireplaces must sit on noncombustible surfaces such as concrete or tile. There are requirements from manufacturers that limit how close the appliance can be to a wall or ceiling. The pipe that exhausts the smoke outside of the house is called a chimney vent. Usually this vent will pass through an attic. Because it is near combustible materials such as wood, the chimney usually has several concentric vents within the one visible from the outside. These are either double or triple wall chimney vents. A zero-clearance, manufactured fireplace sits in a frame wall.

Chapter 14

PRE-INSULATION PICK-UP

Now the Work Begins

Okay, so you've made it this far, you're pretty confident about your achievements, and you're ready to begin your exterior and interior finish work. The plumbing, mechanical, and electrical systems are all roughed-in, and you're ready to insulate and drywall after the required inspections. There are some very important details to check before you order insulation and drywall. The following are very important considerations to address now in order to avoid significant problems later. You see, it's easier to fix a problem in the open-framing stage than it is to remove drywall and then fix the same problem. This chapter can be considered as your own personal inspection checklist. It is based on each building trade. These checklists will represent some of the aspects which an inspector will use to evaluate your project. Read the language in the code for each aspect and understand the purpose to better understand the standards enforced by your inspector.

Before you begin to check these items, be sure and completely clean up your project. Remember that you must call for inspection on these installations before you cover, and the cleanliness of the project will correlate to the results. Extend the cleanup area to your entire site. Make this an event that marks the next step in construction. If necessary, order a roll-off dumpster. If you have collected a lot of superfluous building materials that you thought you would use and didn't, now is the time to extricate yourself from them.

Electrical System*

___ Verify the required clearances required for overhead service drop. *Figures E3504.1 and E3504.2.1*

___ Verify electrical layout per plan. Doublecheck that the height and location of outlet and switch boxes are correct.

___ Boxes, wiring, conduit to be identified with manufacturer name. *Section E3303.3*

A building inspector may have questions about your project. Be on hand to answer those questions.

___ Verify panel boards not allowed in clothes closets or bathrooms. *Sections E3605.7*

___ Verify clearances for lights in closet. *Figure E3903.11*

___ Verify kitchen outlets are fed with #12 wire, check for a required island or peninsula outlet. *Section E3603.2*

___ Smoke alarm outlets installed as required by code. *Section R317*

___ Separate branch circuit for laundry (#12 wire). *Section E3603.3*

___ Separate branch circuit for bathroom lavatories (#12 wire). *Section E3603.4*

___ Separate branch circuit for kitchen appliances (two required with #12 wire). *Section E3603.2*

___ No pendent fixtures planned in bathtub area. *Section E3903.10*

___ Verify free conductor length at boxes is at least 6 inches. *Section E3306.10.3*

___ Verify exterior outlets front and back and lighting at doors. *Section E3801.7*

___ Verify approved ceiling fan boxes. *Section E3806.8*

___ Metal water bond required clamp must be accessible. *Section E3509.3*

___ Verify receptacles in lavatory base if two sinks are used. *Section E3801.6*

___ Verify working clearances for equipment, a floor area: minimum 30 inches wide, 36 inches deep. *Section E3305*

___ Verify staples on wire for support on studs at boxes. *Section E3805.3.2*

___ Verify that conductor crimp rings and compression fittings are approved and installed properly. *Section E3306.9*

___ Verify that conductor fill of outlet boxes does not exceed the maximum allowable. *Section E3805.11.1*

___ Verify that conductors in plenums or air ducts comply. *Section E3804.7*

___ Conductors not permitted in concrete, adobe, or cinder block without conduit protection. *Article 336-5 NEC'99*

___ Motors must be ventilated and accessible. *Article 430-14 NEC'99*

___ Neutral wires to be "pig tailed" on multi-wire branch circuits. *Section E3602*

___ Metal plates installed on studs to protect wiring within stud cavity if cable is within 1¼ inches. *Article 300-4 NEC'99*

___ Verify that all ground wires are tied together properly. *Article 250-130(C) NEC'99*

___ Verify that all neutral wires are tied together properly. *Article 300-13 NEC'99*

___ Metal boxes must be grounded. *Section E3805.2*

___ Proper support for fixtures such as ceiling fans. *Section E3805.6*

___ Mark the home runs at service panel with circuit name to save time later. *Section E3606.2*

___ Doublecheck that the planned circuit breaker size matches the wire. *Section E3605.5*

** Note that where the electrical provisions of the International Residential Code do not cover certain aspects of electrical installations, the provisions of the National Electrical Code prevail as outlined in Section E3301.*

Mechanical Systems*

___ Verify the approved listing on vents and ducts. Check that they are installed per listing. *Section M1307.1*

___ Verify that the duct material is installed properly for size, material, location. *Section M1601.3*

___ Verify adequate support for duct work and distribution boxes per manufacturer's specifications. *Section M1601.3.2*

___ Verify adequate air flow for flexible duct work (in other words, straight, not kinked or bent). *Section M1601.2*

___ Verify that return air is installed per plan; check the size, routing and material. *Section M1602*

___ Check for adequate connectors between joints of duct material and distribution boxes. *Section M1601.3.1*

___ Verify exhaust fans in bathrooms installed and vents to the outdoors. *Section R303.3 (exception)*

___ Verify that 26 gauge material passes through fire wall between dwelling and garage. *Section R309.1.1*

___ Verify that HVAC equipment in garage is protected from damage by car. *Section M1307.3.1*

___ Verify condensate drain; materials, size, route. No PVC pipe is permitted in return air plenum. *Section M1412.2*

___ Verify that access will be available to replace or work on mechanical equipment. *Section M1305*

___ Verify clothes dryer vent length does not exceed the maximum permitted. *Section M1501.3*

___ Verify that a working platform is provided for heating or cooling equipment in attic. *Section M1305.1.3*

___ Verify that the attic has switched light and electrical service outlet. *Section M1305.1.3.1*

** Assumption is that no gas appliances are installed as recommended for first-time home builder.*

Plumbing

Rough-In or Underground DWV Plumbing

___ Verify drain-waste-vent piping; materials, size and placement. *Section P3002*

___ Check that plumbing vents meet the minimum size. *Section P3113.1*

___ Check that piping is properly supported. *Table 2605.1*

___ Check vent pipe fittings for proper usage. *Chapter 31*

___ Check horizontal pipe for proper slope. *Section P3005.4.2*

___ Verify no leaks at pipe joints. *Section P3003.1*

___ Check for clean-outs where required or per plan. *Section P3005.2*

___ Verify trap arm sizes and maximum lengths. *Table 3201.7*

___ Verify for proper listed pipe materials. *Section P2608*

___ Backwater valves installed if required. *Section P3008*

___ Verify under-slab water supply pipe installed with proper materials. *Section P2904*

___ Verify water test for DWV pipe. *Section P2503.5.1*

___ Verify building clean-out fitting is installed in building sewer, near the building line. *Section P3005.2.7*

Plumbing Top-Out

___ Verify water test on two-story portion of DWV piping. *Section P2503.5.1*

___ Verify proper size and type of all DWV piping; look for label with proper listing. *Section P3002*

___ Verify that the height of vents is above fixture rims. *Section P3102.2*

___ Check that termination of plumbing vents are at least 2 feet above or 10 feet from the side of windows. *Section P3103.6*

___ Verify required plumbing vent termination height of 6 inches above roof. *Section P3103*

___ Check exhaust ventilation of toilet rooms. *Section R303.3 (exception)*

___ Verify that location for water closet is correct for wall frame and future cabinetry. *Section P2705.1 (Item #5)*

___ Verify all required clean-outs installed per plan. *Section P3005.2*

___ Verify water test on water supply lines. *Section P2503.6*

___ Verify nail plates installed on stud walls to protect plumbing pipes within stud wall cavities. *Section P2603.2.1*

___ Verify proper trap sizing and trap arms lengths. *Table 3201.7*

___ Verify water supply pipe materials, size and adequate support. *Section P2904.5*

___ Check for cross connections between potable water supply and sewage system. *Section P2902.1*

___ Verify the water lines are under test conditions. *Section P2503.6*

___ Verify hot water is on left of each fixture. *Section P2722*

___ Verify water heater location, access and support. *Section P2801*

___ Verify that pressure and temperature pipe materials discharge to daylight between 6 inches to 24 inches above finish floor. *Section P2803.3*

___ Verify shower minimum size: 1024 sq in and 30 inches diameter. *Section R307.2 and Figure R307.2*

___ Verify water test on shower pans. *Section P2503.5.2*

___ Verify adequate pipe support both horizontal as well as vertical runs. *Table 2605.1*

___ Doublecheck that each fixture's exact location is correct per plan.

Wood Structural Framing

___ Verify that all earlier utility inspections (electrical, plumbing and mechanical) have passed. *Section R109.1.2*

___ Check for excessive cutting or notching of framing members. *Section R602.6*

___ Verify that wall intersections and corner construction are adequately connected. *Figure R602.10.5*

___ Verify that lateral bracing, such as shear walls, are installed where required and with proper nailing. *Section R602.10*

___ Verify that proposed wall openings and windows have the required natural light and ventilation. *Section R303*

___ Minimum widths for stairways and hallways of 36 inches. *Section R311.4*

___ Minimum stairway headroom height of 6 feet 8 inches. *Section R314.3*

___ Verify stair requirements, 7¾ inches maximum rise, 10 inches minimum run, with no more than ⅜ variance. *Section R314.2*

___ Verify that stair width is at least 36 inches. *Section R314.1*

___ Verify that handrails are installed between 34 inches and 38 inches above the nosing of the tread. *Section R315*

___ Support columns and posts are installed per plan and anchored to foundation and beams. *Section R407.3*

___ Verify that sizes of headers, lintels and beams are per plans. *Tables 502.5(1) and (2)*

___ Verify that wood stud spacing is per plans. *Table R602.3(5)*

___ Verify that attic access is according to plans and a minimum of 30 inches by 24 inches. *Section R807*

___ Verify that sill plates are pressure treated or foundation redwood. *Section R323*

___ Verify that anchor bolts for sill plates are no more than 6 feet apart. *Section R403.1.6*

___ Verify that each individual piece of sill plate has at least two bolts per piece. *Section R403.1.6*

___ Metal connector clips are installed between walls and roof trusses per plan with proper nails. *Section R802.3.1*

___ Verify maximum bored holes and notches in structural lumber. *Figure R502.8 and Figure R602.6*

___ Interior frame walls to be anchored to masonry walls w/ 3 bolts embedded into grouted cell. *Section R602.11.1*

___ Check fire blocking throughout wood wall frame. *Section R602.8*

___ Verify structural blocking: solid blocking at points of bearing. *Section R502.7.1 and Section R802.8.1*

___ Install attic ventilation vents. *Section R806*

___ Install foundation ventilation vents. *Section R408*

___ Install structural blocking for wall cabinets in kitchen, bathroom, laundry and other areas. Good practice.

___ Install structural blocking for any wall-hung pictures that weigh over ten pounds. Good practice.

___ Seal the base of all exterior walls on the inside with a good quality exterior grade caulk. *Section N1102.1.10*

___ For servicing attic equipment, build a platform with plywood. *Section M1305.1.3*

___ To isolate *quiet* rooms from noisy rooms, install insulation in wall frame. Good practice.

___ Consider installing a skylight; better now than later! Good practice.

___ Install any optional exhaust vents that may be required for range or oven. *Section M1502*

___ Install doorbell button, bell and low voltage wiring. Manufacturer's specifications.

___ Install caulking between windows and wood frame to prevent air infiltration and heat loss. *Section N1102.1.10*

___ Photograph interior frame to help remember location of studs, plumbing, and wiring.

Chapter 15

ENERGY CONSERVATION

Heat Loss

In Chapter 3 you learned about heat flow. Heat flows in only one direction—toward the lack of heat. Heat moves this direction in any combination of three ways: conduction, convection, and radiation. Conduction is the heat transfer you experience when you grab a hot frying pan. Convection is the movement of heat through the air, such as the heat you experience from the frying pan when it is inches away. Radiation is the sensation you experience from the sun while sunbathing.

Convective heat flow is the most common method of heat transfer to overcome in trying to maintain a comfortable environment inside your home. In order to maintain a comfortable interior, you must adequately keep adverse weather outside. You can achieve this with quality windows and doors, heavy insulation, weatherstripping, and caulking. The expense of these simple materials is minimal compared to the savings they will help you to realize over the years. The size, placement, quality, and quantity of windows affect the flow of heat, since the ability of glass to resist heat flow is inferior to insulation in a wood frame wall. However, in Chapter 2, you made certain decisions about the aesthetics provided by windows. A large picture window may provide the look you want in a particular room. Although it will give you the view you desire, it will also serve as an energy loser. These are some of the compromises you reached as a family.

You can compensate for heat losses of this nature. Windows are available with multiple panes, separated with a vacuum (or even inert gas) which helps resist heat flow. Large picture windows can be provided with night walls, which slow down the loss of heat. These are thermal blankets placed against the window during the evening hours of the heating season, when you do not need a view. Heavy drapes are a good example of a decorative night wall.

Insulation: Resistance to Heat Flow

The purpose behind insulation is to restrict or at least retard the flow of heat. There are numerous materials that are available for use as insulation, including fiberglass, paper, wood fibers, cellulose, pulp, and synthetic polymers. Insulation material may be installed in a variety of methods such as batts, blankets, rigid panels, blown-in, or even loose fill. The value for resistance to heat flow is referred to as R value. The larger the R value, the more the insulation should restrict the convective flow of heat. The value is the coefficient of heat flow for a particular material. It is the inverse of the R (1/R) value.

Blanket or batt insulation is the most common type of insulation used in wood frame houses. It is designed to lay neatly inside the cavity made between stud framing members, which are usually either 16 inches or 24 inches on center. The thickness of the insulation is also designed to fit the most common studs used for wood framing. Thickness of batt type insulation is usually either 4 inches or 6 inches thick, which corresponds with the thickness of either the 2 x 4 or 2 x 6 wood stud.

However, don't assume that the R value will be maintained if it is compressed into a cavity too small. The R value is based on the density and thickness of the material in its optimum compression. If batt or blanket type insulation is compressed more than it's designed to, its R value will diminish. Therefore, a 2 inch by 6 inch wall may use a 6 inch thick batt, and a 2 inch by 4 inch wall may use a 4 inch batt. Insulation in the attic area is similar in function to that of walls. Usually, trusses are placed on 24 inches o.c. and a 22½ inch wide fiberglass batt insulation fits perfectly in between the bottom chord of each truss.

Blown cellulose is an excellent way of completely filling the stud wall cavity and sealing any heat or air leaks. There are numerous products available that may involve renting equipment to complete the installation.

Air Barrier

An air barrier is the shield between your conditioned space and an unconditioned space. The location of the air barrier is important in deciding how to efficiently install your insulation. The insulation should be installed so as to make firm contact with the air barrier. The reason for this is that voids or spaces between the air barrier and the insulation allow for heat to migrate through, leaving the conditioned space. Once out of the conditioned space, the heat will slowly dissipate out of the wall or attic of the building.

An exterior wall with siding defines an air barrier during the cooling season. The most efficient thermal barrier to heat flowing into the house occurs when the insulation is thoroughly engaged with every surface area of the exterior siding inside the stud cavity. To prevent heat from leaving the conditioned space during the heating season, the side of the air barrier is reversed. The surface of the drywall becomes the air barrier. Because of this, the insulation must completely touch the inside of the drywall. This means that when you install batt insulation, you must be extremely careful to make sure the insulation is completely touching both surfaces and yet still not over-compress the insulation.

Gaps in insulation, voids in wall spaces, and compression of insulation are among the most common mistakes that diminish the insulation's ability to resist heat flow. Insulation must completely fill all of the voids inside the wall cavity. It also must be completely in contact with the drywall from within the attic. Do not let it extend above the finish ceiling drywall in the attic. Be sure to install the insulation in both the ceiling and wall frame in a smooth, workman-like manner. Notice that the manufacturer's instructions on most batt insulation call for you to allow the batts, which are compressed for packaging, to expand or fluff up for a brief period of time before you install them. You should not be able to see anything in the attic from within the room after you have installed these batts properly. The inside of the exterior siding should not be visible. There should be no light entering the room except from the windows.

These blankets or batts are usually covered on one side with a paper, which serves as a moisture barrier. This moisture barrier prevents the insulation from absorbing water vapor and settling, thus reducing the R value of the batt. This moisture barrier should be placed toward the "warm" or living side of the wall or roof frame. This prevents moisture inside the house from migrating through drywall and infiltrating the insulation, thus causing it to compress. In warm, dry climates, the moisture barrier is not as important, and some batts are sold without the moisture barrier.

Notice that a folded tab of paper on the moisture barrier is provided on the sides of the batt. Use this extra paper to staple the insulation to the wood frame. The paper tab on the next cavity should overlap on the previous paper tab to form a moisture lock.

Installing fiberglass insulation is a unique experience. Fiberglass insulation is made from fine fibers that are very caustic and irritating to skin and dangerous to interior parts of your body. Although you may be tempted to work in shorts and a tee shirt, I strongly recommend otherwise. Wear clothing that you can afford to discard (or wash several times to extract fibers). No portion of skin should be exposed. You should wear a breathing mask over your mouth and nose. It would be desirable to wear goggles as well. I am not kidding! If you choose to ignore these warnings, at least pass my warnings to your family and give them a chance to protect themselves. You will be breathing finely spun glass fibers that are harmful to human tissue.

Alternatives

Other common insulation materials include blanket type, rigid, or loose fill type products. Blown-in insulation is extremely efficient since it fills cavities in both wall frames and attic areas very thoroughly. The industry is filled with different versions of each of these types of insulation. Research and use the best product for your particular home.

Infiltration

Because the objective is to keep the weather out, you must seal all the air leaks you can find. Infiltration will occur in even the most tightly built houses. Look for them at joints between dissimilar materials such as wood and concrete or metal windows and a wood wall frame. One very common source of infiltration is the joint between the bottom plate in a wood frame wall and the wood floor or concrete slab. Seal this leak with a healthy, double layer of caulking under the bottom plate before you erect the wall. Then later, before you install drywall, add an extra layer of caulk at the inside base of the plate.

Gaskets and Weatherstripping

Heat will escape through the most convenient path of all the thermal holes within the heated envelope of your home. Some experts consider the most common loss of heat to be through windows and doors and the openings caused by their installation within the wall frame. These openings occur at joints between exterior doors and door jambs or windows and their wall frames. They can be sealed with caulking and tape covering the joint on both interior and exterior. Don't worry about the appearance, since drywall and exterior siding and trim will provide cover.

Doors must be provided with weatherstripping and gaskets, which form a seal when the door is in a closed position. The type of weatherstripping should be of high quality to afford longevity. Make sure the fit is tight. Test at night with the door closed and the porch light on. Look for light leaks, which are synonymous with heat leaks. Adjust the weatherstripping as needed.

Additionally, exterior doors are usually provided with a threshold (over which you will carry your bride). This threshold serves as the door's bottom weather seal. Thresholds come in a variety of lengths, heights, and shapes. There are minor adjustments permitted in the weatherstripping of the threshold—be sure to measure carefully before you buy.

Windows

Depending upon the quality, a little less than half of all heat loss in a house may be through the windows. The quality of windows is critical in preventing this. Search for windows with the lowest U value. (The U is simply the inverse of the R value.) The larger the R value, the smaller the U value. Double pane or thermal windows are desirable if you can afford them. If you have less than 20 percent of your wall area attributed to windows, a value less than .60 would be a good target value for windows.

Doors

Exterior doors are most commonly made from either solid wood, which is at least 1⅜ inch in thickness, or from metal clad doors with an insulated foam core. Either will provide a good method for resisting heat loss. Be sure to add weatherstripping and a threshold to prevent infiltration. Joints between different materials represent a common means of heat loss. A door frame or jamb may not perfectly align with the rough opening. Sometimes even wood shim material must be installed to help adjust the door. Be sure and add insulation in this area to seal the leak. Some expanding foam products are very appropriate for this use. However, some door or window manufacturers will cancel their warranty if certain of these products are used due to the harm they may do to their product.

HVAC Efficiency

The efficiency rating of HVAC equipment is important for consideration in the economic use of energy. Heating your home is made possible by converting one form of energy into another. In the case of electric heating, you are converting electricity into thermal energy. The efficiency is based on the amount of energy loss in this transfer process. Losses due to the transfer increase with older or poorly maintained heating equipment. Some modern equipment such as heat pumps have very high efficiency ratings. They are measured in SEER ratings (Seasonal Energy Efficiency Rating). A rating of 10 exceeds a rating of 8 and so on.

The proper size for a heat pump is critical in the efficiency of energy usage. A unit that is too small, will be running constantly, trying to do more work than it is capable of doing. An oversized unit will deliver too much conditioned air for the volume available and waste energy. It will run for shorter periods than it should and therefore remove very little moisture from the air. Proper sizing along with proper balancing of conditioned air within the building increases efficiency. Seal supply air ducts on ends or connections to supply boxes with mastic in addition to duct tape so as to achieve a substantially airtight vessel.

Plastic Sheeting Over Insulation

Moisture barriers prevent insulation from becoming damp and collapsing and therefore becoming more dense, which would decrease its ability to resist heat flow. Applying a plastic sheet over wall and ceiling insulation will serve as an additional barrier. It also adds another infiltration barrier or drape to the exterior walls, keeping dust and wind at a minimum. It is cheap and labor intensive—but it's only your labor, so if you are in a moist climate, consider it!

I recommend a 6 mil thick plastic material. You can use thicker paint drop cloths as long as you overlap them. Be sure to be careful around electrical openings so as to maintain the integrity of the plastic enclosure. You may need to add extra insulation at these locations and tape around electrical or telephone outlet boxes.

Framing the Energy Efficient House

Carpentry affects a house's energy efficiency. The wood members connect with each other in such a manner as to retard or accelerate heat transfer. Exterior wall frames are built with a bottom plate and two top plates that are at the ceiling height. Interior walls are built in a similar way. These top plates serve as a barrier between the heat that escapes from within the wall frames. However, when electrical outlets and wiring as well as plumbing pipes and fixtures are installed, holes are bored and notches are made in this pristine wall frame. These cuts are necessary to add electrical and plumbing facilities to your home, but they should be limited to the size and shape needed. Avoid excessive cutting and notching in all framing members. You need a 2 x 6 wall frame for a 3 inch pipe. The hole should be just slightly larger than 3 inches in diameter. Stuff insulation in the annular space around these penetrations. Use a smaller drill bit for a single wire passing through a top or bottom plate. Do not permit shafts or large openings to transcend from wall space into the attic. Even with insulation, these spaces will be energy losers and sources of infiltration.

Living in a Tight House

Of course, there are those who contend that all of the tightening up in construction and material assembly creates the effect of living inside a plastic bag. An air conditioner and a heat pump are closed systems, meaning that they recondition the air already in the building and do not bring in any outside air. So, it is possible to create a stagnant air environment if you don't use your windows and doors. Opening windows may increase natural ventilation, but it will also cause loss of conditioned air from within the house.

The cost of wasted energy is significant. During the life of your house, the energy costs may exceed the cost of its construction. In tight houses, there is less air movement. Less outside air comes in and less inside air is lost outdoors. Houses do need to be tight enough to avoid energy loss, while allowing for the replenishment of natural ventilation. Without proper natural ventilation, the indoor air will become stagnant and even nurture and carry disease. Air filtration is critical in a tight house. Some filters will collect dust and particles smaller than you can see. Removal of these will improve the quality of the air you breathe. Change the filter on a regular basis. This improves the efficiency of the HVAC unit as well.

Discuss this problem with your mechanical equipment dealer or professional installer. If the equipment is adaptable, consider modifying the duct system to add a small portion of outside air. The return air plenum cycles the stale, conditioned air back into the air handler portion of the heat pump or air conditioning unit for re-conditioning. Tap into the return air plenum with a 4 inch or 6 inch diameter flexible duct and place near an outside location. This will ensure that a portion of the recycled air is fresh, outside air.

Attic Vents

If you live in an area that is very hot in the summer months, your attic will be a source of heat for your home. It's true that heat rises, but if heat has nowhere else to go, it will also descend into your cooler house. Remember, heat seeks the easiest path toward a lack of heat! In certain desert climates, some attics may reach 160 degrees in the summer, even with attic ventilation.

In order to decrease the heat in your attic, you must remove it elsewhere, outside for instance. There are a variety of ways to achieve this result. Vents in an attic must be at least 1/150 of the overall roof area. In other words, a 1500 square foot house (single story) must have attic vents equal to 10 square feet. That may be 5 square feet on each gable wall end.

Chapter 16

INTERIOR WALL FINISH

Gypsum Board Should Be Kept Dry

A simple, obvious reminder. Drywall is a factory-made product. It is assembled from gypsum dust particles that are bonded together thermally and chemically. They disassociate and turn into their constituent parts when subjected to moisture. Therefore, don't leave drywall where it can get wet. Store on a dry, flat surface, covered with plastic until ready to install.

Types

There are a variety of manufacturers of drywall. There are also a wide variety of gypsum products, shapes, and sizes. They all have unique uses and applications and methods for installation. Some drywall is made with a certain fire-resistive quality. Other drywall is designed to be installed in bathtub or shower surrounds to resist water penetration or moisture decay. Plaster sheathing boards are even designed to be installed on outside walls. Drywall varies in thickness between ¼ inch and ¾ inch. The ¼ inch is usually only for covering other solid, hard wall surfaces. A ¾ inch wall board is permitted to be installed in some circumstances depending upon the center spacing of the stud support and whether the drywall will be applied vertically or horizontally.

The most common thickness used in houses is the ½ inch type. This is allowed for both vertical and horizontal assemblies. That is, it may be applied on walls as well as ceilings, with certain nailing or screw patterns. The second most common thickness used in homes is ⅝ inch drywall. This is used mainly as a rated separation between the residential portion of your home and the garage. One layer of ⅝ inch Type X drywall applied over stud framing that is 16 inches o.c. establishes a fire resistive separation between the two occupancies. In the IRC, the separation between a dwelling and its garage is limited to regular ½ inch drywall.

The Beginning

Okay, you have followed the methods outlined in Part One of this book and have ordered adequate drywall material. You have enough drywall nails or screws on hand. You have both a drill with screw bit and hammer ready. You have a tape measure, marking pencil, large tee square, utility knife, and keyhole saw available. You have even rented a drywall jack for ceiling installations. What's next?

You have carefully followed the advice in Chapter 14. All of the pickup items are checked off your list. Your intermediate inspections have passed. Any additional agency needing to see inside your walls have done so and are satisfied. Now what? Remember that when you put drywall on your walls, you will never be able to see inside again without removing the drywall. Why take chances with memory? Get the camcorder or camera out and record everything you may need to know about later. Believe me, knowing where a plumbing drain, waste and vent pipe runs may become essential. You may forget the layout or center-spacing for studs in a certain wall section. Where you branched off with electric wiring may really become important. Its so simple to take a picture—before you install drywall—and so hard afterward!

Another last minute thing you may consider is to install insect or vermin repellent. This may be the last, best chance to get at this very inaccessible spot. Use your imagination. Try to imagine where you might intend to hang a heavy mirror or picture. Doing so where there is no stud or wood framing member will destroy the drywall and result in the heavy picture frame falling. You may need to install some wood blocking in between studs for added support.

Ready to Rock

Begin hanging your first effort in a room that will bear the least scrutiny, like a bathroom or laundry room. Learn the trade in the least conspicuous room, like a closet. Remember that drywall subject to moisture, as in tub areas, must be a moisture-resistant type. This type of drywall may have some limitations. For instance, some types of drywall may not be permitted to be installed as a horizontal assembly—in other words, it can't go on the ceiling. Read the installation instructions for all different kinds of drywall before you install.

Horizontal Installation

Hanging drywall on the ceiling can be an adventure, which is a modern euphemism for tedious, hard work. Drywall supported on the heads and shoulders of your family, standing on chairs, while you attempt to nail or screw into ceiling members causes short tempers. An invaluable tool, which will improve both the quality and quantity of your work, is the drywall jack. You can even rent this device from most rental stores. In a weekend, you can install drywall on your entire ceiling—without a single headache!

Although it is okay to install drywall in either direction, I recommend that you install it with the long direction, perpendicular to the ceiling joist or the bottom chord of the trusses. This allows more structural members to provide support and hides any imperfection in the framing better. Beginning in one corner, measure and install a row at a time throughout each room. When complete with a row, begin a subsequent row from the opposite end. This will allow the joints in the drywall to be staggered. If for some reason, the ceiling frame is incorrect on layout, you may need to add a wood nailer along one side of the ceiling joist. This nailer will accommodate the poor layout and allow a full length of drywall to be installed without cutting to exact length.

Vertical Installations

Installing drywall on walls is simple and less strenuous work than installing on ceilings. You have two options for the installation. You can install the drywall upright or lengthwise. Drywall must be supported on at least three supports or studs. If you have purchased 12 foot lengths of drywall, it is preferable to install it lengthwise. If you have purchased 8 foot lengths, you can install it in either direction.

Cutouts for Light and Outlet Boxes

Outlet and light boxes must be flush with the surface of the drywall to ensure an even fit. In order to be precise in the box location, use a tape measure and measure both the horizontal and vertical distance to the center of the box. Transfer this dimension to the drywall and make a small hole. Position the drywall onto the surface to be covered, and again verify the exact location for the cut. Then carefully cut out the shape of the box. There may be multiple iterations of this measuring and cutting sequence, in order to not over-cut the hole size. For ceiling outlet and light boxes, use the jack to raise the drywall into position. Use a colored chalk-line powder jar and squirt some of the powder onto the light box. Then move the drywall exactly into position. The outlet box will imprint its outline onto the back of the drywall. Lower the drywall and cut the hole along the outline.

Some professional drywall installers will mark the center location for all outlet and light boxes in the method previously outlined, then actually install the drywall without making the full cut. Later,

Careful measurement is essential when cutting drywall.

they use a small drill with a cutting bit to route the hole using the outside of the outlet or light box as a template for the cut. With experience, this really makes a perfect fit between the outlet or light box and the drywall. However, it takes experience to avoid damaging wiring and devices inside the boxes. If you try this method, I strongly suggest that you use a short drill bit and push all electrical wire or cable deep inside the outlet box to ensure that it is protected and suffers no damage.

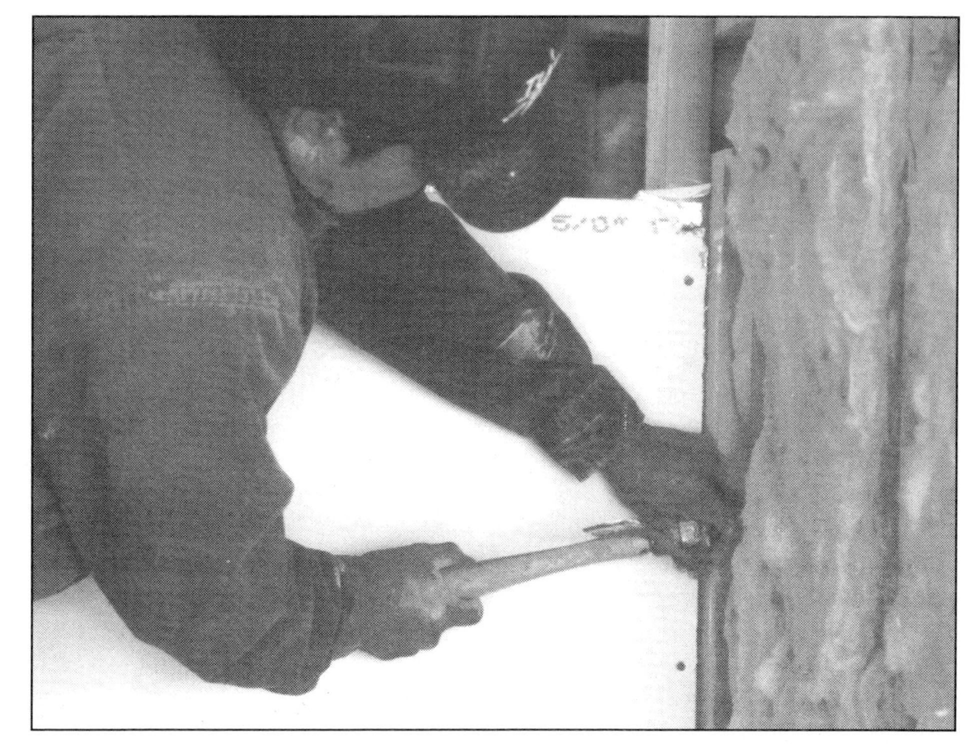

Drywall must have backing on all sides. Nails are applied to the outer perimeter and the inner field.

Nails and Screws

Ends of drywall must be supported by a structural frame such as a ceiling joist or truss chord. The connections that hold the drywall to the wall or ceiling may be either drywall screws or nails. The nails or screws must be set at least ⅜ inches from the edge of the drywall. This keeps the drywall from tearing out of the support.

The nails permitted to be used for drywall attachment are clearly identified in manufacturers' labels. They must be at least #13 gauge nails, and for ½ inch drywall at least 1⅝ inches long. The nails should be spaced no more than 6 inches apart and fastened to a framing member such as a stud or ceiling joist. The hammer should be a drywall hammer. Be careful during the last hammer thrust to avoid any damage to the drywall around the nail. This damage will diminish the nail's ability to hold the drywall onto the frame.

Certain screws are permitted to be used as fasteners for drywall to framing members. Because of the holding power of the screws, less of them are required than nails. They can be spaced farther apart. The screws should be spaced no more than 10 inches on center and fastened to a framing member such as a stud or a ceiling joist. They should be long enough to penetrate at least ⅝ inches into wood framing. For ½ inch drywall, a screw must be at least 1¼ inches long. I recommend using 1½ inch screws for both walls and ceiling installation.

Another seldom used method of connecting drywall to framing is an approved adhesive. Although I do not recommend that this be used by an amateur working alone, it would be an excellent addition to conventional mechanical connections. Different brands of adhesive are approved for wood-drywall fastening. Just install the adhesive according to the manufacturer's installation instructions, and then add screws or nails.

Joints

Where different sections of drywall are installed adjacent to one another, a joint occurs. In order to hide the joint with tape and texture, the two pieces should be as level as possible. Even a wood splinter lodged underneath one piece can cause an unsightly joint, which must be sanded to create a level surface. Sometimes it will be difficult to determine the reason for the unlevel surface. You may be unintentionally covering an outlet box or telephone receptacle. Look for irregularities and debris that will cause a problem, before you lift the drywall into place.

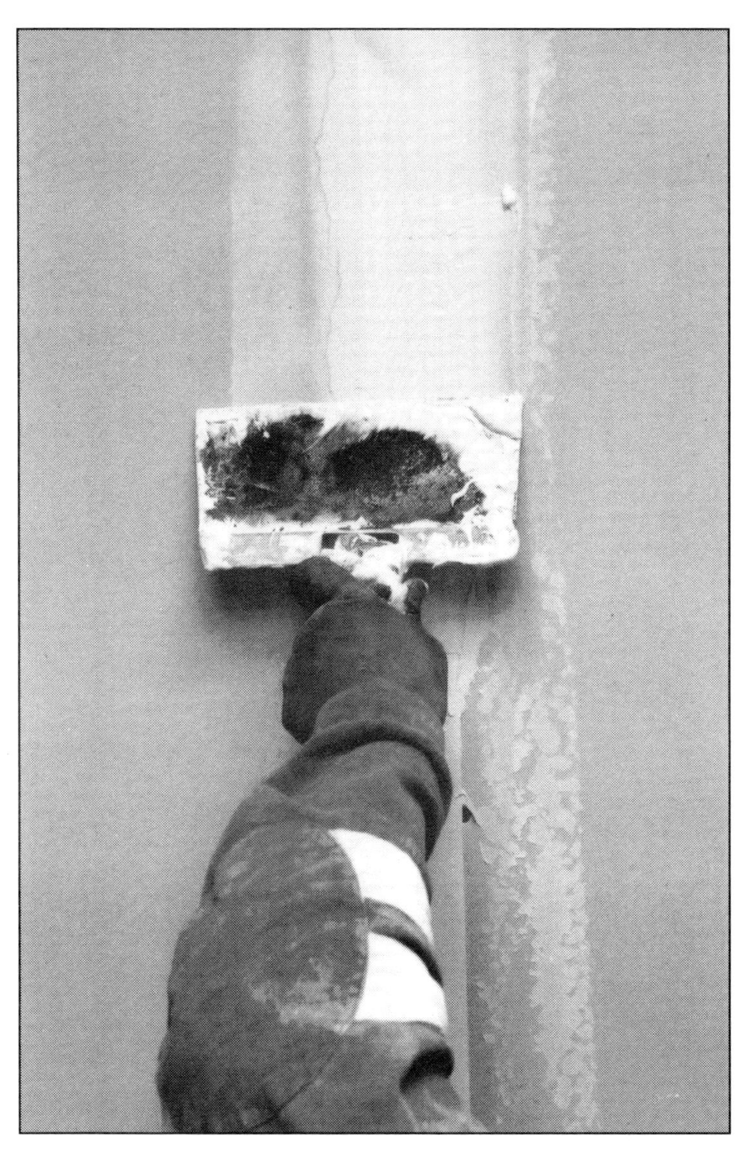

A taping knife is used to apply tape to cover joints in drywall.

Corners and Wall Intersections

It is necessary to add a semi-rigid material to corners that are subject to accidental damage. Corner bead is available either in thin metal or plastic. It forms a 90 degree outside corner, which is applied to all outside corners of walls. Nail the corner bead through the drywall and into a framing member on each of two surfaces. Holes in the corner bead permit nails to penetrate into the supports. Corner bead is also available in a shape that creates a rounded corner instead of a sharp corner. This adds a distinctive look to any wall.

All Purpose Ready-Mix Joint Compound

The material that seals drywall tape and subsequently covers the tape is called joint compound. It is a premixed vinyl base compound. It is made to seal the tape to the drywall and to provide a textured finish to both drywall joints and surfaces. It contains additives that allow it to seal the tape to the surface. It gains strength when it dries. The joint compound embeds the tape and corner bead. It also will allow you to laminate drywall to an existing surface such as a masonry wall.

There are two types of all purpose compound. One kind is thick and is designed to be used in hand applications, and the other is thinner and is designed to be used in mechanical equipment applications

Attractive details and niches can be created with drywall with little effort.

such as a hopper gun powered by an air compressor. The joint compound is water-based, so it can be thinned to increase workability. Store any compound away from extreme heat or cold. All purpose joint compound can also be used to achieve an attractive surface for interior walls and ceilings. All purpose ready mix joint compound can serve another function as well. If you are applying the drywall to another flat surface where nails or screws would not be appropriate, joint compound serves as the adhesive. It allows the drywall to be laminated to other surfaces such as block or plaster walls.

Taping Joints

All joints between drywall should be taped with drywall tape. Essentially, tape hides the joint and successive layers of texture hides the tape. All purpose joint compound is used as a mastic that holds the tape to the drywall. Use a joint taping knife about 3 inches wide, and apply a thin layer of joint compound for the entire length of the joint. Then place the tape over the thin layer of joint compound and press it gently in place with your fingers and palm. Then to smooth it out better, use the taping knife to seal the tape into the bed of joint compound. After installing the tape, add another very thin layer of joint compound to the top of the tape and smooth it into the tape. Repeat this same process throughout every drywall joint in your home.

The result of successful texture finish.

Sanding Joints

When the first layer of joint compound has been applied to the tape and is dry, it should be sanded to remove any roughness or surface irregularities. Use a pole sander to achieve a level surface. A word of caution though—if you have allergies, wear a breathing mask. The dust you generate within the confines of a building may aggravate allergy symptoms.

Floating Joints

In order to completely hide the taped joint it is necessary to apply repeated thin coatings of all purpose texture over the taped and sanded joint. Use a wider taping knife to spread all purpose texture in order to float the joint. This texture will dry and may need to be sanded and floated again in order to completely obscure the joint.

Applying Texture Finish

When all of the joints have been taped, sanded, and floated successfully, it is time to apply a finish texture to the entire surface of the drywall. This textured pattern will turn otherwise flat, dull drywall into a wall pattern that is interesting and stimulating to the eye.

The texture is the same material that is used in applying tape onto drywall joints—all purpose compound. Texture can either be installed by hand using a trowel or mechanically. A hopper gun is a device that uses air pressure to scatter and blow texture onto walls. This gun is usually available from rental stores. If you don't have a compressor, rent one to provide pressurized air to the hopper gun.

Because there are two ways of installing the texture, there are two different viscosities. The manual installation method will rely on a slightly thicker texture. A hopper gun cannot blow this thicker texture. You can select a thinner all purpose compound or add water to the thicker compound to thin it to a consistency appropriate to a hopper gun.

Chapter 17

FINAL TRIM: EXTERIOR AND INTERIOR

Sealing Exterior Siding

Exterior wood panel siding has been installed during framing and maybe even before wall erection. However, you must trim your exterior to seal up gaps in connections and joints between dissimilar materials that are exposed. These and other openings must be sealed to prevent interior heat from escaping.

For instance, windows and doors in a wall frame always leave at least a small gap, which allows air infiltration into your home. Also, wall panels at corners cause a joint opening. If you have trusses that overhang a wall frame, the soffit area under the truss causes an opening into the attic area that must be sealed. Above the top plate in the wall frame, in between the trusses, another gap exists that allows birds and vermin, as well as moisture, to invade.

There will be many such nooks and crannies that you must isolate and seal. Use insulation, expanding polyurethane foam, caulking, or anything that will not provide a source of food for termites or other vermin. Exterior siding usually requires studs to be no more than 16 inches apart. Where joints in the siding occur, a nail should be driven through two pieces of siding. The lap portion of each piece of siding is about half the normal thickness and subject to breaking very easily. Check with the manufacturer about gluing the joints prior to nailing.

For siding with tongue in groove joints, cover the vertical joints with dimensional lumber such as a 1 inch by 4 inch board. This provides an architectural decoration as well as a weather seal. It also provides a more secure attachment for the siding at these joints. Any other type exterior wall covering must be installed over an approved weather barrier such as asphalt-saturated felt paper. This paper must be free from holes or breaks. It must be applied horizontally, and joints must be lapped.

Facia

The ends of pitched trusses extend over the exterior wall to provide drip protection for exterior walls. The ends of these truss legs should be capped with an exterior-grade material such as facia. This facia should be applied prior to roofing material, so that drip edge material can cover the facia. The bottom side of the truss legs must be covered as well. Facia material is adequate here as well. Facia installed here prevents birds and bees from using your attic as their hotel. Of course, you may need to provide ventilation in the attic at this point. To do that and still maintain protection against insect infestation, install vented frieze boards. These allow the free-flow of air in the attic and obstruct anything larger than 1⁄64 of an inch.

Exterior Painting

Whether you select exterior wood siding or masonry veneer such as stucco as your finish exterior, it must be protected from weather damage. Paint performs two functions; not only does it add color to your home, it also provides a weather seal. The type of paint you select depends upon the surface material. Exterior wood such as window trim, facia, or soffit board withstands weather better when coated with an oil based alkyd or acrylic latex primer paint product. Exterior wood paneling or stucco veneer can be effectively protected with an exterior grade acrylic latex house paint.

Interior Painting

Your drywall is installed, taped, and texture has been added. Now, you must protect the interior walls from deterioration. Paint serves to protect the wall as well as provide an attractive color. Paint not only adds the look and feel you want, it seals the surface and blocks dust and other deleterious materials. It also preserves the texture from falling into decay. Paint all areas, even if they will be concealed from view by the fixture.

Paint can be divided into two distinct categories: oil-based and water-based. The water-based paint is always identified as a form of *latex* paint. Oil paint is usually identified as either enamel or alkyd. Latex paint is available in different sheens: flat, eggshell, satin, semi-gloss or gloss. Each of these sheens has a different resistance to abrasion or wear. There are two general categories of latex paint: acrylic and enamel. Acrylic latex forms a very hard surface to resist weathering and is usually used as exterior house paint. There is also a version of latex enamel available that simulates oil-based paint. This type of paint is usually used in bathrooms and kitchens to allow for resistance to moisture, smoke and grease and abrasive cleaning. Oil-based paint is durable, long lasting and is thinned and cleaned up with paint thinner. Latex paint is easy to install, reasonably durable, and thins and cleans up with water.

Paint is available in a variety of different sheens. Each of these sheens has specific qualities and purposes. Flat sheen has a nonreflective surface. It is best for walls and ceiling surfaces. It may be easier on the eyes but is harder to wash, due to its durability. Satin sheen has a very soft reflective surface with limited durability for scrubbing. It is suitable for walls and some trim work. Semi-gloss paint has more sheen and high durability. It is suitable for trim and surfaces subjected to moisture or frequent cleaning. Gloss sheen has a highly reflective surface and is suitable for trim and surfaces subjected to moisture accumulation, such as in bathrooms or kitchens. As sheen increases, any imperfections in wall surface are magnified.

Initial interior wall painting requires that a base coat or primer be applied before the finish coat is applied to the interior surface. The primer will seal the drywall surface so as to improve coverage of the finish coat. Primer installation will improve the adhesion of the finish coat and prevent an uneven finish. It will prevent peeling, flaking and the appearance of bleed-through. Apply the primer to walls and ceiling. Apply the finish coat after thoroughly dry, usually 24 hours.

Selection of the color of walls and ceiling is part of the interior design process. The mood and look of the room is set partially by the selection of color. A bold color will seem intense when the surface area is large, such as every wall within a small room. Using an accent color on one wall within a room could provide the look and avoid the intensity. Try painting a sample color on a section of the wall within the room, let it dry, and then evaluate your color choice. Be sure and look at the color in different lighting conditions, direct and indirect natural sunlight as well as artificial light.

What to Use?

For drywall surfaces of interior walls and ceiling, add primer or a pigmented wall sealer, then use a flat or semi-gloss acrylic latex. Most surfaces in living rooms, bedrooms and hallways can be painted with a flat finish. Bathrooms and kitchens and other rooms subject to moisture or intensive abrasive scrubbing should use either a semi-gloss or even high-gloss latex enamel. Interior trim, such as baseboard and doors, should use the same semi-gloss latex enamel. Using nonblocking paint on doors and jambs prevents painted surfaces from sticking together.

How to Paint

Carrying the paint from the bucket to the wall or ceiling surface is the object of the roller or brush. The paint roller is good for covering large surface areas, and the brush is good for applying paint to edges or joints between two surfaces. Rollers are available in varying thicknesses of nap and texture. If the texture is very thick, a thicker nap is needed to carry paint to recessed areas of the texture. For very thin texture, a smaller nap will carry enough paint to the smooth surface. Brushes are divided into synthetic and natural bristles. The natural bristle brush is usually the choice of the professional painter. If maintained properly, it lasts forever (almost). It also produces a very smooth finish without losing bristles onto the painted surface. The natural bristle brush is versatile. It is capable of spreading both oil-based and latex paint equally well. The synthetic bristle is good if you select a quality brush. Some of the cheaper brushes may tend to lose bristles onto the painted surface after a short time. Some brushes have bristles with an angle shape. These are cut-in brushes and are used to apply paint in tight edges or corners so as to avoid painting the adjacent surface.

Begin your painting project with the toughest chore—the edges or corners. Use a small cut-in brush to paint on the wall surface adjacent to the baseboard or door trim. Extend the paint onto the wall approximately 4 to 6 inches away from the trim. This will provide a safety margin when applying paint with a roller. Also, paint to the adjacent surface of every inside corner for both walls and ceiling. Use the same 4 to 6 inch width. Use a single directional stroke to apply the paint with the cut-in brush. Apply paint to the end of the bristles only. The brush should carry only about ½ inch of paint at the tip. Think about it. What good would paint further down the bristles do for your wall surface? It can't possibly touch the painting surface. Painting is a job that requires patience. Just like trim, a good job may take a while, but it will be worth it.

When the edges are complete, clean the brush and grab the roller. The roller cover slips onto a roller cage. The roller can reach tall walls or ceilings with an extension handle. If you are painting an entire house, you will be better off buying paint in five gallon buckets. Use a bucket grid that fits neatly inside a five gallon bucket to screed excess paint from the roller. Too much paint on the roller will spill on the floor or apply unevenly onto the wall surface. Do not dip the roller into the bucket any more than is necessary to collect paint. Paint on the roller cage or handle do the wall surface no good. As you can see, preparation is important in achieving a successful painting project

If flooring or wood trim is already installed (bad move), be sure to install drop cloths and masking tape to avoid a serious cleanup chore. Set the bucket in a central location to promote ease of access. Be aware of the bucket's location when painting overhead so as not to tip it over or step into it. Begin applying gentle roller strokes in a back and forth motion. Next, the ceiling provides some interesting challenges. For those who are of average height, an extension tool will allow you to reach the ceiling without using a ladder. Painting with a roller overhead will become tiresome quickly. Take your time with overhead strokes. Continue with the ceiling until the project is complete.

Creative Alternatives

After selecting a look for your room, you will inevitably choose color to help achieve that look. While paint really helps to establish a mood, some very simple techniques can improve the overall appearance to the room. Instead of relying on the finish surface, try adding texture to the finish paint job. There are a number of ways to use paint to add texture and even simulate the appearance of wall paper. After the primer and finish coat of paint are applied and have dried for 24 hours, use a sponge or even a rag to add a mottled texture with an additional color or the same color to the finish surface. Let a sponge (or rag) soak in water for a few minutes, wring it out and use like a handheld roller to apply the texture coat. Experiment with other tools, like a hard surface brush, whisk broom, or even a steel scrub pad to create the look you desire. When you have achieved the desired texture and it dries thoroughly, seal the textured finish with a varnish.

Wood Trim: Haste Makes Waste

Adding decorative trim to the joints between drywall and floor surface or to cover the door jambs is the next task. This part of the project is where you should back up and take a good, long look at the task ahead. You must discipline yourself to slow down and use patience now more than ever. You may be working with expensive hardwood where an incorrect cut can waste $20 worth of trim. Careful, slow measuring is the key. In fact, if you can, avoid using a tape measure at all during trim work. Use the actual board itself to determine the cut location. Position a length of trim in the location intended and use a sharp pencil to make the cut mark. It is more accurate.

You must allocate several hours, or even a full day to trim out each room. Avoid taking your cell phone or beeper during these jobs. Avoid all distractions and pay attention to what you're doing. You may think that you can trim out a room in an afternoon, and maybe you can. However, it's important to be methodical and careful in trim work. Every good finish carpenter I know has one common personality trait—that of being careful, meticulous, and slow. It may be hard to shift gears from rough framing, utility installation or even drywall, which can be accelerated compared to this job. Rough framing will never be seen again. Trim will be the first thing everyone notices when they enter your home. If you are not satisfied with the joinery of the wood trim, tear it out and do it again!

Door Installation

I recommend that you buy prehung type swinging doors. These are easy for the novice to install. These come in a variety of sizes, swing directions and types. Most interior doors are hollow-core type. That means that they are built with two exterior surfaces clad together and the interior is filled with cardboard spacers. Closet doors that are either bypass or folding are also very easy to install.

Door Passage Set and Lockset Installation

A passage set opens and closes a door without a key. A lockset requires a key, whereas a passage handset cannot be locked from one side. A privacy handset, on the other hand, can be locked from one side. The most common privacy passage set is the bathroom door, while a lockset is most commonly installed on exterior doors.

Wood Trim Types

A variety of decorative baseboard and door trim is available at a reasonable cost. Be sure and pick a trim that is readily available in resupply. You don't want to trim your house in three different styles of trim just because the first two styles you originally selected were on closeout and aren't available anymore. A really great alternative to decorative trim is solid dimensional lumber. I bought 1 inch by 3 inch pine for baseboard as well as door frames. I ran them through both a jointer and a planer. They are as straight and good looking as expensive trim. Everyone loves the idea. The cost is about 10 percent of standard decorative baseboard and looks every bit as good.

Miter Cuts

Links of both baseboard and door trim must be joined. Door jambs must be covered to hide the wall frame. These pieces may be joined with 45 degree joints at the corner. Even baseboard along a wall must have proper joinery to look nice. A manual jig for miter joinery is just as good as a power saw, just slower. There are compound miter saws that allow angle cuts in three dimensions. The quality of the miter cut is inversely proportional to how much caulking is required to fill the gaps between the pieces. If you have to caulk, it is better to paint the trim. If you are careful and avoid the need for caulk, you can stain and varnish the trim, which provides a beautiful wood look. Both are okay, but the latter demonstrates craftsmanship. Learn to make better joints by starting your trim work in a closet or some other hidden location.

Cabinet Installation

Kitchen and bathroom cabinets are available at a cost that almost precludes the need for the owner-builder to develop an additional skill. Your selection of cabinet sizes should have been worked out earlier and within the space limitations of your kitchen or bathroom. I will assume that you acquire cabinets on your own. Installing them is simple. Use wood screws that are large enough and long enough to withstand the weight to be imposed upon them. This is a three-man job. Two will lift and position the cabinet, the third will install the screws. The screws must connect the cabinet onto a wood framing member in the wall frame.

Because you planned ahead, you installed solid wood blocking in between studs at the strategic height of the wall cabinets to provide superior backing behind cabinets in wall frame. Install wood screws according to the cabinet manufacturer's recommendations, usually above and below the cabinet top and bottom surface, through the cabinet and into the blocking. The base cabinets are easier since they do not need to be lifted into position. However, if the wall is not exactly square with the floor, shims may need to be installed to create a plumb surface. Attach to wall frame in a manner similar to the wall cabinets.

Kitchen cabinets installed, awaiting the plumbing fixtures.

Countertops in the kitchen and vanity base top in the bathroom must be installed before the sink or lavatory can be installed. Kitchen countertop could be a manufactured, laminated top, or even a particle board base for ceramic tile. Irrespective of the material, a hole in the counter top must be made for the sink above the appropriate cabinet base. Of course, the size sink you select must match the cabinet base. When the sink is purchased, it comes with installation instructions. Follow these to determine the size of hole to cut into the countertop.

Wall Tile

There are areas around your home where you may want or need to install ceramic tile. If you have a shower or tub/shower combination, you should install wall tile or some other smooth, hard, nonporous surface to prevent water damage on the drywall. Since they are subject to grease build-up, kitchen walls between the countertop surface and the undersides of wall cabinets are good areas to install tile.

Ceramic tile is available in a variety of sizes and shapes, the most common being 4¼ inches by 4¼ inches. In addition to this shape, there are edge, corners, and trim tiles available. Tile is very easy to clean and has excellent water-resistant characteristics. Sometimes, the grout between the tiles will absorb dirt and require scrubbing.

Ceramic tile is made from fired clay with a patina on the surface. The smooth, hard, non-absorbent surface repels moisture. Sizes and colors vary, but the method of installing tile is similar. Tile is attached to a clean wall surface, free of deleterious material, with ceramic tile mastic. The mastic is a special glue that bonds the tile to the wall surface. It is applied with a special screed that has teeth of varying sizes that leave rows of mastic. Tile is installed in a checkerboard (or other) pattern with a space in between each piece, which is filled with grout.

If needed, the tile can be cut with a tile cutter, sanded smooth, then gently placed into position. Cuts will need to be made around the valves, shower head and tub spout. Of course, there is more to gaining a professional look. First, you must measure the wall space to establish a layout for the tile. Then, you must determine where different cuts or pieces will rest within the layout. Be sure to allow for the width and length of the tile plus the width and length of the grout joint. The width of the grout joint should be between ⅛ inch to ¼ inch.

Bathroom lavatory base is moved into place.

Wall tile is applied using spacers to control pattern.

Wall tile after grouting.

Wait 24 hours after the tile is set before applying grout. Grout is available in various colors that will match your tile. Specific kinds of grout are made for heavy water usage areas, such as a tub surround or even areas subjected to a buildup of grease, such as in a kitchen area. Mix the grout according to the instructions and apply to the spaces between tiles. Be sure to wash the grout from the surface of the tile before drying. Let grout dry thoroughly, then seal with a grout sealer.

Another possible alternative to tile is full slab sections of cultured marble. While cultured marble is very expensive, some manufacturers will sell *seconds* at a substantially reduced price. The advantage of the marble is that it is applied in contiguous sections with no need for grout. Also, it installs in less than half the time of ceramic tile. It also looks magnificent!

Cultured marble can be cut with a skill saw using a masonry blade. Fine cut outs, such as the valves and shower head, can be made with a drill bit and then a jigsaw. Bond the cultured marble to the wall surface with a construction adhesive. Use caulking to seal the joints.

Cultured marble provides a professional look without a lot of effort.

Touch Up Paint

After everything else is complete, you will have some touch-up painting to perform. Inevitably, you will need to add touch-up paint to cover smudges or abrasions caused by installing cabinets, doors, or wood trim. Be sure and use a small brush and use the same batch of paint used in that particular room. Install touch-up paint using a cut-in brush by painting X's in a cross over motion. This permits the new paint to blend better into the existing paint. Be sure to save a portion of paint for each room. A quart of each color would be sufficient.

Electrical Trim

Your interior walls and ceilings are painted, doors are installed, cabinets and trim are complete. Now is the time to begin connecting electrical outlets, switches and lights. Outlets and switches are installed in outlet boxes using methods outlined in Chapter 12. Lights are installed according to the installation instructions for the light. Remember that outlets in kitchen countertops and bathrooms must be ground fault circuit interrupter type. For consistency, install the ground post down. Also, when installing a light switch, the light should be on when the switch is in the up position. Cover plates can be installed over light switches and outlets. A good energy savings technique is to install a foam insert over the outlet or switch behind the cover plate to keep infiltration out. A very small amount of caulking inside the cover plate can help seal the plate to the wall.

Next, install smoke detectors onto the outlet base by following manufacturer's directions. In some cases, smoke detectors must be interconnected with other smoke detectors to carry the signal from the particular smoke detector that senses the smoke. In this case, the signal wire is prohibited from being inside the outlet box, so it must enter the smoke detector from outside the box. Read manufacturer's instructions about how to achieve this interconnection.

Electrical outlet is trimmed by applying a face plate.

Ranges, ovens, and similar appliances may be directly wired to wiring inside outlet boxes instead of by cord and plug. If an appliance is permitted to be directly wired, the manufacturer's instructions will explain methods. If the appliance requires a cord and plug method of connection, you may need to purchase a pigtail separately. Ask the appliance dealer for guidance.

Now that all of your outlets, switches, lights and appliances are installed, it is time to install breakers in your service panel. The circuit breaker connects the circuit phase conductor (black or red wire) to the phase buss bar inside the service panel. Select the circuit breaker according to the wire size (see panel schedule). With the circuit breaker in the open position, connect the circuit phase conductor to the circuit breaker with a compression screw, which is inside the circuit breaker. Then carefully install the circuit breaker onto the service panel. This will complete your electrical trim work. Be sure and identify each circuit breaker as to the appliance or room or area it serves. Be sure to avoid turning on any circuit breaker for which an appliance or equipment has not yet been installed, such as a range, oven, heat pump or water heater.

Plumbing Trim

Plumbing fixtures and water cut-off valves are set when the interior painting is completely finished. Before drywall was installed, the bathtub and shower drain pan should have been installed. Start installing the other plumbing fixtures such as toilet and lavatory in the bathroom. Angle-stop water supply valves are installed on the wall and serve as cut-off valves for each fixture. The back of the lavatory base cabinet must be cut to allow the valves to project into the cabinet and connect with the faucet, and the waste line to connect with the trap.

The kitchen sink normally is installed over the top of the countertop surface. Therefore, you will need to install the countertop onto the cabinet, carefully measure and cut the countertop, then install the sink according to the manufacturer's installation instructions. Install drain strainer(s) in the sink. Assemble drain trap and waste arm to the plumbing drain. Then, install hot and cold water cutoff valves, and connect water supply to faucet. Food waste disposers are installed between the sink and the trap. The food waste disposer has a cord and plug connection to an outlet. The outlet should be activated by a switch near the kitchen sink. The dishwasher is installed in its under cabinet location. The dishwasher has a cord and plug connection to an electrical outlet. The waste line from the dishwasher is installed either to the tail piece or disposer (if you have installed one).

In 1992, Congress passed legislation that mandated that certain water-saving fixtures be required. The 1.6 gallon per flush toilet was one of the results of this action. There are similar limitations on water flow rates in lavatory and kitchen sinks. This water conserving act was passed before there were an adequate amount of any of these fixtures available. Their research and invention largely occurred after 1992. In newer fixtures, the toilets will scour and flush better than earlier models. For the most part, this type of fixture is all that is available. It makes good sense to use them, and they do a good job conserving a precious resource.

The water heater can be installed now. Because you made precautions before the drywall was installed, this should be a snap. Position the water heater in its place and begin by connecting the cold water supply pipe, which has a cut-off valve, to the cold inlet on the water heater and hot outlet pipe from the water heater. To avoid accidentally reversing the connection, during the rough-in you have marked each pipe as either hot or cold. The connection between the cut-off valve and the water heater

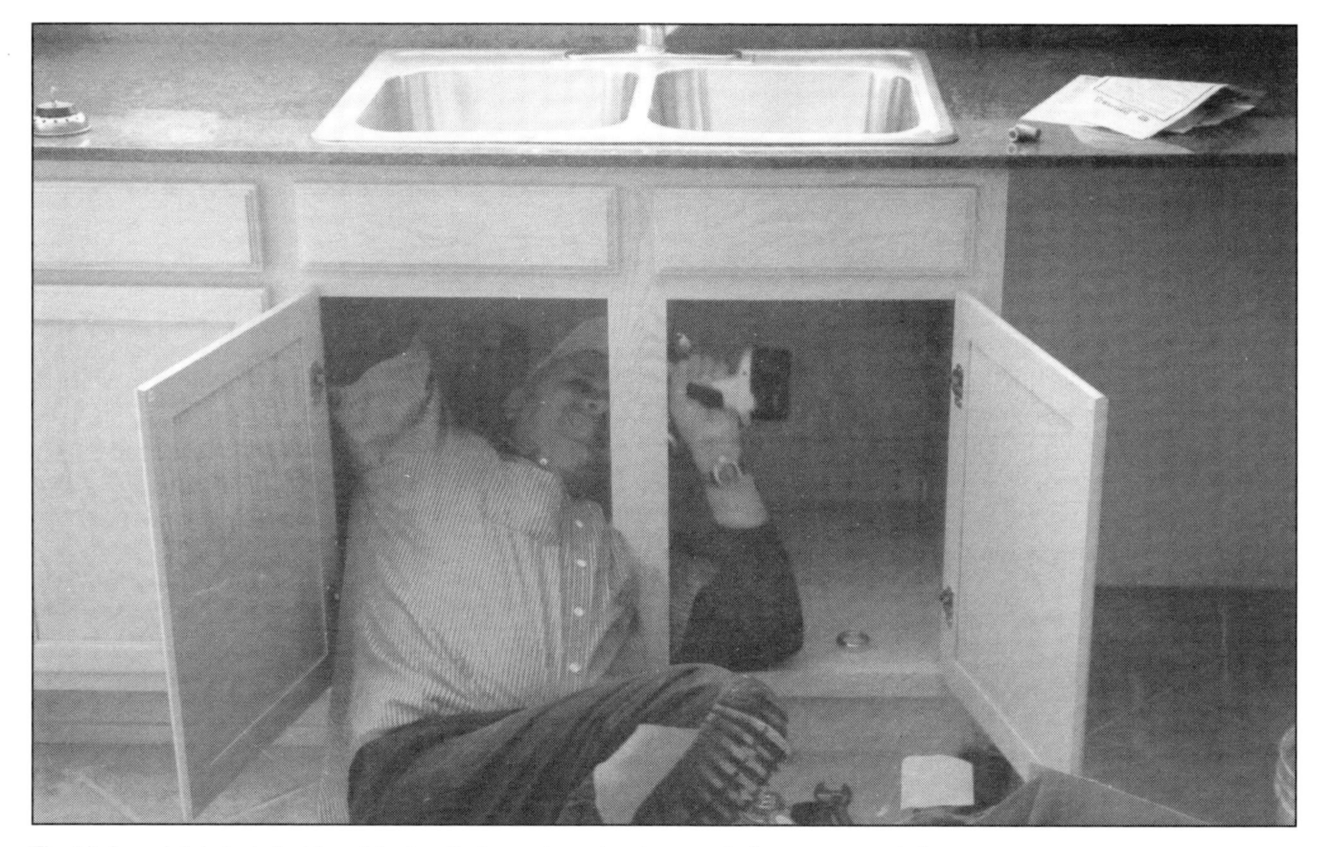

The kitchen sink is installed in cabinetry. Both waste and water supply is now connected.

is made with a flexible copper pipe. Next, connect the electrical wiring to the water heater. The wires should be color coded. The cable from the electrical box should be installed inside a flexible metal conduit. This prevents the cable from being exposed and damaged.

The water heater must be secured to prevent falling if you live in an area subject to seismic activity. The method of securing a water heater to a wall is with a metal strap near the top and bottom of the water heater, which connects to a framing stud. Be sure to place the straps far enough away from the controls of the water heater.

Bathrooms will be trimmed with towel bars and toilet paper dispensers. Remember, alternatives may be just as aesthetic as the manufactured products. You could craft these with wood or even use steel pipe to create an individual look.

After the sink is installed, the kitchen comes to life.

Mechanical Trim

Interior mechanical equipment such as a heat pump should already be in place and connected. The exterior equipment can now be set in place and installed. Have your mechanical professional assist you in making the final connections for equipment and carefully read the manufacturer's installation instructions.

Registers that open, close or divert conditioned air must be installed on the duct terminal. These registers look like grillwork and are able to be closed to restrict air flow in certain rooms. They can be installed with wood screws into framing. Finish grillwork on the return air duct terminal must not be capable of being closed. This would seriously harm the mechanical equipment.

Flooring

Installing flooring is the final thing you do, just before you begin moving in your furniture and setting up the house. You have arrived! This is the stage you have dreamed about for several months or years. I am in favor of using wood or tile for a finish floor. Ceramic floor tile can be installed on either concrete or wood subfloor. There are manufacturers' instructions for installation on either surface. Read these thoroughly before beginning.

The water heater is connected with copper pipe. Note the water supply is equipped with a cut-off valve.

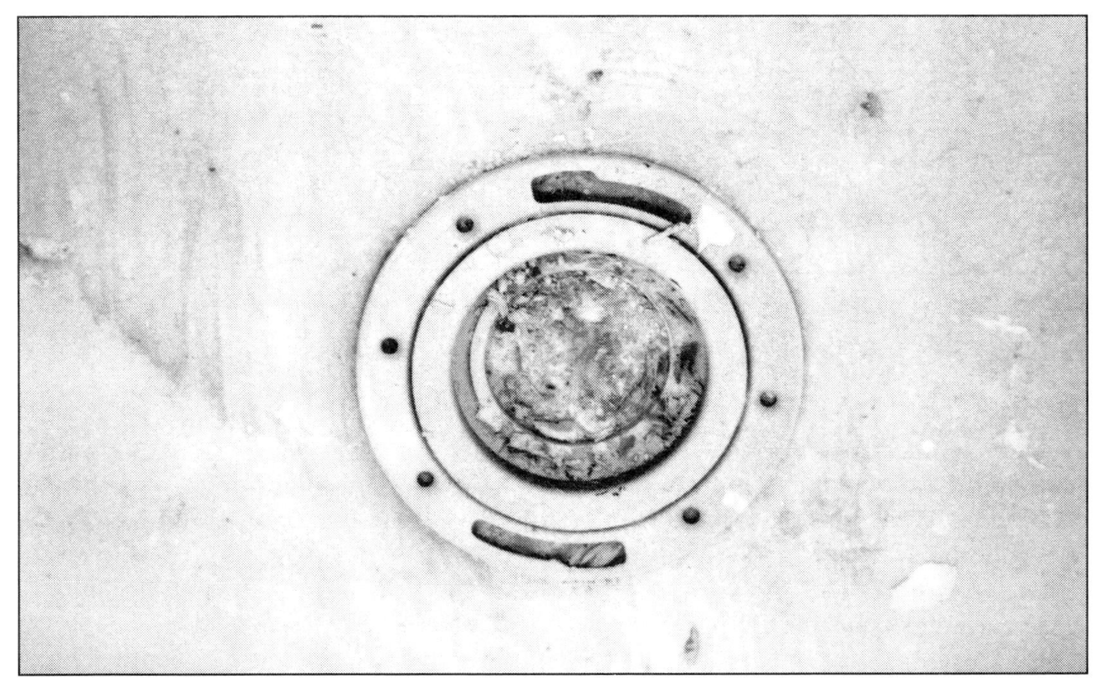

The water closet is installed following the manufacturers' instructions. Normally a closet flange extending from a plumbing fitting called a closet bend allows for easy connection to the base of the water closet.

A heat pump condensing unit is installed on a substantial base above grade. Electrical connections must be according to the manufacturer's recommendations.

Material Choices

As always, your choices will be based on your personal needs and desires. Carpeting is cheap, can be installed quickly and adds warmth to a home. It also collects and hides dirt, and with heavy traffic, wears out relatively soon. Tile is more expensive, takes longer to install, but adds great beauty and elegance to a home. Hardwood flooring is more expensive and takes even longer, but it adds an element of charm to your castle. Softwood pine boards are a cheap alternative to hardwood flooring.

If everyone cannot agree on a material, try to work out the decision in a creative manner. For instance, use large area rugs over tile floor. Install different materials in different rooms: carpet in bedrooms, tile in entryway, living room and other heavy traffic areas, and hardwood in family room or den.

Registers and vent covers seal supply ducts and exhaust vents.

Carpet and tile side by side.

Concrete Floor

If your sub-floor is a concrete slab, a cost-effective way of providing an elegant floor is simply to paint the concrete. As always, there are a variety of ways to do this. If you thought ahead, you could add concrete colorant when you pour your slab. This adds the color in a very uniform manner. I have seen a good cement finisher score the slab in a grid fashion to resemble a tile floor. If you have waited until the end of your project to consider flooring, you can still use paint. You can paint the slab with an exterior or concrete enamel paint. If you're interested in a really exotic look, after this paint is dry, smear wood stain in a series of random, circular patterns. When the stain has thoroughly dried, add polyurethane varnish topcoat. It will look like an exotic solid slate tile.

Wood Floor Options and Alternatives

Tongue and groove hardwood flooring is the most attractive and durable of all floors. Tongue and groove hardwood flooring is available in random lengths that are usually about 2 inches in width. Overlay the subfloor with a single layer of 15 lb felt building paper. Special nails are necessary to drive through the hidden corner of each plank. The subsequent plank hides the nail when installed. The hardwood flooring planks are installed a row at a time. Most hardwood flooring is milled with a manufactured tongue and groove on opposite sides of the plank. Because it is a manufactured product, a set of instructions and additional installation advice will accompany the package.

If you have a wood subfloor and cannot afford hardwood tongue and groove flooring, another very attractive substitute is a softwood floor. Use 1 inch by 10 inch dimensional pine boards. Lay them in a staggered pattern and connect to the plywood subfloor with wood screws. As with hardwood, install building paper on top of the subfloor before you install the pine boards. Set screws into the pine boards sufficient to prevent their withdrawal. If the pine boards are not perfectly straight, there will be a gap with which to contend. It may be necessary to by or rent a wood jointer. The jointer will provide a very smooth edge for both sides of the pine boards before installing them. Be sure and select boards which are flat, so as to avoid cupping of the finish floor.

Carpet

Carpet provides a quick and immediate finish floor. It is also very luxurious. There is nothing like the feel of walking barefoot on carpet in your new home to convince you this was all worth it. The drawback to carpet is that it tends to collect and retain dirt. Without a very good vacuum cleaner, it will remain there. Just like dirt in the oil of an automotive engine, dirt will cause wear and tear on the carpet.

Carpet is distinguished by its thickness and height of pile. If the weave of the carpet is low and tight, it can withstand lots of wear and tear and clean up well. This class of carpet is regarded in the industry as commercial grade. If the weave is thin and has a high pile, it may be subject to more wear and tear.

Carpet is sold by the square yard, but usually only in 12 foot widths. For example, if your bedroom is 11 feet by 14 feet, you must purchase a piece that is 12 feet wide by 14 feet long. This will represent 3 yards by 4⅔ yards (14 square yards).

Final Inspection

General

___ Verify that all earlier inspections have passed and building permit is still valid.

___ Verify that approved plans are on-site for inspector's review.

General Exterior and Garage

___ Verify that final grade slopes away from the building. *Section R401.3*

___ Verify that building is generally complete in appearance.

___ Verify that the lot is generally clean and free of debris. *Zoning code and subdivision rules*

___ Verify that exterior guards are installed 36 inches high and that openings are less than 4 inches in diameter. *Section R316*

___ Verify that safety glazing is installed in any hazardous location according to *Section R308.4*

___ Verify that fireplace chimney extends 24 inches above roof. *Section M1804.2*

___ Verify that the rated separation between house and garage is one layer of ½ inch gypsum drywall. *Section R309.2*

___ Verify that any door between house and garage is 1⅜ inches thick solid-core door. *Section R309.1*

___ Verify that garage floor is non-combustible, such as concrete slab. *Section R309.3*

___ Verify that roofing is installed and meets the requirements for slope. Chapter 9

___ Verify that exterior wall covering meets the requirements for weather protection. *Section R703*

___ Verify that protection from termite damage is complete and documented. *Section R324*

An approved final inspection certifies that your home meets safety standards.

___ Verify that any exposed wood is protected from decay. *Section R323*

___ Verify that address is installed on-site to identify property. *Section R325*

___ Verify that flood protection is provided. *Section R327*

General Interior Exits and Stairs

___ Verify that the proper rise and run of stairs is 7¾ inches maximum rise and 10 inches minimum run. *Section R314*

___ Verify that minimum stair width is at least 36 inches. *Section R314.1*

___ Verify that handrail height is between 34 inches and 38 inches above the nosing of the tread. *Section R315*

___ Verify that there is adequate headroom height of 6 feet 8 inches in the stairwell. *Section R314.3*

___ Verify that any guards are installed 36 inches high and that openings are less than 4 inches in diameter. *Section R316*

___ Verify that the house has at least one exit door that is 36 inches wide and 6 feet 8 inches high. *Section R311.3*

___ Verify that exit doors do not have any special latches or keyed locks. *Section R311.2*

___ Verify that hallways are at least 36 inches in width. *Section R311.4*

___ Verify that all sleeping rooms have emergency egress windows that meet minimum size. *Section R310*

___ Verify that basements have emergency egress windows that meet minimum size. *Section R310*

___ Verify that all exits have acceptable landings according to *Section R312.1.*

___ Verify that doors do not swing over steps. *Section R312.1.*

___ Verify that any ramps meet the maximum slope requirement of 1:8 and other criteria. *Section R313*

General Interior Architectural

___ Verify that windows meet the design requirements for energy conservation. *Table N1102.1*

___ Verify that interior wall covering such as drywall meet the requirements of *Section R702.*

___ Verify that ceiling height in habitable rooms meets the minimum requirements of *Section R305.*

___ Verify that windows are provided for required natural illumination and ventilation. *Section R303*

___ Verify that heating facilities provide sufficient heat as required. *Section R303*

___ Verify that rooms meet the minimum dimension and area requirements as specified in *Section R304.*

___ Verify that bathroom and kitchen facilities are provide and operable. *Section R306*

___ Verify that insulation certificate from supplier matches the requirements for *Section R320 and Table N1102.1.*

___ Verify that any foam plastic meets materials standards and is protected from exposure. *Section R318*

___ Verify that ceramic tile is installed in shower surrounds to a height of 6 feet above drain inlet.

___ Verify that fireplace or wood stove is installed according to manufacturer's instructions and *Section R1002.*

General Interior Plumbing

___ Verify that plumbing fixtures are installed properly and are operable. *Section P2701.1*

___ Verify that water heater is provided with a drain pan and drain to prevent damage. *Section P2801.5*

___ Verify required location, access and clearance for water heater. *Section M2005*

___ Verify that water heater's temperature, pressure relief valve is installed properly. *Section P2803*

___ Verify that water supply pressure is at least 40 psi and not greater than 80 psi. *Section P2903.3*

___ Verify that fixtures meet the requirements for maximum flow rate and water consumption. *Tables P2903.1 and 2*

___ Verify that all fixtures for water supply are provided with cut-off valves. *Section P2903.9.3*

___ Verify that any drinking water treatment units meet the requirements of manufacturer and *Section P2907.*

___ Verify that hose bibs have backflow prevention devices. *Section P2902.3.3*

___ Verify that flush tanks are equipped with anti-siphon ballcock. *Section P2902.3.1*

___ Verify that drain, waste and vent clean-outs are installed where needed. *Section P3005.2*

___ Verify that no drain, waste and vent piping is subjected to freezing temperatures. *Section P3001.2*

___ Verify that provisions for future fixtures are capped according to *Section P3005.1.6.*

___ Verify that fixtures with flood rims below manhole covers are protected with backwater valves. *Section P3008*

___ Verify that any sump pump or sewage ejector is installed according to *Section 3007 and Table P3113.4.1.*

___ Verify that vent piping is flashed and extends through the roof to a required height as specified in *Section P3103.*

___ Verify that traps meet the minimum size of *Table P3201.7* and maximum distance as specified in *Table P3105.1.*

___ Verify that any air admittance valves are installed per *Section P3114.*

___ Verify that fixture tail pieces meet the minimum size requirements. *Section P2703*

___ Verify that discharge line for dishwashers is protected with air gap or backflow protection. *Section P2717*

___ Verify that whirlpool bathtubs are equipped with an access panel door. *Section P2720*

___ Verify that sinks and laundry tubs are provided with tailpieces at least 1½ inches in diameter. *Section P2714 and P2715*

___ Verify that bathtubs are equipped with waste and overflow at least 1½ inches in diameter. *Section P2713*

___ Verify that hot water valves for sinks, lavs, bathtubs and showers are on the left. *Section P2722* (see I told you!)

___ Provide final plumbing test by filling traps with water to test for leaks. *Section P2503.5.2*

General Interior Electrical

___ Verify that electrical service disconnecting means are provided with access and clearance. *Section E3305*

Washing the windows may be the final task before you move in.

___ Verify that all electrical equipment is approved, listed and labeled for the use intended. *Section E3303.3*

___ Verify that all electrical equipment is in working condition and is not damaged. *Section E3304.6*

___ Verify that equipment that is exposed to physical damage is protected by guards. *Section E3304.9*

___ Verify that artificial lighting is provided for service equipment or panelboards installed indoors. *Section E3305.6*

___ Verify that conductor and terminal identification meets the requirements of *Section E3307.*

___ Verify that wiring in circuits is protected by overcurrent devices (breakers) according to *Table E3503.1.*

___ Verify that electrical overhead service drops meet the height and clearance requirements of *Section E3504.*

___ Verify that service entrance conductors are installed properly and protected from damage. *Section E3505*

___ Verify that electrical service equipment meets the installation standards of *Section E3506.*

___ Verify that electrical service equipment is grounded to earth as previously inspected. *Section E3507*

___ Verify that metal parts of the service equipment is bonded to ensure electrical continuity of fault. *Section E3509*

___ Verify that branch circuits for bathrooms, kitchens and outdoors are protected with GFCI. *Section E3802*

___ Verify that branch circuits for general lighting and outlets are protected according to *Table E3605.5.3.*

___ Verify that outlets are installed in habitable rooms according to *Figure E3801.2.*

___ Verify that outlets for counter spaces are installed according to *Figure E3801.4.*

___ Verify that a convenience outlet is installed for HVAC equipment servicing. *Section E3801.11*

___ Verify that switched light or outlets are provided in habitable rooms. *Section E3803*

___ Verify that switched light is provided at stairs and garage. *Section E3803.3*

___ Verify that outlets installed in outdoor locations are equipped with a weather-proof cover. *Section E3902.8*

___ Verify that lights installed in closet meet the clearance requirements of *Figure E3903.11.*

___ Verify that outlet and switch boxes are covered and protected. *Section E3904*

___ Verify that appliances are installed according to their listing and Chapter 40.

___ Verify that all metal face plates are grounded. *Section E3806.10 and E3902.4*

General Mechanical

___ Verify that all appliances are labeled. *Section M1303*

___ Verify that appliances are installed according to manufacturers' instructions. *Section M1307*

___ Verify that appliances are accessible for repair or replacement. *Section M1305*

___ Verify that appliances meet the requirements for clearances from combustibles. *Section M1306*

___ Verify that portions of heat pump installed outdoors are elevated at least 3 inches above grade. *Section M1403.2*

___ Verify that condensation is disposed according to provisions of *Section M1411.3.*

___ Verify that evaporative cooling is installed to manufacturer's instructions and *Section M1413.*

___ Verify that fireplaces or wood stoves are installed according to manufacturer's instructions. *Section M1414*

___ Verify that clothes dryer exhaust is installed according to the provisions of *Section M1501.*

___ Verify that range hoods are installed according to the provisions of *Section M1502.*

___ Verify that supply air and return air is connected properly and operates according to Sections *M1602 and 1603.*

Chapter 18

SEPTIC TANKS

Purpose

Unless you buy it in the store, most drinking water comes from underground. The hydraulic cycle describes how water transitions from the atmosphere via condensation to the earth's surface and via evaporation back into the atmosphere. Our use of the water inevitably adds impurities. Where a public sewer system is not available, a private sewage disposal system is the logical alternative. The septic tank is a small, on-site sewage treatment plant. The objective of the tank is to allow nature to disintegrate impure biological particles, which otherwise would pose a health hazard when added to ground water. In residential use, those impurities are usually biological and result from human excretion and food waste.

Septic Tanks

Septic tanks are water-tight tanks, and made of substantial materials such as concrete, grouted block, metal or even plastic or fiberglass. Permeable materials, such as concrete and block, must be water-proofed with any of a variety of materials such as fly-ash, or even an emulsified asphalt coating. This waterproofing isolates the wastewater inside the chamber and prevents premature absorption into the ground.

The septic tank is provided with an upper baffle that separates the tank into two chambers. This baffle prevents solids, which float, from entering the secondary chamber. In the first chamber, these solids float to the top of the liquid level where bacteria consume and dissolve most of these biological solids by *digesting* them. The bacteria are then heavier than the ambient water, and they drop in the tank beneath the baffle and enter the second stage of the septic tank. There, the bacteria biologically reproduce. The offspring rise into the first chamber and consume more biological waste, then drop, continuing the cycle. In the second chamber, the older bacteria, which have died, are carried out into leach fields that are connected to the second stage of the septic tank.

Filtration

In order to avoid drinking water that is soiled with your neighbor's biological waste, modern sanitary rules and regulations have led to a very time-tested method: ground filtration. The purpose of a private filtration system is to contain any residual waste and prevent living or dead bacteria from entering the ground water. The specific design of filtering these remaining impurities involves the use of a leach line.

Leach lines are absorption fields that further strain pollutants from waste water expelled from the septic tank. The leach line(s) allows liquid from the second stage of the septic tank to enter a long hollow chamber dug in the soil—essentially, a wide trench. The trench is usually filled with rock to prevent collapse. On top of the rock, a perforated pipe conveys and distributes the liquid along the top of the trench. The polluted liquid then fills the trench and is leached or filtered before ultimately entering the groundwater.

A septic tank must be accessible from grade. If it has to be deep due to elevation changes, a manhole must be installed to gain access to the tank for cleaning.

Tank and Leach Line Design

Septic tanks are usually built in standard sizes based on the number of people living in the home. The most common capacity for a tank is between 1000 and 1200 gallons. Usually, the regulating jurisdiction will establish the size based on the number of bedrooms. The size of the leach field is a little more complex.

As you have learned in Part One, different soils have different properties, which define capacities for supporting a load. Well, these same characteristics that define a soil's structural capacity also affect its ability to drain liquid. In order to determine the rate at which your leach field will adequately filter and drain effluent from your septic tank, a test must be performed.

This field test, called a *percolation test* is usually made by a soils engineer. Percolation is the flow of ground water in the direction of flow toward the ground water table. The method of testing is usually established by the state or local health or environmental enforcement agency. Usually, a *perc test* is an in-situ test at the site of the proposed leach lines. A hole is dug to the design width and depth of the leach line. A measured amount of water is poured into the hole, and the absorption time of the water is measured. Sometimes, more than one hole is required in order to establish the average absorption capacity of the soil.

Classification of the soil, rate of absorption in the perc test and other factors are involved in establishing the design absorption rate of the leach field. With this estimate of the predicted absorption rate, the width and depth of the leach field may be determined. An engineer or qualified professional must make this determination. Following is a simplified method of determining the length and depth of field.

Example:

Consider, after a percolation test, that the required absorption field was determined to be 1000 square feet. What length and depth of leach line must be dug?

Solution:

If you plan to dig two separate trenches, each trench must be 500 square feet each. Each side of an open trench is considered a separate surface for waste water absorption. Since each trench is assumed to have two sides, each side must have 250 square feet. A nominal depth for a backhoe is about six feet below grade. If the leach line begins 12 inches below grade, the *effective depth* of the leach line is five feet. The required length for 250 square feet of leach area is 250 divided by 5 or 50 feet. Therefore, you must dig two trenches, about 14 inches wide, 50 feet long.

Installation

I recommend that you purchase a tank from a qualified professional. They can usually build tanks, bring a backhoe and dig a hole for the tank and the leach lines, and install the gravel bed considerably cheaper than you could. You might make a deal with them to assemble the distribution chamber, install the perforated pipe, cover with asphalt paper and backfill with soil to save some expense. Here's what to expect from the septic tank installation professional.

A truck carrying your new tank will probably be towing a backhoe. The backhoe will dig both an oversize hole for the tank and two or more trenches for the leach lines. Another dump truck will arrive with a load of washed gravel. The gravel will be dumped near the trenches. The backhoe will use the front end loader to scoop gravel and fill the trenches after they have been dug. Then, your septic tank will be moved in place with a crane and hoist and lowered into position in the ground. The tank should be within a few feet of the ends of the leach trenches. If you have made arrangements to do some of the labor in exchange for a lower bill, here's where your work will start.

Distribution Box

You probably will have two (or more) leaching trenches, which must carry waste water from the septic tank. A distribution box connected to two more leaching trenches leads to the outlet side of the septic tank. This is a fancy name for a concrete box with at least three circular openings for pipe. One pipe leaves the secondary chamber of the septic tank and deposits waste water into the distribution box. Two others connect to pipes leaving the distribution box carrying the waste water to the beginning portion of the perforated pipe in the leach field. This is a simple task. It's important to ensure that the distribution box is level and rests on a substantially compacted bed of soil. This prevents its collapse. The line to the distribution box must be level. The lines from the distribution box and the beginning of the perforated leach lines should have a moderate slope of ¼ inch per foot.

Perforated Pipe

Perforated pipe is usually schedule 40, four inch PVC and is sold in almost all larger hardware stores. The perforated pipe on top of the leach lines should also have a moderate slope of ¼ inch per foot. Install the perforated side down and lay against the washed gravel in the center of the trench. The pipe comes in 10 foot segments with a female and male end. Connect segments of pipe with an approved PVC pipe solvent. Arrange the "run" of the pipe so that water flows over the male end at the joint. At the end of the trench, install a pipe cap. Some jurisdictions require a testing tap be installed at the end of the trench to test whether the leach field is doing its job. This is simply a solid pipe installed vertically to the bottom of the trench before gravel is placed in the trench. This pipe extends upward above the ground. It is capped and available to provide sampling of filtered waste water at some future time.

Other Considerations

A septic tank is designed to last for the lifetime of the house it serves, assuming only biological waste products are deposited. However, our modern sewage system has become a sort of dumping ground for all of our human waste products. We flush all sorts of things down the toilet with total disregard as to the implications. Other modern impurities we arbitrarily discard into our sewage disposal system include soap products, soil and petroleum debris, chemical by-products, poisons, and a host of other material. Just look around your home now and take an inventory of all the products in your pantry and storage closet. Now consider the likelihood of those products finding their way into your plumbing drain.

Out of Sight, Out of Mind

As a society, we have become very complacent about dumping our waste down the drain. We barely consider the implications when we flush the toilet or run water and wash our hands after changing our car's oil. Take a moment and actually inventory just two rooms in your home and list the ingredients of a sample of products. Here's an inventory of just my bathroom.

PRODUCT	CHEMICAL(S)
Toilet cleaner	Hydrogen chloride
Mouthwash	Alcohol, benzoic acid
Powdered cleaner	Calcium carbonate, bleach
Bleach	Sodium hypochlorite
Medicated oil	Methyl salicylate, camphor
Hand lotion	Demithicone, petrolium, glycerin, magnesium, aluminum silicate, sodium hydroxide
Tile cleaner	No ingredients, but a warning not to get on skin
Bubble bath	sodium lauryl sulfate, sodium chloride, glycol stearate
Aftershave	Glycol, alcohol
Toilet paper	Colorant

This is not even considering the medicines, shower splashes, perfumes, window cleaner, baby powder, ointments, salves, shampoo, sun screen, eye drops, and myriad other chemicals we casually use and then discard into the drainage system. Nor is it considering products and chemicals in other areas of your house including the garage, kitchen, pantry, office, storage rooms, or yard. Nor do we consider those things we bring home from work. Many workplaces include interaction with hazardous materials or chemicals, which are washed down the drain when you shower. All of these chemicals adversely affect the sewage disposal system. Some chemicals will kill the very bacteria that maintain a healthy sewage treatment system. Tobacco or cigarette filters will clog a septic tank. The use of solid or powdered laundry soap will congeal at the baffle inlet to the septic tank and create a "dam," which will be noticed when a toilet will not flush fast enough.

Regular Cleaning

Because of the nature of contaminants, the action of the productive bacteria may be impaired or even stopped. Therefore, it may be necessary to empty a septic tank on a regular basis. Depending upon how many people are served by the septic tank, I recommend that it be cleaned about every three years. Every 18 months, check the depth of sludge in the first chamber. If the depth of sludge is greater than the depth of the baffle (24 inches), the tank needs to be cleaned.

Chapter 19

CASE STUDIES: ADVICE FROM OTHERS

From Those Who Have Been There and Done That

If you plan on doing something you've never done before, you always consult with someone who has. You have some specific questions and you know you need advice, but sometimes you don't know all of the right questions to ask. Because I know what you will be aspiring to achieve, I wanted to include thoughts, suggestions and advice from others who have been successful in their aspirations. Let me start with a very brief summary of my story:

My wife, Cynthia Underwood, and I, along with our two adolescent children, bought a neighbor's travel trailer and moved onto our newly purchased acre lot. Utilities were extended to the home site and from there, to the RV. We began by digging and building a basement. Some progress ensued, including interior bearing walls in the basement, but it was slow. After a year, the RV began to shrink as the kids grew older. So, a guest house was built within two months. The RV was sold. We moved into the guest house and worked on finishing the main house.

Our family built our home together. You may consider our list of advice to be the contents of this book, but let me summarize some aspects you may wish to strongly consider:

1. Be considerate. This means to everyone, your family first! Then be considerate to your boss for funding this project, your neighbors for enduring it, your helpers and your educators, salesmen, building department staff, and everyone you meet. It may sound like the golden rule, but it works as well now as it did in kindergarten. Being patient and considerate means just that—consider how your actions are adversely affecting others and try to avoid those actions.

2. Be patient. Your project will take several years. Do not grow weary or impatient. You will be like the turtle in the race—you will win by being consistent. Remember, slow and steady wins the race. Work to maintain morale among your family. Get some rest away from your project. Keep the faith. Keep your eye on the prize. Remember for whom you are doing this.

3. Be careful. Your project should not cause injuries to you or your family. Keep a clean project. Maintain safety rules. Be methodical about your actions. Look before you leap. Watch each other. Above all, parents, watch your kids!
4. Be frugal. No matter how dedicated you are, you cannot build your home unless you save money in a disciplined manner. Just put the money into a building fund and pretend it is gone—like rent money! First-run movies and the new suit will have to wait. You're on track toward an achievement. Obviously, there are exceptions to this rule including illness, accidents and the inevitable car repair. Just try to keep deviations from your building fund contribution to a minimum.
5. Be open. Accept new ideas, concepts and ways of doing things. Listen to advise of others who are in a position to know. Alter your paradigm to reflect reality when it faces you. Accept set-backs or compromises. Respond with renewed optimism. Remember, character building is like house building. It's supposed to be tough, otherwise everyone would be doing it!

John and Pat Cooper

John and Pat Cooper have done what you plan. They built a home with no mortgage while living across the street in a mobile home. Despite both being over the age of 50, John and Pat engaged their dream with the enthusiasm needed to be successful. They bought a lot with a mobile home and an attached room addition for a decent price and moved the mobile home off and proceeded to build a house onto the attached addition. While they easily could have lived in the addition, they chose to live across the street in a mobile home.

Their plan was to build a CMU block foundation, slab on grade, 2 x 6 wood stud wall frame, wood trusses, stucco over OSB siding, architectural grade shingles and double glaze windows. They have included an on-demand water heater and other custom features, including outlet at desk height in sewing room and secondary refrigerator. Here are some words of wisdom from John and Pat:

One of the main advantages of building your own home is that you can make changes as you build. You can customize your home to your furniture and add new ideas as they develop, such as lighting needs. These minor changes would cost a fortune if a builder was working for you. Being able to move a light switch just a few inches may allow your cabinets or furniture to fit. It is convenient and easy if you do the job.

- We found that we wanted to do a better job and exceed the minimum code requirements.
- Ask a lot of questions. Our building inspectors have given us lots of invaluable advice and information that helped.
- Hire for work for which you do not feel competent. If you can't handle certain work or do not have the equipment, you will need some help anyway. We had help with foundation, concrete, shingles and excavation.
- Use trades where you need help. We hired a mechanical and plumbing contractor, but we were able to do the electrical wiring.
- Work in an orderly and systematic fashion. There is a system to these things
- Keep a running to-do list and a supplies and materials list. Add to it every time. Then, use it to follow up on loose ends or to be on the lookout for something from the store. When wrapping up a phase of construction, make a "punch list."
- Take your time and do each thing in order and completely if possible. Put any loose ends on a punch list.

- Clean up on a regular basis and do dump runs. It's much nicer and safer to work in a clean environment.
- Be prepared for delays and changes in schedules.
- Take time off and get away every now and then. This keeps you fresh.
- It is great to own a home with no mortgage. Maneuver finances to accumulate funds from which to build. Live on less to pay everything off.
- Take advantage of sales, closeouts, and discount days. We have beaten contractor prices by just being patient and methodical.
- Barter and trade labor with friends.
- Combine dump runs with supply pickup runs.
- Take advantage of contractor moonlighting hours.
- Set a budget and track your expenses. We use a columnar pad.
- Keep your eye out for good buys or freebies along the way regardless of whether you're ready for them or not. Storage of materials is a small price to pay for free materials.

Art Boehm and Jeff Ramos

Art Boehm and Jeff Ramos are life partners and have actually undergone the owner-builder process twice. Art brought lifelong building and remodeling skills to the project, while Jeff added physical energy and a quick knack for learning. Each had a desire to create a quality home. I met with them while performing inspections on their first home. Their story follows.

We built our first home with a third person providing financial backing, while the two of us furnished much of the labor and funds to complete the project at the end. An important lesson learned was to know your investing partner(s) as well as your own self. Despite knowing the third party for years, it became obvious before the house was complete that the arrangement was not going to work out. Fortunately, because of sound legal written agreements, all involved received equitable proceeds from the venture. The house, built largely on the same principles outlined below, has since become a home to the satisfied buyers.

To assure satisfaction in the building process, planning is the key. Think about what you have enjoyed in past residences as well as what has annoyed you. Ask questions such as, "Why does hot water take so long to reach the bathroom?" Sketch out rough diagrams of the proposed house and live with those designs for a while. Imagine a day and a night in the house you have designed, and you may discover that large bedrooms aren't as important as a great room for family and friends to gather. An angled area included for aesthetics could serve as much needed space for built-in storage or even a bookshelf. We often annoyed our friends with our ever-present tape rules measuring features in their homes and endless questions about how they liked or disliked various aspects of their design. We found that what worked best for us were simple, clean designs with an emphasis on functionality.

Using our home as collateral, we secured a 1.25 acre desert lot for building. We noted its physical features and decided where we could locate the house to maintain the majority of the natural vegetation, especially any native trees. To be able to enjoy the beautiful Southwestern views and sunsets, we oriented our house accordingly. Bearing in mind the potential for heat gain on that side of the house, a covered patio runs a majority of the length of the home. Maximum insulation in ceilings and exterior walls was utilized to minimize heating and cooling costs. In addition, insulation was added to interior walls for its sound dampening ability. Operable skylights were added to bathrooms and kitchen areas to add natural lighting and ventilation in these locations. As an afterthought, a skylight provided much needed lighting to a dark walk-in closet. A split bedroom plan provided privacy for guests and/or children.

Largely unseen, but highly desirable, is choosing to use the best quality materials your budget will allow. As an example, we chose to use copper pipes over PVC supply lines for their durability as well as for keeping supply and waste lines as short as possible. Have the foresight to place stub-outs where you may add plumbing fixtures or equipment such as a water softener or a solar heater. This facilitates their addition at a future point. While you're at it, add sufficient electrical outlets, phone lines and television cable where you might want them. In addition, make these runs easily accessible in a crawl space or attic, should the need arise for repair.

Finally, a key to surviving the actual building process is to take each day as its own separate job rather than being overwhelmed mentally with the thought, "I'm building a whole house." Often, we began our day by saying, "Today we're working on plumbing (or electrical or shopping for materials)." Keep a list of upcoming supplies that will be needed so as to eliminate extra time-consuming trips to the store. Take pride in each step completed. Acknowledge your limitations in areas of construction where you will need help. There will be some aspects of construction that you will need to subcontract. There will be labor-intensive jobs where you will need help. Some jobs will require unique skills. With the exception of poured slab on grade foundation, the exterior stucco and the HVAC system, we built every aspect of our home. From the drawings to the luscious berry-colored front door, we built our home ourselves. We are very satisfied and appreciate our ability to accomplish our goal.

Dale and Nancy Fish

Dale and Nancy Fish and their family decided to build their home while living on-site. Their plan was to build a residence with an attached garage. They designed and built the garage first, then moved into it. They planned the garage to have the amenities of a bathroom and kitchen sink. They lived in the garage portion while using their spare time to finish the home. Here is their advice.

When we first discussed building a house of our own, I thought my husband had lost his mind. It seemed to be an endeavor way out of our league. We began by looking for property by driving all over town. We discussed the aspect of travel time to and from work for various sites and how it would affect our commute. We even discussed the effect of the sun shining in our eyes, driving to and from work. We finally found a reasonably-priced acre of land approximately half an hour from town, which was easily accessible from a major freeway, yet far enough away from the noise of town. There were no utilities at the site, but we reasoned that we could install them before we were ready to build a home. As luck would have it, by the time our plans were complete, then approved by the county, we found an adjoining acre that had utilities installed. We decided on an alternate location for our home. That is where our home now sits.

After finding the area we wanted to live, the first step we took was to decide what our house should look like. We began at the public library, looking at different floor plans for the size of house we wanted. Then we discussed what each of us wanted in a house. I wanted a garage entry into the kitchen for easy access to bring in groceries. We finally reached consensus on a floor plan. We looked around for an architect to draw our plans. After considering the costs, we discovered a building designer and draftsman with an excellent reputation who understood our needs. With his help, we reviewed some more plans, modified our design slightly and had our plans drawn. We then took a crosscountry trip to visit relatives. Of course, we took our plans with us to discuss with our family and friends. We became more excited about the project, as we shared our goals with others. The more we thought about it, the more anxious we became to get out of the city and begin our new home. That is when we came up with the concept of building our home in two phases. The design of the house lent itself very easily to the idea, and it would help us out financially. We asked our draftsman to help by dividing the proposed plan into two phases. He redrew them as phase one and phase two housing projects. We submitted the proposal to the county and were approved for this staged project. The plan was to live in the garage for three years and complete the house during the second stage.

We lived in the garage with our young family for those three years. It was very comfortable. The garage was 1014 square feet and we had partitioned it into two bedrooms, 1 bath, a laundry, kitchen and living area. We now have a completed house that is 3800 square feet, which includes 6 bedrooms, 3 bathrooms, formal dining room, family room, and living room. This is the house we had dreamed about and at the place we loved. Here are our recommendations for others who want to do what we have done:

1. Keep the home construction money in a separate account from other personal funds. Get a three-ring binder to keep all check stubs in one place for future record keeping.
2. Watch the quality of work done by outside contractors. Visit the project every day!
3. Select all plumbing fixtures before starting the house, so the plumber knows how to assemble the plumbing drain, waste and vent pipe as well as water supply pipe.
4. If you have any skills that you can trade with others, consider bartering with professionals for those services that you cannot perform yourself. If you have a car you don't need, your electrician may want to trade for his labor.

Jim Herbig and Debbie Haas

Jim Herbig and Debbie Haas own 56 acres in a remote part of the desert. After nine months of research and study, a conventional construction approach was selected for their home. Here is their advice.

Jim spent hours walking the desert to find the appropriate site for house placement, which would afford the best views while taking full advantage of the sun. Prior to that, with the help of an inexpensive CAD program, he also spent weeks designing and enhancing the proposed house layout, attempting to minimize damage to the environment. Further, propane was also needed for heating, water, and cooking.

Jim, at this writing has spent almost 11 months in the construction of the property and has experienced most of the rewards and pitfalls, challenges and successes, of being an owner-builder. This is a dream Jim has always had, and he found, as his dream came to fruition, that he could and did save approximately 35-40 percent on overall construction costs. These savings were the result of conscientious bidding and the use of family and friends who assisted in various stages of construction, from framing through porch building. Through this work he has developed a series of fundamental suggestions, which would have made the construction activities progress more smoothly, had he been aware of them at the outset. Here they are.

The initial phase of the construction involved contacting and interviewing a prospective architect, who would be charged with drawing up the official house plans for submission to approving authorities. This phase taught us one valuable lesson, namely, that an architect is not a prerequisite for construction plans. In fact, after receiving the bid, we discovered that the talents of a draftsman, who charges per square foot, were just as acceptable to the county. Furthermore, the draftsman's plans were priced substantially lower than the architect's.

Next, and very important, is to maintain open lines of communication with people who have experience, such as county employees, contractors and others in the industry. Listening to various approaches from individuals with backgrounds in this work does aid in clarifying direction, ability, and goals. Remember, the construction will generally take longer than you planned, and there will be cost overruns in some instances.

Be somewhat familiar with the zoning requirements and building regulations. When in doubt, ask. This will keep costs down and minimize mistakes. The old adage "measure twice, cut once," is appropriate. You will get frustrated. When the job seems to become too overwhelming to manage, it's best to walk away for a short period of time, knowing that you'll go back with a fresh outlook and heightened concentration. Home construction is a laborious process that requires time and concentration.

Take advantage of discounts. Utilize qualified weekend and off season workers, and once you receive an acceptable bid, don't be afraid to ask for "best and final," which is lower than that received. Jim would consistently present the successful contractor with a counteroffer for a lower amount, and he normally succeeded in reducing the final contract price. Get bids from reputable individuals and negotiate from there. Although, as an owner-builder, you are allowed to do the work yourself, it pays to have qualified help. Additionally, if you hire licensed contractors, you have recourse if the work winds up being substandard or fails inspection. Also, shop at stores that will meet and beat competitor's prices. Many of the larger home improvement stores will do this.

Building off the grid as we were, we needed to be especially cognizant of energy efficiency. We were including a rather large solar energy system in our plans; however, we still could not afford to be wasteful. As a result, we initially contacted the US Department of Energy for information on efficient appliances. Their web site, however, had minimal information. Further research led us to a web site maintained by the Canadian government (http://www.energuid.nrcan.gc.ca/), which publishes extensive information on the energy usage of all types of appliances. Certainly, some of these items are not available in the United States; overall however, we found the data to be very helpful as we shopped for refrigerators and washers. Also bear in mind when comparing efficiency numbers that they are usually based upon usage by a family of four. That means, approximately 16 loads of wash per week, and numerous refrigerator and freezer door openings. If you do not have children or have a smaller family, these figures can generally be adjusted downward.

Remember to try to build the house you want. Some upgrades may indeed be affordable as you save on labor costs by doing various portions of the construction yourself. Neighbors and relatives make good workers—just be sure to maintain control over the process. County officials are very helpful and a wonderful source of information and direction. Keep hallways to a minimum to minimize wasted space, and use the environment to help with heating and cooling.

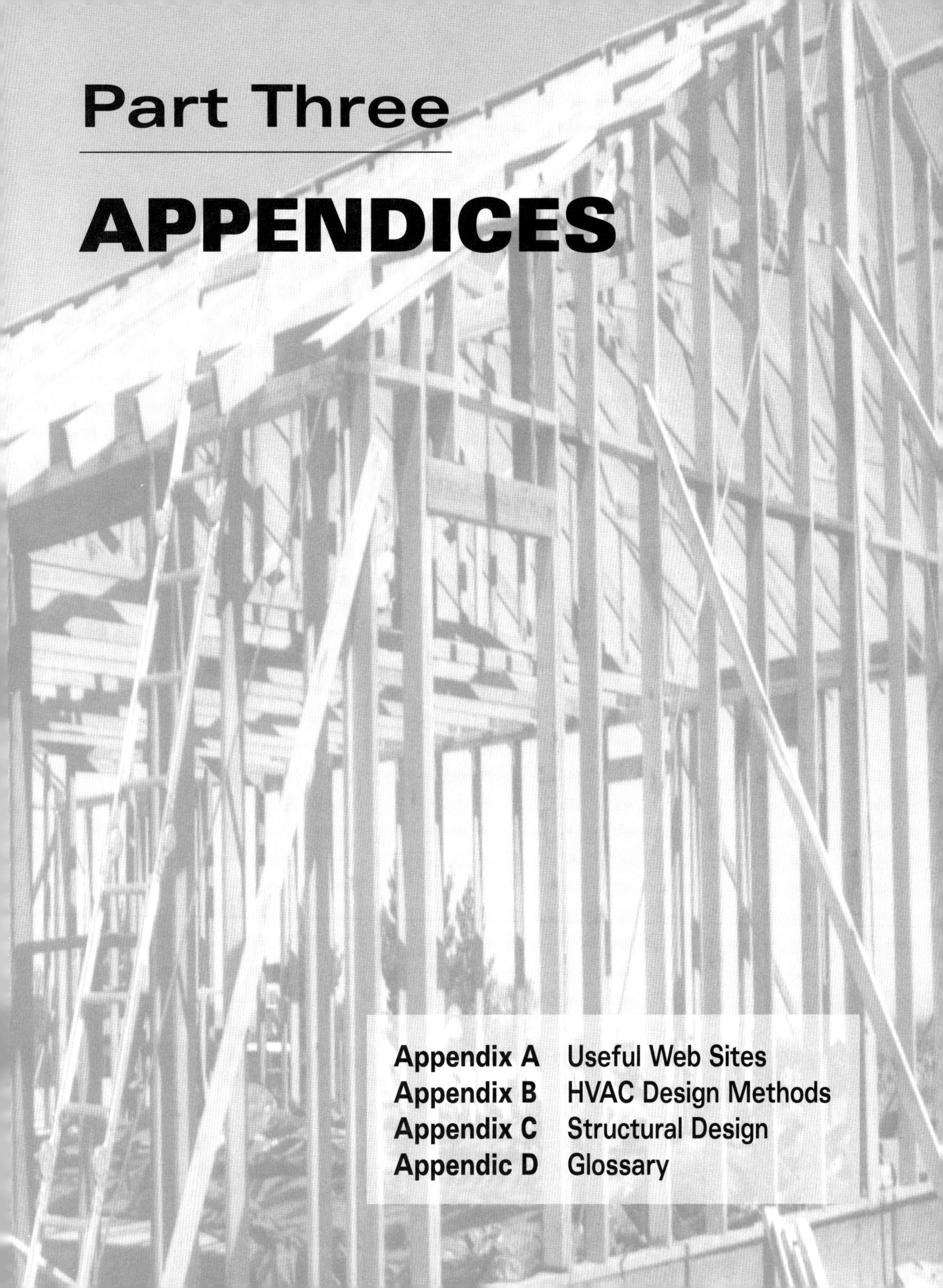

Part Three

APPENDICES

Appendix A

USEFUL WEB SITES

American Concrete Institute	http://www.aci-int.org/
American Timber Industries	http://www.arcat.com/
APA: The Engineered Wood Association	http://www.apawood.org/
Building Officials and Code Administrators	http://www.bocai.org/
Energy Codes	http://www.energycodes.org/
Energy Codes	http://www.21Design.com
Fine Homebuilding Magazine	http://www.taunton.com/
Gypsum Association	http://www.gypsum.org/
Hardwood/Plywood and Veneer Association	http://www.hpva.org/
Hometime Television	http://www.hometime.com/
International Code Council	http://www.intlcode.org/
International Conference of Building Officials	http://www.icbo.org/
Masonry Institute of America	http://www.masonryinstitute.org/
Mother Earth News	http://www.motherearthnews.com/
Soil Science	http://www.soils.org/
Southern Building Code Congress International	http://www.sbcci.org/
Western Wood Products Association	http://www.wwpa.org/
Wood Truss Council of America	http://www.woodtruss.com/

Appendix B

HVAC DESIGN METHODS

Design Considerations

The required heating capacity is a function of how cold it is outside and how warm you want to be inside. Begin by establishing what temperature you want to maintain inside. Most codes require that an HVAC system be capable of maintaining a temperature of 68 degrees inside habitable rooms within the dwelling.

Next, determine the outside average winter temperature. The outside temperature where you live is recorded in weather data. Use the dry bulb winter temperature. Call your local weather bureau for this information. The difference between these two temperatures is used to calculate the rate of heat flow away from a given structure. This difference in temperature is commonly referred to as **ΔT** (delta T or change in temperature).

The formula for calculating heat flow is:

$Q = UA\Delta T$

Q = Rate of heat flow in Btu/hour × ft^2
U = Coefficient of heat transfer (no units, but equal to 1/R value)
A = Area such as wall section or ceiling (square feet)
ΔT = Temperature difference between interior design and outdoor temperature (degrees F)

The coefficient of heat transfer: U is the most complicated and troublesome part of this equation. U is the inverse of the R value of a given material within a floor, wall or ceiling area.

$U = 1/R$

A wall section has many different materials such as drywall, insulation, wood stud, windows, doors, and siding that must be analyzed as a composite unit to determine an accurate total R value as follows:

U Total = (U wall × A wall) + (U window × A window) + (U door × A Door) ÷ Area of the Wall Section

Example:

A wall section 40 feet long, 8 feet high with three 4/0 by 4/0 windows and no doors. The wall section has sufficient insulation and siding to develop a combined value of R-25. The windows have a U-60 rating.
Calculate the rate of heat flow out of the wall section if delta T = 30 degrees.

Solution:

1. U wall = R-25 = 0.04
2. U (total) = [0.04 × (40 × 8)] + [(3 × (4 × 4)) × (.6)] /320

   = ([.4 × 320] + [48 × .6]) /320

   = (128 + 28.8)/320

   = 156.8/320=.49 Btuh ft^2 degree F
3. $Q = UA\Delta T$

   Q = .49 × 320 × 30 = 4704 Btu/hour

This means that every hour 4704 Btus leave this wall section and need to be replaced with some sort of heating to maintain the desired temperature.

Heat Flow Calculation of Typical Wall Section

A whole house heat loss calculation is a matter of finishing this procedure for the entire building and adding each of the Qs for each wall or roof section to determine the whole house heat loss. This represents what must be provided by heating equipment to maintain the 68 degree temperature. There is more involved in the accurate calculation of heat loss. Analyzing such things as infiltration from crack areas around doors and windows, the general tightness of the house, geographic orientation to collect solar gain, foundation heat loss, below grade heat losses, and bare surface heat losses all add to an accurate depiction of total heat loss.

Appendix C

STRUCTURAL DESIGN

The principles of engineering follow standard structural analyses whenever designing wood beams, rafters, joists and columns. They analyze the following conditions:

1. Bending
2. Deflection
3. Compression perpendicular to grain
4. Horizontal shear

For wood studs, joists, and rafters, you will use dimensional lumber (2 x 4s, 2 x 6s, 2 x 8s, 2 x 10s and 2 x 12s), whose allowable loads are the subject of an engineered design and are tabulated in model building codes. To determine the proper selection of a joist or rafter, you must know some of the physical properties of the wood. Modulus of Elasticity (*E*) is a measure of the relative stiffness of the wood. This elastic strength in wood varies between species of trees. Fiber Stress (*Fb*) is a measure of the force needed to bend the wood beyond its capacity to resist and also varies according to species. Tables in design manuals list species of trees along with their grading to establish their respective *Fb* and *E* values. Use these to determine allowable span in other tables. Use the tables on the next sheet to estimate sizes. Another term used in this genre is the *section modulus* (S). The section modulus is a number that is entirely dependent upon the shape and size of the cross section. The section modulus for rectangular members is:

$S = bd^2/6$
Where
b = width of member in inches
d = depth member in inches

The standard tables, which size joists and beams, are not esoteric. They are based upon empirical data from testing and engineering formulae. Here in a nutshell is some background for the armchair engineer.

Bending Moment Calculation for Wood Joist or Beam

Under a load, wood experiences flexural bending. The allowable limit to this bending is defined by *maximum allowable moment*. The manner in which a moment of a joist is calculated is widely dependent upon the way a load is applied. The most common condition is a uniform load applied to a simply supported joist (supported on each end). The formula for moment under these conditions is:

$M = wL^2/8$

M = Moment developed from load in ft•lbs
w = Unit loading in pounds per lineal foot (per joist)
L = Span of joist in feet

Example:

Calculate the moment of one simply supported floor joist that is 24 inches o.c. carrying a square foot load of 55#/ft^2. The joist is 12 feet long.

Solution:

$w = (55\ \#/ft^2) \times (2ft) = 110\#/ft$
$L = 12$ ft
$M = 110 \times 12^2/8 = 1980$ ft•lbs

Now, how useful is this number? Well, you can solve required sizes for joists the way the guys who wrote the tables did. You see, the moment is equal to the fiber stress of a wood times the section modulus.

$M = Fb \times S$

M = Bending moment in fiber stress
Fb = Allowable fiber stress of material
S = Section modulus of bending member

Assume $Fb = 1450$ psi
Convert feet to inches: (1980) (12) = 23,760 in•lbs

Solution:

$M = Fb \times S$, or $S = M/Fb$
23760 in•lbs/1450 psi = 16.4 in^3
$S = bd^2/6$
$16.4 = (1.5 \times d^2)/6$
$16.4 \times 6 = (1.5 \times d^2) = 98.4$
Try a 2 x member ($b = 1.5$ inches)
$98.4/1.5 = (d^2) = 65.6$

$d = \sqrt{65.6} = 8.1$ inches

The solution is 8.1, so in this case, you would use a 2 x 10. Always round up to the next available size.

Joist Sizing Using Deflection Calculation

There are two other very common engineering formulae to reckon with in designing wood joists. First, deflection is the maximum amount of distance a bending member may deflect before risking safety. The way to calculate allowable deflection is a little more complicated. The formula for deflection, based on a simply supported, uniformly distributed load, is:

$d = 5wL^4/384EI$
where
w = weight on joist or beam in pounds per foot
L = length of joist in feet
E = Modulus of elasticity
I = Moment of inertia
$I = bd^3/12$

Most codes establish that $L/240$ is maximum allowable deflection for a roof member. That means that the L or length in inches divided by the actual deflection in inches may not exceed the dimension-less number: 240. Let's run through a typical calculation of deflection and verify if the condition meets or exceeds the requirement for maximum allowable deflection.

Determine if 2 by 8 on 24 inches o.c. spanning 12 feet long carrying roof load is within limits established by code.
Given: E = 1,600,000 psi, live load = 40 psf, dead load = 15 psf

Solution:

w = (40 psf + 15 psf) × 2 ft = 110 lb/ft
L = 12 ft
$I = bd^3/12 = (1.5) \times (7.5^3)/12 = 52.7^4$ in
correction factor for feet to inches = 1728 (1728 cubic inches = one cubic foot)

$d = [(5) \times (110) \times (12^4)/(384) \times (1{,}600{,}000) \times (52.7)] \times 1728$
= .60 inch (5/8 inch)
$144/.6 = 240 \therefore L/240$

Calculation of Horizontal Shear in a Bending Member

The last analysis check is for horizontal shear. When you examine a joist or beam with a load, it is under two opposite kinds of pressure. On the top, the joist is under compression. On the bottom, the joist is under tension, In the middle, these forces resolve themselves. The resolution is a force called horizontal shear (Fv). In common terms, the joist is trying to come apart along this axis. The maximum allowable shear is established by code requirements specifically according to the species and grade of lumber. Generally, for a Number 1 grade Douglas fir construction framing lumber, the maximum allowable shear is 85 psi. The allowable horizontal shear for other species is located within tables in the building code. The way of determining horizontal shear is as follows: First determine the reaction of total load at one end of the joist or beam. It is half of total load on the joist. This is known as shear, and abbreviated V.

Now, use the following formula to determine *Fv*:

$Fv = 3V/2bd$
where
V = reaction at one end only in pounds
(This can be a post supporting the end of a beam)
b = width of joist in inches
d = depth of joist in inches

Example:

Determine if joist is within limits for *Fv*

Solution:

Find reaction. $V = \frac{1}{2}$ total load, simply supported.
Load = (110#/ft) × (12 feet) = 1320#
V = 1320/2 = 660#
$Fv = 3V/2bd = (3) \times (660)/(2) \times (1.5) \times (9.5) = 69.5$#/sq in
69 psi is less than 85 psi for Douglas fir joist horizontal shear value

Now, use this value to size the joists or beam you need. This particular design criteria is based on a simply-supported beam or joist with a uniform load applied. Other loading conditions require consultation with a structural engineer.

Column Design

Wood columns or posts can be designed with some simplified analysis methods. There are two criteria to analyze: compressive stress and buckling. The first method of column design is allowable compression stress. Compressive stress is simply a material's ability to resist a squeezing or crushing effect without rupturing. Wood, as a building material, is able to resist compression load extremely well. This ability to withstand compressive stress is referred to as

$F'c$ = Allowable compression design value

The value varies subject to the species of tree from which the wood member is hewn. The values listed below are not necessarily accurate but are presented for the purposes of example only. $F'c$ is presented in terms of pounds per square inch (#/in^2). The following list provides the design values for some common species.

Species	*F′c (*Compression perpendicular to grain*)**
Douglas fir	625 psi
Hem fir	370 psi
Ponderosa pine	535 psi
Southern pine	375 psi
Spruce-pine-fir (SPF)	425 psi

Western cedar	425 psi
Western white pine	375 psi

** Design assumed design value for select grade of each species*

The useful translation can be best understood if you know the cross sectional area of the structural member that you are trying to analyze or design. For instance, a 4 x 4 post has a cross sectional area as follows:

3½ inch x 3½ inch = 12¼ in^2

Apply the load uniformly to the cross sectional area and you have the allowable unit uniform load. For instance, consider that your 4 by 4 inch post is from a Douglas fir tree and is of a select grade. What is the maximum allowable load permitted to be applied to the post?

(625#/in^2) × (12¼ in^2) = 7,656#

The second method of column design to analyze is the ability of a column to resist buckling or bending away from a true vertical direction. This analysis method is a little more complex, in that more variables must be considered. If a column is too slender, a massive load may cause it to buckle. Its ability to withstand that load is then compromised because of this buckling. The criteria that defines a column as too slender is the equation:

le/d

This is an imaginary number that tells us something about the ratio of length to thickness of the column. The effective span of a column in inches is referred to as *le*. The least dimension of a compression member is referred to as *d*. For instance, if our 4 x 4 inch post is 8 feet tall, its *le/d* would be as follows:

[(8) × (12)] / 3.5 = 27 (columns are normally limited to an *le/d* of less than 50)

Remember that columns support beams, which in turn support floor joists or roof trusses. The analysis for column design is not accurate unless all of the loads for a particular column are taken into account.

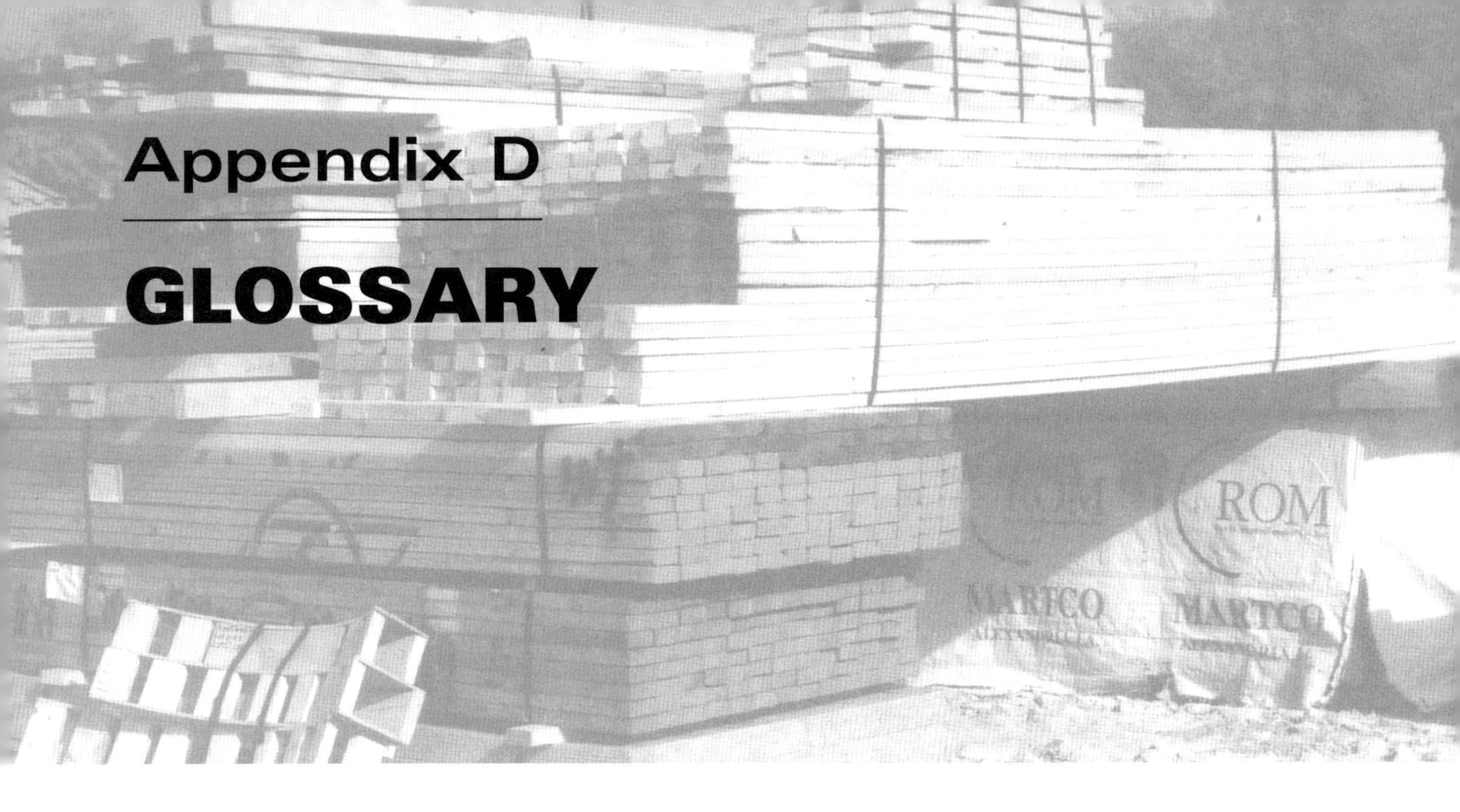

Appendix D

GLOSSARY

ABS is an abbreviation for acrylonitrile butadiene styrene, which is a plastic pipe commonly used for drain, waste and vent uses.

accessible (plumbing) means that access to a device or fixture must be possible with minor effort.

air admittance valve is a local one-way air valve used on drain waste and vent systems, which avoids the use of a vent.

air gap is a space between the faucet and the flood rim of a fixture such as a sink, which prevents syphonage.

allowable soil bearing capacity is a measure of the strength of the soil to resist imposed load.

alluvial soil is sedimentary material deposited in a delta or a riverbed.

amperage is the unit that measures the amount of electrical current flowing in a circuit.

ASHRAE is an abbreviation for American Society of Heating, Refrigeration, and Air Conditioning Engineers.

anti-syphon valve is a valve that prevents syphoning of waste into the fresh water system. A common type is a vacuum breaker, which is on hose bibs on outdoor locations.

approved agency means an agency charged with the testing and certification of a product for use.

backfill is an earth material that is used to infill a cavity created by excavation.

base course is a layer of graded earthen material placed on sub-grade or sub-base under a footing or slab.

beam is a horizontal structural member used to support rafters of joists and is supported by posts or walls.

bearing wall is a wall that supports a load, such as a floor or roof.

bending moment is a measure of the load imposed upon a structural member in terms of bending.

board foot is a unit of measure of dimensional lumber. A 1 inch thick board 12 inches wide and 10 feet long is said to be 10 board foot.

bonding is the electrical connection between any metal part and the building ground.

branch drain is a part of the drain system other than the main drain.

branch vent is a vent that serves as a vent for more than one fixture. It connects to a vent through the roof.

building drain is the name for a main drain within the perimeter of a building. It does not include various branch lines, trap arms or venting associated with the DWV pipe.

building sewer is the name for the drain outside the perimeter of a building that conveys sewage from the building drain to a septic tank or a sewer.

caisson is a concrete foundation cast in a circular excavation.

camber is a slight curve in a beam that helps resist vertical load.

canales is a channel to divert rain off low slope roofs with a parapet.

caulk is a joint sealing compound used around window or door frames.

certificate of occupancy is legal permission from a jurisdiction to occupy a building. It is issued only after a series of inspections have been approved.

clean-out is an access to the DWV piping system, which allows for physical cleaning of the pipe.

closed circuit is a circuit that is carrying current.

closet bend is the 90 percent plumbing fitting specially designed to convey soil waste from a water closet (toilet) to the building drain.

closet flange is a transition fitting between the closet bend and the water closet (toilet).

collapsible soil is soil that, when under a load, will collapse and cause general failure in the supported structure.

column is a vertical supporting member.

common (wire) is a term for a neutral conductor within an electrical system.

concrete masonry unit (CMU) is a manufactured block designed for construction.

concrete slab on grade is a concrete surface layer cast directly on subgrade.

conduit is a pipe or tube in which electrical wiring is installed.

cripple studs are smaller studs above and below openings in a stud wall frame.

crown is a term that refers to the slight natural curl of a wood framing member, which occurs in the logitudinal dimension.

dead load is the weight of the structure, including wall, floor and roof framing, roof decking, insulation, siding, windows, doors, drywall, stucco, and roofing.

deflection is the maximum distance a structural member contorts from its original shape due to loads imposed.

developed length is the total running length of a plumbing pipe.

dimension is a distance between points, which may be reduced to a specified scale for a plan, such as ¼ inch per foot.

distribution box connects two or more leaching trench fields to the outlet side of the septic tank.

drainage fitting is a fitting that connects sections of drain pipe to allow for changes in direction or pipe sizes.

DWV is an abbreviation for drain, waste, and vent piping system.

dried-in refers to the condition of the interior of the building when enough roofing is installed to protect it from adverse weather.

drywall refers to materials used as interior wall covering, usually in sheets, normally composed of gypsum products.

easement is land that you own but have deeded exclusive use to another for the purposes of access to repair essential services such as water service, electrical power lines, and sewer tap.

eaves is the lower part of a roof that projects over an exterior wall.

elevation views are a type of plan drawing of a building's exterior, much like a picture from all sides.

engineered fill is a fill material that is specified by an engineer for density and bearing pressure and other conditions.

existing grade is the ground surface prior to any excavation.

expansive soil is a soil with a potential to increase its volume due to absorption of moisture.

facade is the main or front elevation.

feeder is a conductor supplying electricity from one panel board to another, as in a sub-panel.

fiber stress is a measure of the active stress imposed upon a structural bending member, such as a wood beam or floor joist.

fixture refers to a plumbing device such as a bathtub, sink, shower, or lavatory, which is designed to accumulate and dispose of waste.

fixture unit is the unit of measure for drainage and water supply calculations and is specific to each fixture.

footing is the lowest structural support of any bearing wall, usually concrete poured into trenches.

foundation is the structural support of exterior or interior bearing walls. A foundation is supported by a footing.

framing is the act, product or building of a floor, wall or roof structure and usually includes wood studs, joists, rafters, trusses, beams and posts.

free conductor length is a term that describes the distance a cable assembly must extend from an electrical outlet or switch box to facilitate equipment installation.

general contractor is a person who manages the overall building project including framing, plumbing, mechanical and electrical trades.

girder is a large or principal bearing support such as a beam or a truss that holds other framing members.

glazing is another word for glass or glass products such as windows or glass doors.

grade (building) refers to exterior soil location.

grader is the title of a person who decides on and marks lumber with its grade, thereby classifying its design strength.

gray water is waste water from sinks, showers or tubs, and washing machines, which may be used for gardening.

gypsum board is the broad name for a variety of gypsum sheet products with a paper covering.

HVAC is an abbreviation for heating, ventilation, and air conditioning equipment.

headers are solid or built-up beams within a wall frame that support loads over a door or window opening. Headers are supported by trimmer studs from the bottom.

heating degree days are the cumulation of the differences in the outdoor temperature and the indoor design temperature throughout the heating season.

hertz is a term for cycles per second and is the measure of frequency of alternating current.

home run is the name for the wire between the circuit breaker and the first outlet or device in a circuit.

horizontal shear is a measure of certain forces resolving themselves in the center of a horizontal structural member.

individual vent is a pipe that conveys air to the drain waste vent system to a single fixture trap.

insolation is solar radiation exposed to a certain area.

insulation is any material, placed in a wall frame crawl space or attic, that will resist heat transfer.

jack stud, also known as a trimmer stud, is the supporting stud under the end of a header in a wall frame.

joinery is a term for any method of joining wood, usually in finish work.

joint is a connection between two similar building materials, such as pipe joint or masonry joint.

joist is a solid wood dimensional member used as one of a series of parallel framing members, such as a rafter.

kicker is long piece of framing lumber used, temporarily, as a brace to support and shore up a wall frame prior to its permanent bracing.

king stud is the full-height stud adjacent to a trimmer or jack stud in a wall frame.

lateral bracing is a term used to describe the bracing required for a wall frame to resist lateral loads such as wind or earthquake forces. Lateral bracing helps to create a *shear wall.*

lath is attached to exterior wall siding to serve as a base of attachment for plaster. It can be any approved material and is most commonly chicken wire.

leach lines are absorption fields that further strain pollutants from waste water expelled from the septic tank.

ledger is a horizontal structural element that serves as a means of support for a joist or truss from the side of a wall.

lien is the right to claim a portion of some property until a bill or charge is satisfied.

light construction usually includes wood stud framing, wood rafters and floor joists. It is usually limited to houses and smaller buildings.

listed means that a material or product has received an approval from a listing agency, which is an independent testing agency that reviews the product for efficacy.

live load is the load imposed upon a structural member by all other forces other than dead load, including atmosphere, rain, snow, wind, seismic as well as furniture and human traffic.

let-in bracing is a form of lateral bracing in which the studs are cut to allow for a 1 inch dimensional board to be *let-in* to the stud.

lift is a layer of material such as sand or fine aggregate used to backfill an excavation. It is a strata of materials such as soil installed sequentially to allow for uniform consolidation.

light-frame construction is a term that refers to simple, standard wood framing and includes wood studs, floor and ceiling joists, and roof rafters or trusses.

lintel is another term for a header. Usually it has come to define a horizontal element over an opening in a masonry wall.

load is the weight that must be supported by the structural elements.

Dead Load is the weight of all of the building materials.

Live Load is the load anticipated from nature and occupants, such as people, pianos, books, and atmosphere.

Load Bearing capacity is the ability of the soil to resist loads applied by a structure before collapsing.

load (electrical) the demand in watts by a particular motor, outlet or other device.

member is any portion of a system, such as a post being a structural member supporting a beam.

modulus of elasticity is a number unique to a structural bending member, which describes its elastic behavior prior to rupture.

monolithic slab describes an all-in-one concrete pour that includes the footing, foundation and slab.

neutral wire is a wire that carries the unbalanced current back to the transformer in a circuit.

nominal size refers to the design size of a sawn wood framing member; i.e. a 2 inch x 4 inch stud is only 1½ inches x 3½ inches.

nonbearing partition is a wall that does not support any load except itself.

open circuit is a circuit that is not carrying current due to a break in the circuit.

on center (abbreviated o.c.) refers to the distance between the center of a framing member, such as 2 inch x 4 inch @ 16 inches o.c.

panel board is an electrical panel that distributes electrical current through circuits.

partition is a wall that separates one room or space from another and can be a bearing or nonbearing wall.

passive solar is a method of obtaining and using solar energy without mechanical or electrical devices.

permit is the legal permission from a jurisdiction to begin construction on a proposed project.

pier is a small masonry or concrete column that is used as structural support.

pitch is the slope. It is used to define roof pitch and is usually specified in units of 12ths, such as 3:12 means 3 vertical units for every 12 horizontal units.

plan is a drawing that clearly delineates the intent of a proposed construction project.

plates are longer pieces of framing lumber in a wall frame that secure studs, trimmers and headers together. Generally, there is one bottom plate and two top plates in a wall frame.

PB refers to Polybutylene, which is a plastic material used for certain pipes.

PE refers to Polyethylene, which is a plastic material used for certain pipes.

potable water is water fit for human consumption.

power (electrical) is expressed as watts and is the product of volts times amps.

prism (masonry) is a sample masonry wall built with similar mortar and grout and is used to test the design strength.

rafter is one of a series of repetitive structural members that create a roof structure.

rake wall is a wall with different height studs within its frame. Its shape may define the slope of the roof.

readily accessible means that you can walk right to the object without any obstruction or access panel.

retaining wall is usually a masonry or concrete wall that retains soil, such as a basement wall.

rim is the lip or edge of a fixture, such as a lavatory bowl.

riser (building) is one of a series of vertical elements in a stair between treads.

roughing-in is the work of installing the initial plumbing or electrical components.

rough opening is the length and width of an opening such as a window or a door frame.

run wild is a term for any part of a wall, floor or roof system that is unsupported except for a cantilever.

scaffold nails are double headed nails that allow for easy removal. They are usually used as connectors for temporary shoring.

scarify is action that mechanically loosens or breaks down existing soil structure.

section modulus is a number unique to a structural bending member that describes its strength as a function of its cross sectional shape. For rectangular members, the Section Modulus = width x depth squared ÷ 6 (in inches).

septic tanks are water-tight tanks made of substantial materials such as concrete, grouted block, metal or even plastic or fiberglass.

service (electrical) is the equipment that receives electricity from the power company and distributes it to various uses inside the building. The service is the location where power can be disconnected.

setback is the distance from a property line in which no building may be erected.

sewage is liquid waste from a drainage system.

sewage ejector is a device that lifts sewage to a point where it can enter the gravity drainage system. It uses a sewage pump.

shear panel is a section of a wall frame that is designed to resist wind and seismic forces by transferring loads from the roof through the wall to the foundation.

sheathing can refer to exterior siding or roof decking material. It also refers to exterior building paper that is used to wrap around exterior siding before the application of stucco.

shoring is any means of bracing a structural element to resist collapse.

siding is the exterior covering of a wall, such as plywood, aluminum, vinyl, etc.

soil pipe is drainage pipe that carries human waste.

soil series or *profile* is a description of the type of soil that is generally found at your site as determined by a soil survey for your county.

span is the total distance between supporting framing members.

stringers are the support structure for stair treads.

studs are vertical framing lumber within a frame wall, which support the floor, roof and siding.

subpanel is a remotely located, secondary electrical panelboard extending current from the main service panel.

sump pump is a pump used to eject sewage from a basement or lower floor bathroom or other plumbing fixture to an upper level sewage discharge pipe.

survey is an exact map of your property with dimensions, angles and distances established by a registered land surveyor.

tailpiece is a pluming pipe that connects the outlet of a fixture to the fixture's trap.

toenailing means to drive a nail at an angle to achieve a connection.

topographic map is a map that portrays elevation differences within a defined area.

trap is a fitting that uses water to prevent odor or gasses in the drainage system from entering the living area of a home without adversely affecting the flow of waste.

trap seal is the section inside the trap that, when filled with water, prevents sewage gases from entering the room or space.

trap arm is a section of pipe between the vent and the fixture trap.

tributary area is that area of a span that is being supported by a particular structural member.

tributary width is that width of a span that is being supported by a particular structural member.

trimmer studs are pieces of framing lumber within a wall frame shorter than adjacent studs that support headers. The top of a trimmer defines the opening height of a window or door frame.

true and plumb describes the perfect positioning between horizontal and vertical members in wall framing.

truss is a structural component of a building, which can serve as either floor or roof support. It is designed by a registered professional engineer.

vent is a pipe that allows air to enter into the DWV system.

volt is the electrical unit that measures the potential difference in electrical force.

wallboard is any of several different sheet materials that are fastened to the building frame as a finish surface.

water closet is a toilet.

watt is a unit of electrical power. It defines the rate at which a device converts electrical current to another form of energy also known as volt-amps.

wet vent is a vent or portion of a vent that acts as both vent for a downstream fixture and as a drain for an upstream fixture.

yard line refers to a water supply pipe that taps into a water meter or well and carries water to the building.

Index